Student Atlas of
World Geography

Fourth Edition

John L. Allen
University of Wyoming

The **McGraw·Hill** Companies

Book Team
Managing Editor: *Larry Loeppke*
Developmental Editor: *Nichole Altman*
Developmental Editor: *Ava Suntoke*
Designer: *Tara McDermott*
Typesetting Supervisor: *Kari Voss*
Typesetter: *Jean Smith*
Typesetter: *Sandy Wille*
Typesetter: *Karen Spring*
Cover Design *Maggie Lytle*
Cartography *Carto-Graphics, Oxford, MD*

We would like to thank Digital Wisdom Incorporated for allowing us to use their Mountain High Maps cartography software. This software was used to create maps 73, 74, 82, 83, 85–90, 92–95, 97–101, 103–106, 108–114, 116–120

The credit section for this book begins on page 226 and is considered an extension of the copyright page.

A Note to the Student

The study of geography has become an increasingly important part of the curriculum in secondary schools and institutions of higher education over the last decade. This trend, a most welcome one from the standpoint of geographers, has begun to address the massive problem of "geographic illiteracy" that has characterized the United States, almost alone among the world's developed nations. When a number of international comparative studies on world geography were undertaken, beginning in the 1970s, it became apparent that most American students fell far short of their counterparts in Europe, Russia, Canada, Australia, and Japan in their abilities to recognize geographic location, to identify countries or regions on maps, or to explain the significance of such key geographic phenomena such as population distribution, economic or urban location, or the availability of natural resources. Indeed, many American students could not even locate the United States on world maps, let alone countries like France, or Indonesia, or Nigeria. This atlas, and the texts it is intended to accompany, is a small part of the process of attempting to increase the geographic literacy of American students. As the true meaning of "the global community" becomes more apparent, such an increase in geographic awareness is not only important but necessary. If the United States has learned any lesson from the tragic events at the World Trade Center and the Pentagon on September 11, 2001, these lessons would surely include the considerations that we are not isolated from events that transpire in other parts of the world; our boundaries do not make us secure; and we ignore the conditions of political, economic, cultural, and physical geography outside those boundaries at our great peril.

The maps in the *Student Atlas of World Geography* are designed to introduce you to the patterns or "spatial distribution" of the wide variety of human and physical features of the earth's surface and to help you understand the relationships between these patterns. We call such relationships "spatial correlation" and whenever you compare the patterns made by two or more phenomena that exist at or near the earth's surface—the distribution of human population and the types of climate, for example—you are engaging in spatial correlation. Like the maps, the data sets in the atlas are intended to enable you to make comparisons between the distributions of different geographic features (for example, population growth and literacy rates) and to understand the character of the geographic variation in a single geographic feature. In many instances, the data in the tables of your atlas are the same data that have been used to produce the maps. At the very outset of your study of this atlas, you should be aware of some limitations of the data tables. In some instances, there may be data missing from a table. In such cases, the cause may represent the failure of a country to report information to a central international body (like the United Nations or the World Bank), or it may mean that the shifting of political boundaries and changed responsibility for reporting data have caused some countries (for example, those countries that made up the former Soviet Union or the former Yugoslavia) to delay their reports. It is always our aim to use the most up-to-date data that is possible. Subsequent editions of this atlas will have increased data on countries like Slovenia, Ukraine, or Uzbekistan when it becomes available. In the meantime, as events continue to restructure our world, it's an exciting time to be a student of world geography!

You will find your study of this atlas more productive if you study the maps and tables on the following pages in the context of the five distinct themes that have been developed as part of the increasing awareness of the importance of geographic education:

1. *Location: Where Is It?* This theme offers a starting point from which you discover the precise location of places in both absolute terms (the latitude and longitude of a place) and in relative terms (the location of a place in relation to the location of other places). When you think of location, you should automatically think of both forms. Knowing something about absolute location will help you to understand a variety of features of physical geography, since such key elements are so closely related to their position on the earth. But it is equally important to think of location in relative terms. The location of places in relation to other places is often more important as a determinant of social, economic, and cultural characteristics than the factors of physical geography.

2. *Place: What Is It Like?* This theme investigates the political, economic, cultural, environmental, and other characteristics that give a place its identity. You should seek to understand the similarities and differences of places by exploring their basic characteristics. Why are some places with similar environmental characteristics so very

different in economic, cultural, social, and political ways? Why are other places with such different environmental characteristics so seemingly alike in terms of their institutions, their economies, and their cultures?

3. *Human/Environment Interactions: How Is the Landscape Shaped?* This theme illustrates the ways in which people respond to and modify their environments. Certainly the environment is an important factor in influencing human activities and behavior. But the characteristics of the environment do not exert a controlling influence over human activities; they only provide a set of alternatives from which different cultures, in different times, make their choices. Observe the relationship between the basic elements of physical geography such as climate and terrain and the host of ways in which humans have used the land surfaces of the world.

4. *Movement: How Do People Stay in Touch?* This theme examines the transportation and communications systems that link people and places. Movement or "spatial interaction" is the chief mechanism for the spread of ideas and innovations from one place to another. It is spatial interaction that validates the old cliché, "the world is getting smaller." We find McDonald's restaurants in Tokyo and Honda automobiles in New York City because of spatial interaction. Advanced transportation and communications systems have transformed the world into which your parents were born. And the world your children will be born into will be very different from your world. None of this would happen without the force of movement or spatial interaction.

5. *Regions: Worlds Within a World.* This theme helps to organize knowledge about the land and its people. The world consists of a mosaic of "regions" or areas that are somehow different and distinctive from other areas. The region of Anglo-America (the United States and Canada) is, for example, different enough from the region of Western Europe that geographers clearly identify them as two unique and separate areas. Yet despite their differences, Anglo-Americans and Europeans share a number of similarities: common cultural backgrounds, comparable economic patterns, shared religious traditions, and even some shared physical environmental characteristics. Conversely, although the regions of Anglo-America and Eastern Asia are also easily distinguished as distinctive units of the earth's surface, they have a greater number of shared physical environmental characteristics. But those who live in Anglo-America and Eastern Asia have fewer similarities and more differences between them than is the case with Anglo-America and Western Europe: different cultural traditions, different institutions, different linguistic and religious patterns. An understanding of both the differences and similarities between regions like Anglo-America and Europe on the one hand, or Anglo-America and Eastern Asia on the other, will help you to understand the world around you. At the very least, an understanding of regional similarities and differences will help you to interpret what you read on the front page of your daily newspaper or view on the evening news report on your television set.

Not all of these themes will be immediately apparent on each of the maps and tables in this atlas. But if you study the contents of *Student Atlas of World Geography,* along with the reading of your text and think about the five themes, maps and tables and text will complement one another and improve your understanding of global geography.

John L. Allen

About the Author

John L. Allen is professor and chair of Geography at the University of Wyoming and emeritus professor of Geography at the University of Connecticut, where he taught from 1967 to 2000. He is a native of Wyoming. He received his bachelor's degree in 1963 and his M.A. in 1964 from the University of Wyoming, and in 1969 his Ph.D. from Clark University. His special areas of interest are perceptions of the environment and the impact of human societies on environmental systems. Dr. Allen is the author and editor of many books and articles as well as several other student atlases, including the best-selling *Student Atlas of World Politics.*

Acknowledgments

Nozar Alaolmolki
Hiram College

Barbara Batterson-Rossi
Palomar College

A. Steele Becker
University of Nebraska at Kearney

Koop Berry
Walsh University

Daniel A. Bunye
South Plains College

Winifred F. Caponigri
Holy Cross College

Femi Ferreira
Hutchinson Community College

Eric J. Fournier
Samford University

William J. Frazier
Columbus State College

Hari P. Garbharran
Middle Tennessee State University

Baher Gosheh
Edinboro University of Pennsylvania

Donald Hagan
Northwest Missouri State University

Robert Janiskee
University of South Carolina

David C. Johnson
University of Louisiana

Effie Jones
Crichton College

Cub Kahn
Marylhurst University

Artimus Keiffer
Franklin College

Leonard E. Lancette
Mercer University

Donald W. Lovejoy
Palm Beach Atlantic College

Mark Maschhoff
Harris-Stowe State College

Richard Matthews
University of South Carolina

Madolia Mills
University of Colorado–Colorado Springs

Robert Mulcahy
Providence College

Otto H. Muller
Alfred University

J. Henry Owusu
University of Northern Iowa

Steven Parkansky
Morehead State University

William Preston
California Polytechnic State University, San Luis Obispo

Neil Reid
The University of Toledo

A. L. Rydant
Keene State College

Deborah Berman Santana
Mills College

Steven Slakey
University of La Verne

Rolf Sternberg
Montclair State University

Richard Ulack
University of Kentucky

David Woo
California State University, Haywood

Donald J. Zeigler
Old Dominion University

Table of Contents

A Note to the Student

Introduction: How to Read an Atlas   x

Part VII World Regions   106

Part VIII Tables

Part IX Geographic Index   223

Introduction: How to Read an Atlas

An atlas is a book containing maps which are "models" of the real world. By the term "model" we mean exactly what you think of when you think of a model: a representation of reality that is generalized, usually considerably smaller than the original, and with certain features emphasized, depending on the purpose of the model. A model of a car does not contain all of the parts of the original but it may contain enough parts that it is recognizable as a car and can be used to study principles of automotive design or maintenance. A car model designed for racing, on the other hand, may contain fewer parts but would have the mobility of a real automobile. Car models come in a wide variety of types containing almost anything you can think of relative to automobiles that doesn't require the presence of a full-size car. Since geographers deal with the real world, virtually all of the printed or published studies of that world require models. Unlike a mechanic in an automotive shop, we can't roll our study subject into the shop, take it apart, put it back together. We must use models. In other words, we must generalize our subject, and the way we do that is by using maps. Some maps are designed to show specific geographic phenomena, such as the climates of the world or the relative rates of population growth for the world's countries. We call these maps "thematic maps" and Parts I through VI of this atlas contain maps of this type. Other maps are designed to show the geographic location of towns and cities and rivers and lakes and mountain ranges and so on. These are called "reference maps" and they make up many of the maps in Part VII. All of these maps, whether thematic or reference, are models of the real world that selectively emphasize the features that we want to show on the map.

In order to read maps effectively—in other words, in order to understand the models of the world presented in the following pages—it is important for you to know certain things about maps: how they are made using what are called *projections;* how the level of mathematical proportion of the map or what geographers call *scale* affects what you see; and how geographers use *generalization* techniques such as simplification and symbols where it would be impossible to draw a small version of the real world feature. In this brief introduction, then, we'll explain to you three of the most important elements of map interpretation: projection, scale, and generalization.

MAP PROJECTIONS

Perhaps the most basic problem in *cartography,* or the art and science of map-making, is the fact that the subject of maps—the earth's surface—is what is called by mathematicians "a non-developable surface." Since the world is a sphere (or nearly so—it's actually slightly flattened at the poles and bulges a tiny bit at the equator), it is impossible to flatten out the world or any part of its curved surface without producing some kind of distortion. This "near sphere" is represented by a geographic grid or coordinate system of lines of latitude or *parallels* that run east and west and are used to measure distance north and south on the globe, and lines of longitude or *meridians* that run north and south and are used to measure distance east and west. All the lines of longitude are half circles of equal length and they all converge at the poles. These meridians are numbered from 0 degrees (Prime or

The Coordinate System

Greenwich Meridian) east and west to 180 degrees. The meridian of 0 degrees and the meridian of 180 degrees are halves of the same "great circle" or line representing a plane that bisects the globe into two equal hemispheres. All lines of longitude are halves of great circles. All the lines of latitude are complete circles that are parallel to one another and are spaced equidistant on the meridians. The circumference of these circles lessens as you move north or south from the equator. Parallels of latitude are numbered from 0 degrees at the equator north and south to 90 degrees at the North and South poles. The only line of latitude that is a great circle is the equator, which equally divides the world into a northern and southern hemisphere. In the real world, all these grid lines of latitude and longitude intersect at right angles. The problem for cartographers is to convert this spherical or curved grid into a geometrical shape that is "developable"; that is, it can be flattened (such as a cylinder or cone) or is already flat (a plane). The reason the results of the conversion process are called "projections" is that we imagine a world globe (or some part of it) that is made up of wires running north-south and east-west to represent the grid lines of latitude and longitude and other wires or even solid curved plates to represent the coastlines of continents or the continents themselves. We then imagine a light source at some location inside or outside the wire globe that can "project" or cast shadows of the wires representing grid lines onto a developable surface. Sometimes the basic geometric principles of projection may be modified by other mathematical principles to yield projections that are not truly geometric but have certain desirable features. We call these types of projections "arbitrary." The three most basic types of projections are named according to the type of developable surface: cylindrical, conic, or azimuthal (plane). Each type has certain characteristic features: they may be *equal area* projections in which the size of each area on the map is a direct proportional representation of that same area in the real world but shapes are distorted; they may be *conformal* projections in which area may be distorted but shapes are shown correctly; or they may be *compromise* projections in which both shape and area are distorted but the overall picture presented is fairly close to reality. It is important to remember that all maps distort the geographic grid and continental outlines in characteristic ways. The only representation of the world that does not distort either shape or area is a globe. You can see why we must use projections—can you imagine an atlas that you would have to carry back and forth across campus that would be made up entirely of globes?

CYLINDRICAL PROJECTIONS

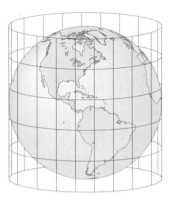

The Mercator Projection

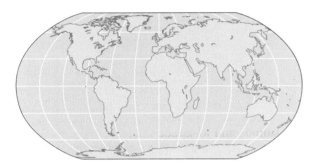

The Robinson Projection

Cylindrical projections are drawn as if the geographic grid were projected onto a cylinder. Cylindrical projections have the advantage of having all lines of latitude as true parallels or straight lines. This makes these projections quite useful for showing geographic relationships in which latitude or distance north-south is important (many physical features, such as climate, are influenced by latitude). Unfortunately, most cylindrical-type projections distort area significantly. One of the most famous is the Mercator projection shown above. This projection makes areas disproportionately large as you move toward the pole, making Greenland, which is actually about one-seventh the size of South America, appear to be as large as the southern continent. But the Mercator projection has the quality of conformality: landmasses on the map are true in shape and thus all coastlines on the map intersect lines of latitude and longitude at the proper angles. This makes the Mercator projection, named after its inventor, a sixteenth-century Dutch cartographer, ideal for its original purpose as a tool for navigation—but not a good projection for attempting to show some geographical feature in which areal relationship is important. Unfortunately, the Mercator projection has often been used for wall maps for schoolrooms and the consequence is that generations of American school children have been "tricked" into thinking that Greenland is actually larger than South America. Much better cylindrical-

type projections are those like the Robinson projection used in this atlas that is neither equal area nor conformal but a compromise that portrays the real world much as it actually looks, enough so that we can use it for areal comparisons.

CONIC PROJECTIONS

Conic Projection of Europe

Conic projections are those that are imagined as being projected onto a cone that is tangent to the globe along a standard parallel, or a series of cones tangent along several parallels or even intersecting the globe. Conic projections usually show latitude as curved lines and longitude as straight lines. They are good projections for areas with north-south extent, like the map of Europe to the right, and may be either conformal, equal area, or compromise, depending on how they are constructed. Many of the regional maps in the last map section of this atlas are conic projections.

AZIMUTHAL PROJECTIONS

Azimuthal projections are those that are imagined as being projected onto a plane or flat surface. They are named for one of their essential properties. An "azimuth" is a line of compass bearing and azimuthal projections have the property of yielding true compass directions from the center of the map. This makes azimuthal maps useful for navigation purposes, particularly air navigation. But, because they distort area and shape so greatly, they are seldom used for maps designed to show geographic relationships. When they are used as illustrative rather than navigation maps, it is often in the "polar case" projection shown below where the plane has been made tangent to the globe at the North Pole.

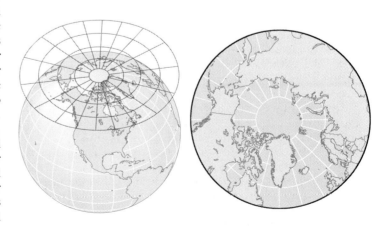

Azimuthal Projection of the North Polar Region

MAP SCALE

Since maps are models of the real world, it follows that they are not the same size as the real world or any portion of it. Every map, then, is subject to generalization, which is another way of saying that maps are drawn to certain scales. The term *scale* refers to the mathematical quality of *proportional representation,* and is expressed as a ratio between an area of the real world or the distance between places on the real world and the same area or distance on the map. We show map scale on maps in three different ways. Sometimes we simply use the proportion and write what is called a *natural scale* or representative fraction": for example, we might show on a map the mathematical proportion of 1:62,500. A map at this scale is one that is one sixty-two thousand five-hundredth the size of the same area in the real world. Other times we convert the proportion to a written description that approximates the relationship between distance on the map and distance in the real world. Since there are nearly 62,500 inches in a mile, we would refer to a map having a natural scale of 1:62,500 as having an "inch-mile" scale of "1 inch represents 1 mile." If we draw a line one inch long on this map, that line represents a distance of approximately one mile in the real world. Finally, we usually use a graphic or linear scale: a bar or line, often graduated into miles or kilometers, that shows graphically the proportional representation. A graphic scale for our 1:62,500

map might be about five inches long, divided into five equal units clearly labeled as "1 mile," "2 miles," and so on. Our examples below show all three kinds of scales.

The most important thing to keep in mind about scale, and the reason why knowing map scale is important to being able to read a map correctly, is the relationship between proportional representation and generalization. A map that fills a page but shows the whole world is much more highly generalized than a map that fills a page but shows a single city. On the world map, the city may appear as a dot. On the city map, streets and other features may be clearly seen. We call the first map, the world map, a *small-scale* map because the proportional representation is a small number. A page-size map showing the whole world may be drawn at a scale of 1:150,000,000. That is a very small number indeed–hence the term *small-scale* map even though the area shown is large. Conversely, the second map, a city map, may be drawn at a scale of 1:250,000. That is still a very small number but it is a great deal larger than 1:150,000,000! And so we'd refer to the city map as a *large scale* map, even though it shows only a small area. On our world map, geographical features are generalized greatly and many features can't even be shown at all. On the city map, much less generalization occurs–we can show specific features that we couldn't on the world map–but generalization still takes place. The general rule is that the smaller the map scale, the greater the degree of generalization;

Map 1 Small Scale Map of the United States

Map 2 Map of the Northeast

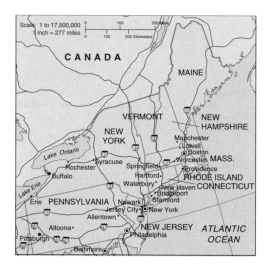

Map 3 Map of Southeastern New England

Map 4 Large Scale Map of Boston, MA

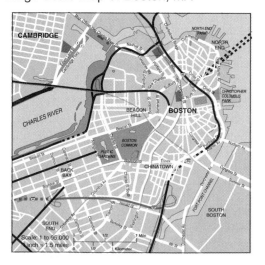

the larger the map scale, the less the degree of generalization. The only map that would not generalize would be a map at a scale of 1:1. and that map wouldn't be very handy to use. Examine the relationship between scale and generalization in the four maps on the previous page.

GENERALIZATION ON MAPS

A review of the four maps on the previous page should give you some indication of how cartographers generalize on maps. One thing that you should have noticed is that the first map, that of the United States, is much simpler than the other three and that the level of *simplification* decreases with each map. When a cartographer simplifies map data, information that is not important for the purposes of the map is just left off. For example, on the first map the objective may have been to show cities over 1 million in population. To do that clearly and effectively, it is not necessary to show and label rivers and lakes. The map has been simplified by leaving those items out. The final map, on the other hand, is more complex and shows and labels geographic features that are important to the character of the city of Boston; therefore, the Charles River is clearly indicated on the map.

Another type of generalization is *classification*. Map 1 on the previous page shows cities over 1 million in population. Map 2 shows cities of several different sizes and a different symbol is used for each size classification or category. Many of the thematic maps used in this atlas rely on classification to show data. A thematic map showing population growth rates (see Map 24 on page 37) will use different colors to show growth rates in different classification levels or what are sometimes called *class intervals*. Thus, there will be one color applied to all countries with population growth rates between 1.0 percent and 1.4 percent, another color applied to all countries with population growth rates between 1.5 percent and 2.1 percent, and so on. Classification is necessary because it is impossible to find enough symbols or colors to represent precise values. Classification may also be used for qualitative data, such as the national or regional origin of migrating populations. Cartographers show both quantitative and qualitative classification levels or class intervals in important sections of maps called *legends*. These legends, as in the samples shown below, make it possible for the reader of the map to interpret the patterns shown.

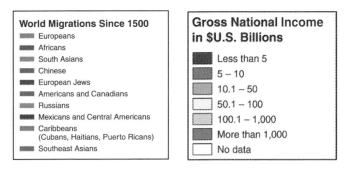

Map Legends from Maps 23 and 37

A third technique of generalization is *symbolization* and we've already noted several different kinds of symbols: those used to represent cities on the preceding maps, or the colors used to indicate population growth levels on Map 24. One general category of map symbols is quantitative in nature

and this category can further be divided into a number of different types. For example, the symbols showing city size on Maps 1 and 2 on the preceding page can be categorized as *ordinal* in that they show relative differences in quantities (the size of cities). A cartographer might also use lines of different widths to express the quantities of movement of people or goods between two or more points as on Map 23 (see page 35).

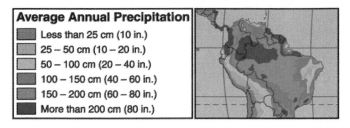

Interval Symbols

The color symbols used to show rates of population growth can be categorized as *interval* in that they express certain levels of a mathematical quantity (the percentage of population growth). Interval symbols are often used to show physical geographic characteristics such as inches of precipitation, degrees of temperature, or elevation above sea level. The sample above, for example, shows precipitation (from Map 3a, page 6).

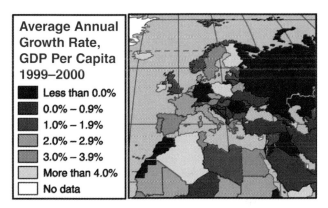

Ratio Symbols

Still another type of mathematical symbolization is the *ratio* in which sets of mathematical quantities are compared: the number of persons per square mile (population density) or the growth in gross national product per capita (per person). The map above shows GDP change per capita (from Map 39, page 53).

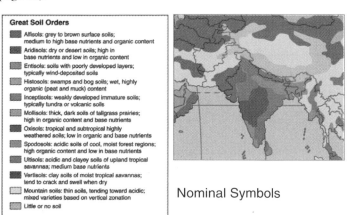

Nominal Symbols

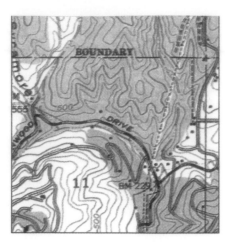

Interpolation

Finally, there are a vast number of cartographic symbols that are not mathematical but show differences in the kind of information being portrayed. These symbols are called *nominal* and they range from the simplest differences such as land and water to more complex differences such as those between different types of vegetation. Shapes or patterns or colors or iconographic drawings may all be used as nominal symbols on maps. The sample map at the bottom of page xiii uses color to show the distribution of soil types.

The final technique of generalization is what cartographers refer to as *interpolation*. Here, the maker of a map may actually show more information on the map than is actually supplied by the original data. In understanding the process of interpolation is it necessary for you to visualize the quantitative data shown on maps as being three dimensional: *x* values provide geographic location along a north-south axis of the map; *y* values provide geographic location along the east-west axis of the map; and *z* values are those values of whatever data (for example, temperature) are being shown on the map at specific points. We all can imagine a real three-dimensional surface in which the *x* and *y* values are directions and the *z* values are the heights of mountains and the depths of valleys. On a topographic map showing a real three-dimensional surface, contour lines are used to connect points of equal elevation above sea level. These contour lines are not measured directly; they are estimated by interpolation on the basis of the elevation points that are provided.

It is harder to imagine the statistical surface of a temperature map in which the *x* and *y* values are directions and the *z* values represent degrees of temperature at precise points. But that is just what cartographers do. And to obtain the values between two or more specific points where *z* values exist, they interpolate based on a class interval they have decided is appropriate and use *isolines* (which are statistical equivalents of a contour line) to show increases or decreases in value. The diagram below shows an example of an interpolation process. Occasionally interpolation is referred to as *induction*. By whatever name, it is one of the most difficult parts of the cartographic process.

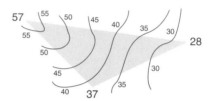

Degrees of Temperature (Celsius)
Interval = 5 degrees

And you thought all you had to do to read an atlas was look at the maps! You've now learned that it is a bit more involved than that. As you read and study this atlas, keep in mind the principles of projection and scale and generalization (including simplification, classification, symbolization, and interpolation) and you'll do just fine. Good luck and enjoy your study of the world of maps as well as maps of the world!

Part I

Global Physical Patterns

Map 1 World Political Divisions

The international system includes the political units called "states" or countries as the most important component. The boundaries of countries are the primary source of political division in the world and for most people nationalism is the strongest source of political identity. State boundaries are an important indicator of cultural, linguistic, economic, and other geographic divisions as well, and the states themselves normally serve as the base level for which most global statistics are available. The subfield of geography known as "political geography" has as its primary concern the geographic or spatial character of this international system and its components.

Scale: 1 to 111,922,000

Note: All world maps are Robinson projection.

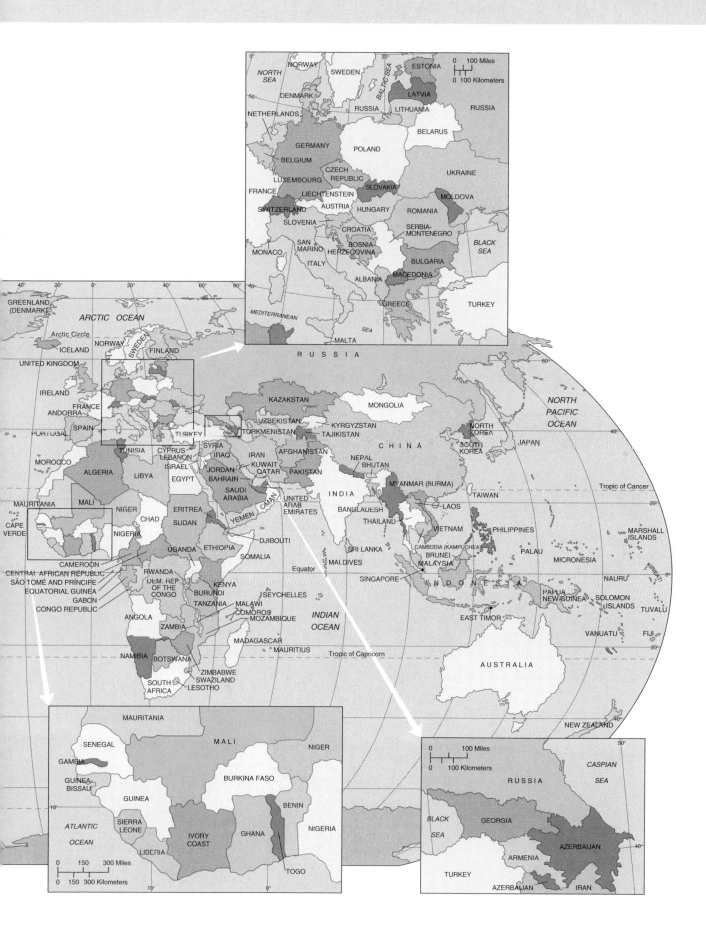

Map 2 World Physical Features

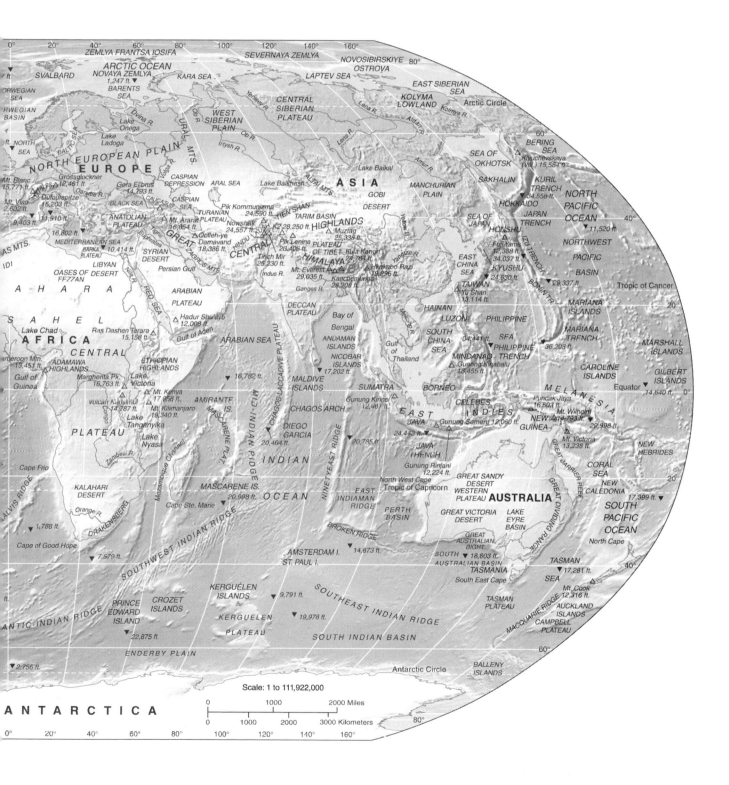

Scale: 1 to 111,922,000

Map 3a Average Annual Precipitation

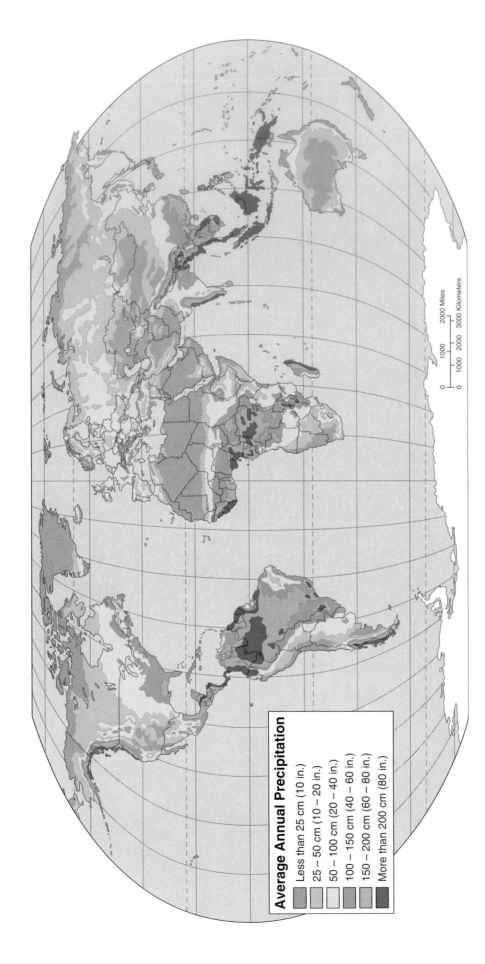

Average Annual Precipitation

- Less than 25 cm (10 in.)
- 25 – 50 cm (10 – 20 in.)
- 50 – 100 cm (20 – 40 in.)
- 100 – 150 cm (40 – 60 in.)
- 150 – 200 cm (60 – 80 in.)
- More than 200 cm (80 in.)

0   1000   2000 Miles
0 1000 2000 3000 Kilometers

The two most important physical geographic variables are precipitation and temperature, the essential elements of weather and climate. Precipitation is a conditioner of both soil type and vegetation. More than any other single environmental element, it influences where people do or do not live. Water is the most precious resource available to humans, and water availability is largely a function of precipitation. Water availability is also a function of several precipitation variables that do not appear on this map: the seasonal distribution of precipitation (is precipitation or drought concentrated in a particular season?), the ratio between precipitation and temperature (how much of the water that comes to the earth in the form of precipitation is lost through mechanisms such as evapo-ration and transpiration that are a function of temperature?), and the annual variability of precipitation (how much do annual precipitation totals for a place or region tend to vary from the "normal" or average precipitation?). In order to obtain a complete understanding of precipitation, these variables should be examined along with the more general data presented on this map. The study of precipitation and other climatic elements is the concern of the branch of physical geography called "climatology."

-6-

Map 3b Seasonal Average Precipitation, November Through April

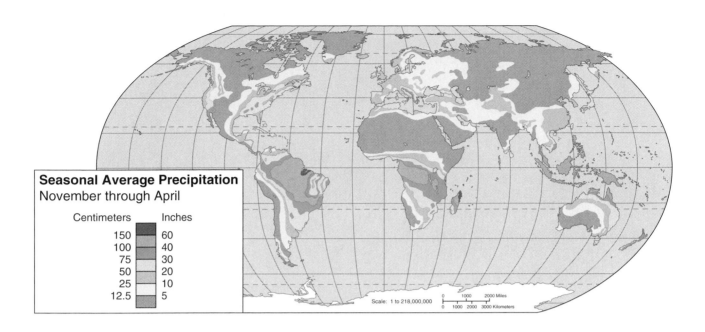

Seasonal Average Precipitation
November through April

Centimeters		Inches
150		60
100		40
75		30
50		20
25		10
12.5		5

Scale: 1 to 218,000,000

0 1000 2000 Miles
0 1000 2000 3000 Kilometers

Map 3c Seasonal Average Precipitation, May Through October

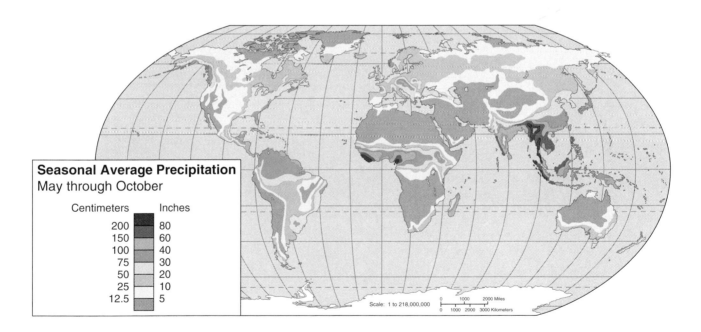

Seasonal Average Precipitation
May through October

Centimeters		Inches
200		80
150		60
100		40
75		30
50		20
25		10
12.5		5

Scale: 1 to 218,000,000

0 1000 2000 Miles
0 1000 2000 3000 Kilometers

Seasonal average precipitation is nearly as important as annual precipitation totals in determining the habitability of an area. Critical factors are such things as whether precipitation coincides with the growing season and thus facilitates agriculture or during the winter when it is less effective in aiding plant growth, and whether precipitation occurs during summer with its higher water loss through evaporation and transpiration or during the winter when more of it can go into storage. Several of the world's great climate zones have pronounced seasonal precipitation rhythms. The tropical and subtropical savanna grasslands have a long winter dry season and abundant precipitation in the summer. The Mediterranean climate is the only major climate with a marked dry season during the summer, making agriculture possible only through irrigation or other adjustments to cope with drought during the period of plant growth. And the great monsoon climates of south and southeast Asia have their winter dry season and summer rain that have conditioned the development of Asian agriculture and the rhythms of Asian life.

Map 3d Variation in Average Annual Precipitation

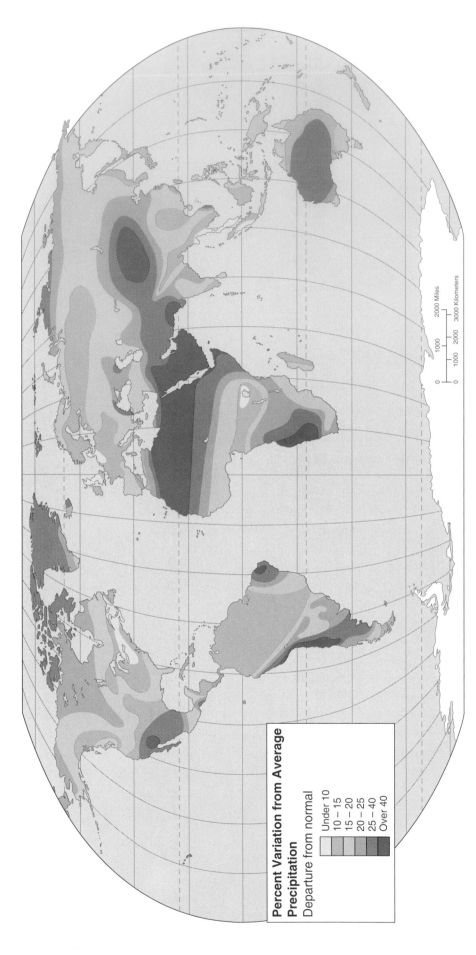

Percent Variation from Average Precipitation

Departure from normal

- Under 10
- 10 – 15
- 15 – 20
- 20 – 25
- 25 – 40
- Over 40

0   1000   2000 Miles

0   1000   2000   3000 Kilometers

While annual precipitation totals and seasonal distribution of precipitation are important variables, the variability of precipitation from one year to the next may be even more critical. You will note from the map that there is a general spatial correlation between the world's drylands and the amount of annual variation in precipitation. Generally, the drier the climate, the more likely it is that there will be considerable differences in rainfall and/or snowfall from one year to the next. We might determine that the average precipitation of the mid-Sahara is 2 inches per year. What this really means is that a particular location in the Sahara during one year might receive .5″, during the next year 3.5″, and during a third year 2″. If you add these together and divide by the number of years, the "average" precipitation is 2″ per year. The significance of this is that much of the world's crucial agricultural output of cereals (grains) comes from dryland climates (the Great Plains of the United States, the Pampas of Argentina, the steppes of Ukraine and Russia, for example), and variations in annual rainfall totals can have significant impacts on levels of grain production and, therefore, important consequences for both economic and political processes.

Map 4a Temperature Regions and Ocean Currents

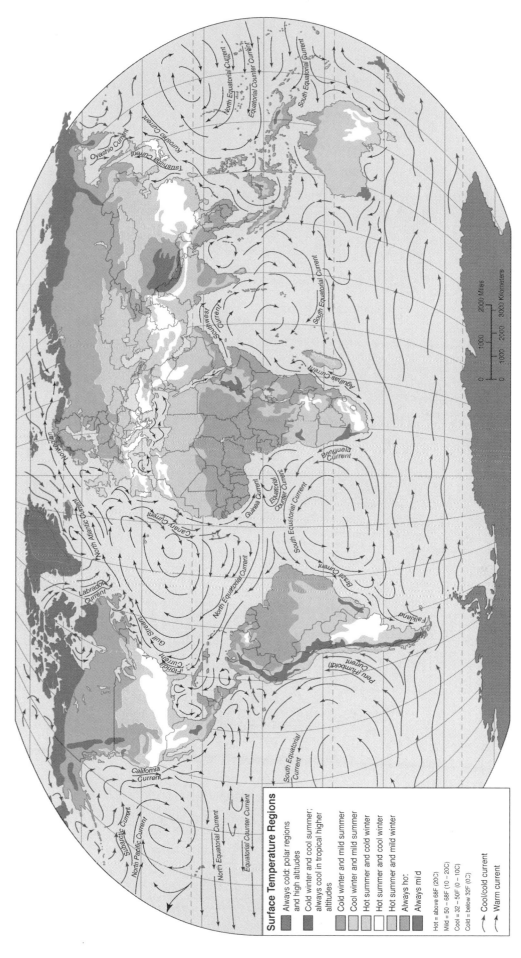

Surface Temperature Regions

- Always cold: polar regions and high altitudes
- Cold winter and cool summer; always cool in tropical higher altitudes
- Cold winter and mild summer
- Cool winter and mild summer
- Hot summer and cold winter
- Hot summer and cool winter
- Hot summer and mild winter
- Always hot
- Always mild

Hot = above 68F (20C)
Mild = 50 – 68F (10 – 20C)
Cool = 32 – 50F (0 – 10C)
Cold = below 32F (0C)

→ Cool/cold current
→ Warm current

Along with precipitation, temperature is one of the two most important environmental variables, defining the climate conditions so essential for the distribution of such human activities as agriculture and the distribution of the human population. The seasonal rhythm of temperature, including such measures as the average annual temperature range (difference between the average temperature of the warmest month and that of the coldest month), is an additional variable not shown on the map but, like the sea- sonality of precipitation, should be a part of any comprehensive study of climate. The ocean currents illustrated exert a significant influence over the climate of adjacent regions and are the most important mechanism for redistributing surplus heat from the equatorial region into middle and high latitudes. Physical geographers known as "climatologists" study the phenomenon of temperature and related climatic characteristics.

Map 4b Average January Temperature

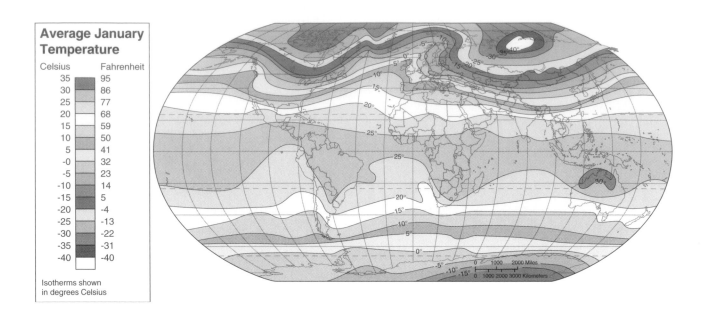

Average January Temperature

Celsius	Fahrenheit
35	95
30	86
25	77
20	68
15	59
10	50
5	41
-0	32
-5	23
-10	14
-15	5
-20	-4
-25	-13
-30	-22
-35	-31
-40	-40

Isotherms shown
in degrees Celsius

Map 4c Average July Temperature

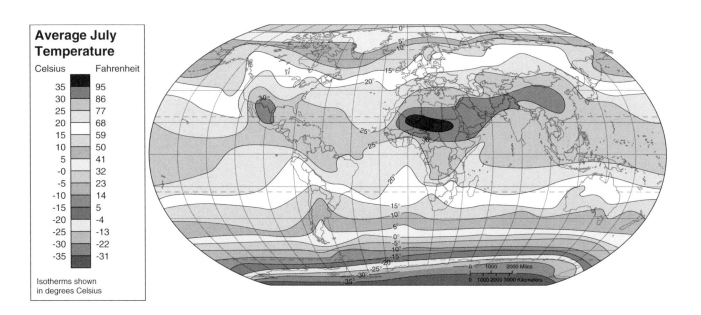

Average July Temperature

Celsius	Fahrenheit
35	95
30	86
25	77
20	68
15	59
10	50
5	41
-0	32
-5	23
-10	14
-15	5
-20	-4
-25	-13
-30	-22
-35	-31

Isotherms shown
in degrees Celsius

Where moisture availability tends to mark the seasons in the tropics and subtropics, in the mid-latitudes, seasons are marked by temperature. Temperature is determined by latitudinal transition, by altitude or elevation above sea level, and by location of a place relative to the world's landmasses and oceans. The most important of these controls is latitude, and temperatures generally become lower with increasing latitude. Proximity to water, however, tends to moderate temperature extremes, and "maritime" climates influenced by the oceans will be warmer in the winter and cooler in the summer than continental climates in the same general latitude. Maritime climates will also show smaller temperature ranges, the differ-

ence between January and July temperatures, while climates of the continental interiors, far from the moderating influences of the oceans, will tend to have greater temperature ranges. In the Northern Hemisphere, where there are both large landmasses and oceans, the range is great. But in the Southern Hemisphere, dominated by water and, hence, by the more moderate maritime air masses, the temperature range is comparatively small. Significant temperature departures from the "normal" produced by latitude may also be the result of elevation. With exceptions, lower temperatures produced by topography are difficult to see on maps of this scale.

Map 5a Atmospheric Pressure and Predominant Surface Winds, January

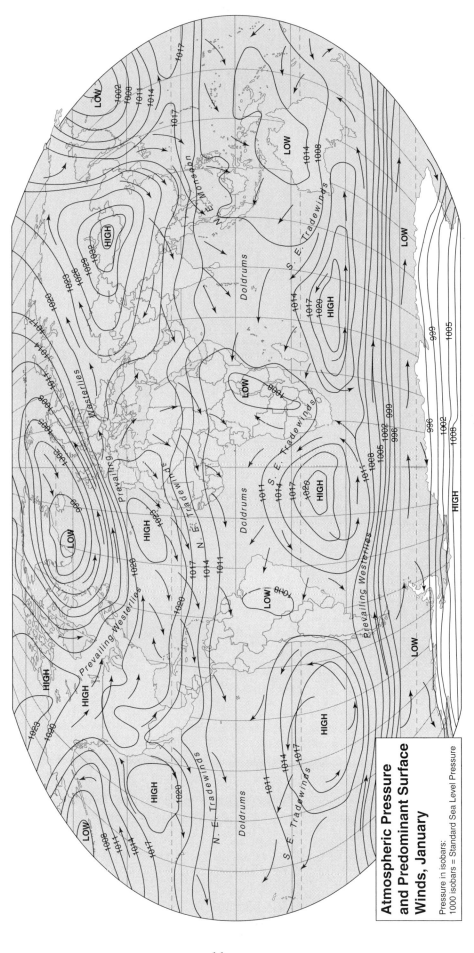

**Atmospheric Pressure
and Predominant Surface
Winds, January**

Pressure in isobars:
1000 isobars = Standard Sea Level Pressure

Map 5b Atmospheric Pressure and Predominant Surface Winds, July

Atmospheric Pressure and Predominant Surface Winds, July

Pressure in isobars:
1000 isobars = Standard Sea Level Pressure

Atmospheric pressure, or the density of air, is a function largely of air temperature: the colder the air, the denser and heavier it is, hence the higher its pressure; the warmer the air, the lighter and less stable it is, hence the lower its pressure. Global pressure systems are the alternating low and high pressure systems that, from the equator north and south, include: the equatorial low (sometimes called the intertropical convergence) centered on the equator for much of the year; the subtropical highs with their centers near the 30th degrees of north and south latitude; the subpolar lows or polar front cen-tered near the 60th parallel of north and south latitude; and the polar highs near the north and south poles. Air flows from high pressure to low pressure regions, and this air flow constitutes the earth's major surface winds such as the tropical tradewinds and the prevailing westerlies. This flow of air is one of the chief mechanisms by which surplus heat energy from the equatorial region is redistributed to higher latitudes. It is also the primary conditioner of the world's major precipitation belts, with rainfall and snowfall associated primarily with lower atmospheric pressure conditions.

-12-

Map 6 Climate Regions

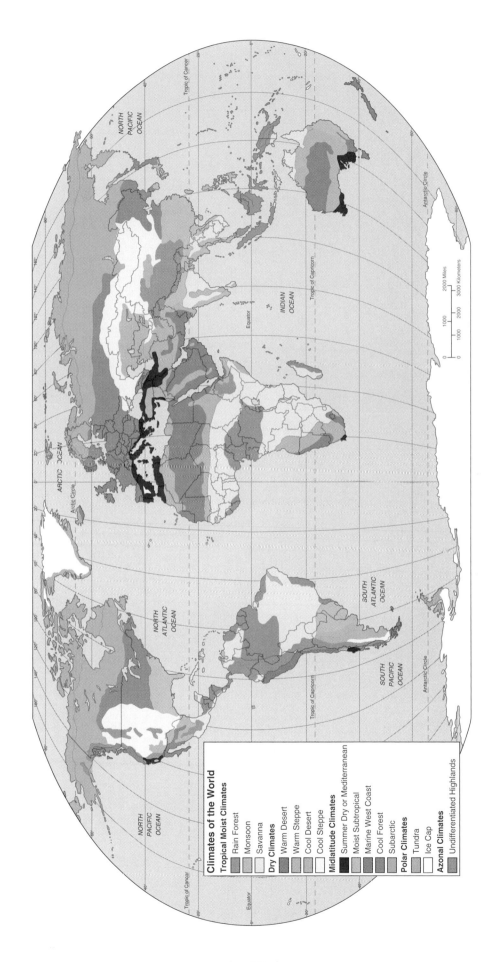

Climates of the World

Tropical Moist Climates
- Rain Forest
- Monsoon
- Savanna

Dry Climates
- Warm Desert
- Warm Steppe
- Cool Desert
- Cool Steppe

Midlatitude Climates
- Summer Dry or Mediterranean
- Moist Subtropical
- Marine West Coast
- Cool Forest
- Subarctic

Polar Climates
- Tundra
- Ice Cap

Azonal Climates
- Undifferentiated Highlands

Of the world's many patterns of physical geography, climate or the long-term average of weather conditions such as temperature and precipitation is the most important. It is climate that conditions the distribution of natural vegetation and the types of soils that will exist in an area. Climate also influences the availability of our most crucial resource: water. From an economic standpoint, the world's most important activity is agriculture; no other element of physical geography is more important for agriculture than climate. Ultimately, it is agricultural production that determines where the bulk of human beings live, and therefore, climate is a basic determinant of the distribution of

human populations as well. The study of climates or "climatology" is one of the most important branches of physical geography.

The climate classification system shown on this map is based on that developed by Wladimir Köppen. To establish his climate regions, Köppen used the climatic parameters of *precipitation*, *temperature*, and *evapotranspiration* as they impacted certain kinds of major vegetative associations. Hence the names for many of the climate regions are also the names of vegetative regions.

Map **7** Vegetation Types

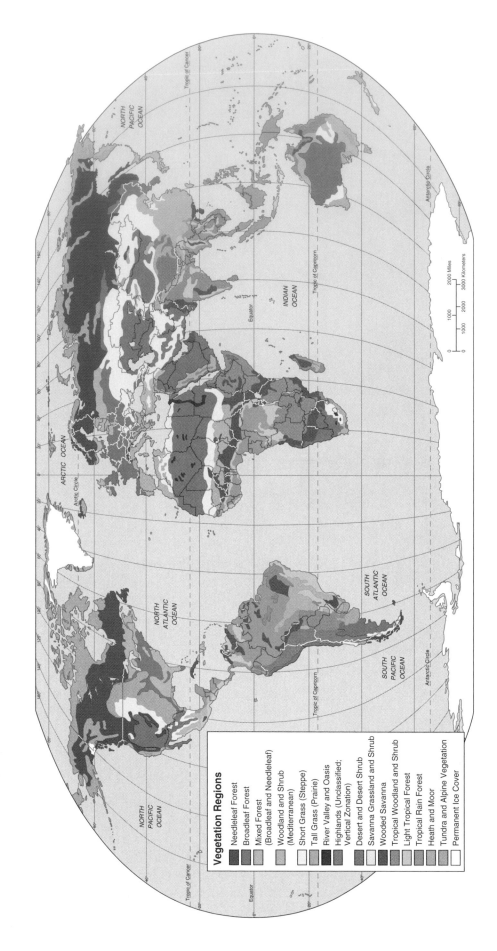

Vegetation Regions

- Needleleaf Forest
- Broadleaf Forest
- Mixed Forest (Broadleaf and Needleaf)
- Woodland and Shrub (Mediterranean)
- Short Grass (Steppe)
- Tall Grass (Prairie)
- River Valley and Oasis
- Highlands (Unclassified; Vertical Zonation)
- Desert and Desert Shrub
- Savanna Grassland and Shrub
- Wooded Savanna
- Tropical Woodland and Shrub
- Light Tropical Forest
- Tropical Rain Forest
- Heath and Moor
- Tundra and Alpine Vegetation
- Permanent Ice Cover

Vegetation is the most visible consequence of the distribution of temperature and precipitation. The global pattern of vegetative types or "habitat classes" and the global pattern of climate are closely related and make up one of the great global spatial correlations. But not all vegetation types are the consequence of temperature and precipitation or other climatic variables. Many types of vegetation in many areas of the world are the consequence of human activities, particularly the grazing of domesticated livestock, burning, and forest clearance. This map shows the pattern of natural or "potential" vegetation, or vegetation as it might be expected to exist without significant human influences, rather than the actual vegetation that results from a combination of environmental and human factors. Physical geographers who are interested in the distribution and geographic patterns of vegetation are "biogeographers."

-14-

Map 8 Soil Orders

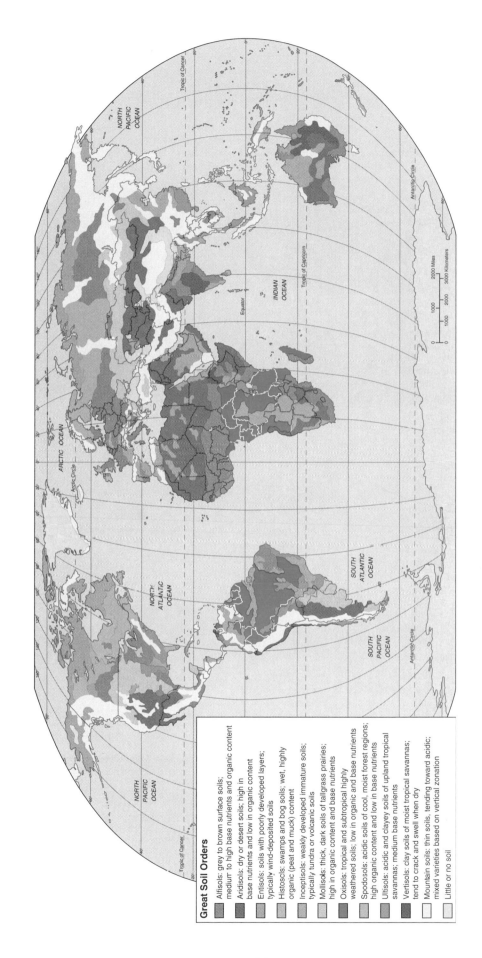

Great Soil Orders

Alfisols: grey to brown surface soils; medium to high base nutrients and organic content

Aridisols: dry or desert soils; high in base nutrients and low in organic content

Entisols: soils with poorly developed layers; typically wind-deposited soils

Histosols: swamps and bog soils; wet, highly organic (peat and muck) content

Inceptisols: weakly developed immature soils; typically tundra or volcanic soils

Mollisols: thick, dark soils of tallgrass prairies; high in organic content and base nutrients

Oxisols: tropical and subtropical highly weathered soils; low in organic and base nutrients

Spodosols: acidic soils of cool, moist forest regions; high organic content and low in base nutrients

Ultisols: acidic and clayey soils of upland tropical savannas; medium base nutrients

Vertisols: clay soils of moist tropical savannas; tend to crack and swell when dry

Mountain soils: thin soils, tending toward acidic; mixed varieties based on vertical zonation

Little or no soil

The characteristics of soil are one of the three primary physical geographic factors, along with climate and vegetation, that determine the habitability of regions for humans. In particular, soils influence the kinds of agricultural uses to which land is put. Since soils support the plants that are the primary producers of all food in the terrestrial food chain, their characteristics are crucial to the health and stability of ecosystems. Two types of soil are shown on this map: zonal soils, the characteristics of which are based on climatic patterns; and azonal soils, such as alluvial (water-deposited) or aeolian (wind-deposited) soils, the characteristics of which are derived from forces other than climate. However, many of the azonal soils, particularly those dependent upon drainage conditions, appear over areas too small to be readily shown on a map of this scale. Thus, almost none of the world's swamp or bog soils appear on this map. People who study the geographic characteristics of soils are most often "soil scientists," a discipline closely related to that branch of physical geography called "geomorphology."

-15-

Map 9 Ecological Regions

Ecological regions are distinctive areas within which unique sets of organisms and environments are found. We call the study of the relationships between organisms and their environmental surroundings "ecology." Within each of the ecological regions portrayed on the map, a particular combination of vegetation, wildlife, soil, water, climate, and terrain defines that region's habitability, or ability to support life, including human life. Like climate and landforms, ecological relationships are crucial to the existence of agriculture, the most basic of our economic activities, and important for many other kinds of economic activity as well. Biogeographers are especially concerned with the concept of ecological regions since such regions so clearly depend upon the geographic distribution of plants and animals in their environmental settings.

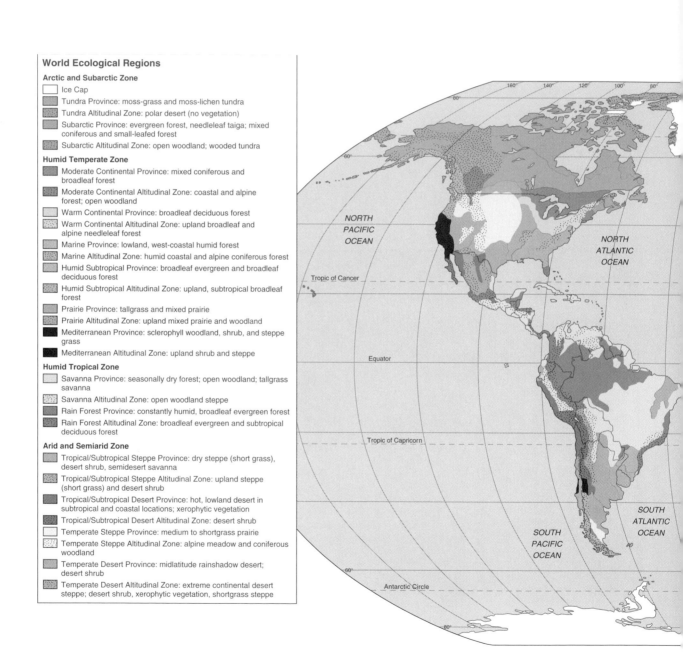

World Ecological Regions

Arctic and Subarctic Zone

- Ice Cap
- Tundra Province: moss-grass and moss-lichen tundra
- Tundra Altitudinal Zone: polar desert (no vegetation)
- Subarctic Province: evergreen forest, needleleaf taiga; mixed coniferous and small-leafed forest
- Subarctic Altitudinal Zone: open woodland; wooded tundra

Humid Temperate Zone

- Moderate Continental Province: mixed coniferous and broadleaf forest
- Moderate Continental Altitudinal Zone: coastal and alpine forest; open woodland
- Warm Continental Province: broadleaf deciduous forest
- Warm Continental Altitudinal Zone: upland broadleaf and alpine needleleaf forest
- Marine Province: lowland, west-coastal humid forest
- Marine Altitudinal Zone: humid coastal and alpine coniferous forest
- Humid Subtropical Province: broadleaf evergreen and broadleaf deciduous forest
- Humid Subtropical Altitudinal Zone: upland, subtropical broadleaf forest
- Prairie Province: tallgrass and mixed prairie
- Prairie Altitudinal Zone: upland mixed prairie and woodland
- Mediterranean Province: sclerophyll woodland, shrub, and steppe grass
- Mediterranean Altitudinal Zone: upland shrub and steppe

Humid Tropical Zone

- Savanna Province: seasonally dry forest; open woodland; tallgrass savanna
- Savanna Altitudinal Zone: open woodland steppe
- Rain Forest Province: constantly humid, broadleaf evergreen forest
- Rain Forest Altitudinal Zone: broadleaf evergreen and subtropical deciduous forest

Arid and Semiarid Zone

- Tropical/Subtropical Steppe Province: dry steppe (short grass), desert shrub, semidesert savanna
- Tropical/Subtropical Steppe Altitudinal Zone: upland steppe (short grass) and desert shrub
- Tropical/Subtropical Desert Province: hot, lowland desert in subtropical and coastal locations; xerophytic vegetation
- Tropical/Subtropical Desert Altitudinal Zone: desert shrub
- Temperate Steppe Province: medium to shortgrass prairie
- Temperate Steppe Altitudinal Zone: alpine meadow and coniferous woodland
- Temperate Desert Province: midlatitude rainshadow desert; desert shrub
- Temperate Desert Altitudinal Zone: extreme continental desert steppe; desert shrub, xerophytic vegetation, shortgrass steppe

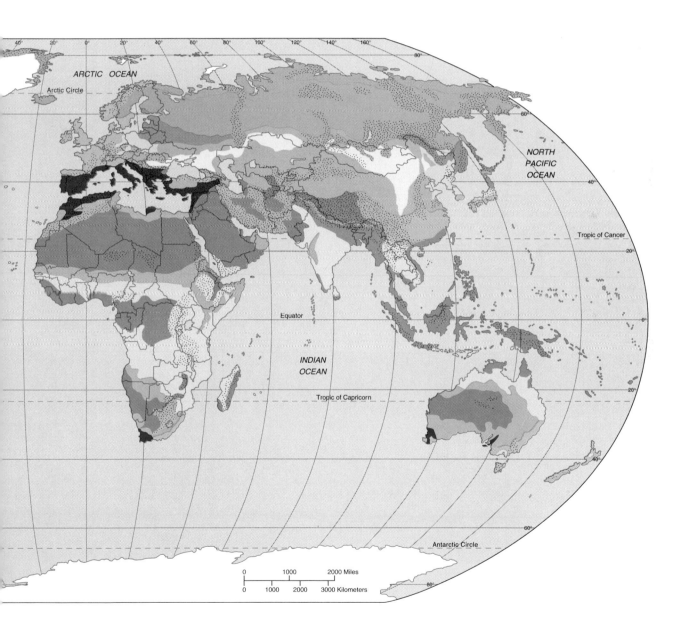

ARCTIC OCEAN

Arctic Circle

NORTH
PACIFIC
OCEAN

Tropic of Cancer

Equator

INDIAN
OCEAN

Tropic of Capricorn

Antarctic Circle

0		1000		2000 Miles
0	1000	2000	3000 Kilometers	

Map 10 Plate Tectonics

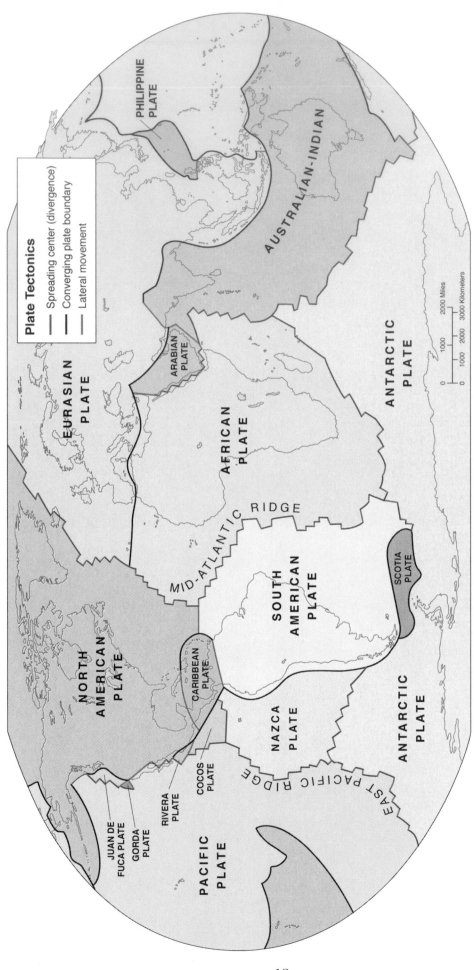

An understanding of the forces that shape the primary features of the earth's surface—the continents and ocean basins—requires a view of the earth's crust as fragments or "lithospheric plates" that shift position relative to one another. There are three dominant types of plate movement: *convergence*, in which plates move together, compressing former ocean floor or continental rocks together to produce mountain ranges, or producing mountain ranges through volcanic activity if one plate slides beneath another; *divergence*, in which the plates move away from one another, producing rifts in the earth's crust through which molten material wells up to produce new sea floors and mid-oceanic ridges; and *lateral shift*, in which plates move horizontally relative to one another, causing significant earthquake activity. All the major forms of these types of shifts are extremely slow and take place over long periods of geologic time. The movement of crustal plates, or what is known as "plate tectonics," is responsible for the present shape and location of the continents but is also the driving force behind some much shorter-term earth phenomena like earthquakes and volcanoes. A comparison of the map of plates with maps of hazards and terrain will reveal some interesting relationships.

Map 11 Topography

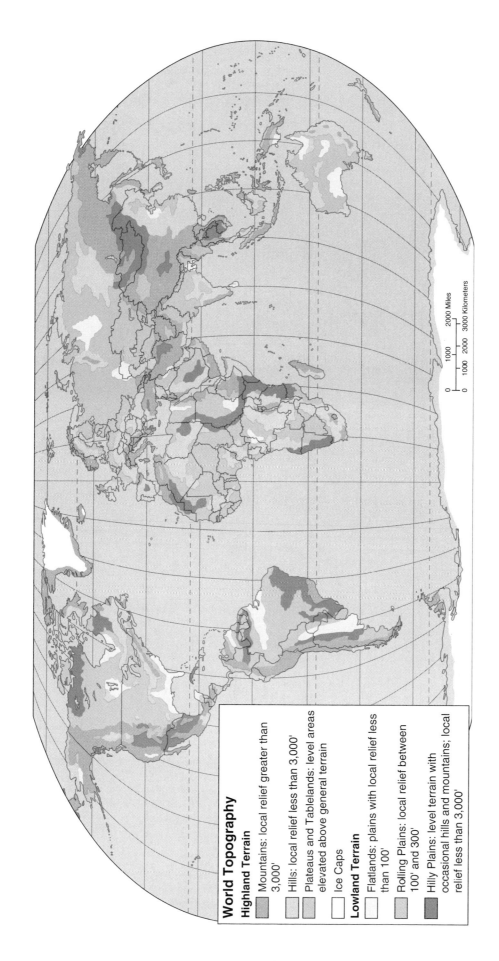

World Topography

Highland Terrain

- Mountains: local relief greater than 3,000'
- Hills: local relief less than 3,000'
- Plateaus and Tablelands: level areas elevated above general terrain
- Ice Caps

Lowland Terrain

- Flatlands: plains with local relief less than 100'
- Rolling Plains: local relief between 100' and 300'
- Hilly Plains: level terrain with occasional hills and mountains; local relief less than 3,000'

0   1000   2000 Miles
0 1000 2000 3000 Kilometers

-19-

Topography or terrain, also called "landforms," is second only to climate as a conditioner of human activity, particularly agriculture but also the location of cities and industry. A comparison of this map of mountains, valleys, plains, plateaus, and other features of the earth's surface with a map of land use (Map 15) shows that most of the world's productive agricultural zones are located in lowland and relatively level regions. Where large regions of agricultural productivity are found, we also tend to find urban concentrations and, with cities, we find industry. There is also a good spatial correlation between the map of topography and the map showing the distribution and density of the human population (Map 14). Normally the world's major landforms are the

result of extremely gradual primary geologic activity such as the long-term movement of crustal plates. This activity occurs over hundreds of millions of years. Also important is the more rapid (but still slow by human standards) geomorphological or erosional activity of water, wind, glacial ice, and waves, tides, and currents. Some landforms may be produced by abrupt or "cataclysmic" events such as a major volcanic eruption or a meteor strike, but such events are relatively rare and their effects are usually too minor to show up on a map of this scale. The study of the processes that shape topography is known as "geomorphology" and is an important branch of physical geography.

Map 12a Resources: Mineral Fuels

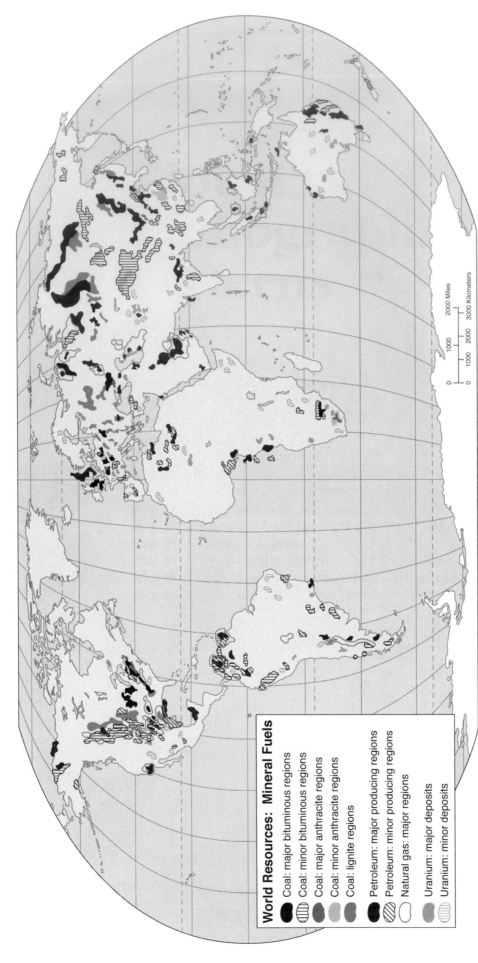

World Resources: Mineral Fuels

- Coal: major bituminous regions
- Coal: minor bituminous regions
- Coal: major anthracite regions
- Coal: minor anthracite regions
- Coal: lignite regions
- Petroleum: major producing regions
- Petroleum: minor producing regions
- Natural gas: major regions
- Uranium: major deposits
- Uranium: minor deposits

0   1000   2000 Miles
0   1000   2000   3000 Kilometers

The extraction and transportation of mineral fuels rank with agriculture and forestry as "primary" human activities that impact on the environment at a global scale. Nearly all of the most highly publicized environmental disasters of recent decades—the Prince William Sound oil spill or the Chernobyl nuclear accident, for example—have involved mineral fuels that were being stored, transported, or used. And the continuing extraction of mineral fuels like oil, natural gas, coal, and uranium produces high levels of atmospheric, soil, and water pollution. The location of mineral fuels tells us a great deal about where environmental degradation is likely to be occurring or to occur in the future. One need only look at the levels of atmospheric pollution and vegetative disruption in central and eastern Europe to recognize the damaging consequences of heavy reliance on coal as a domestic and industrial fuel. The location of mineral fuels also tells us something about existing or potential levels of economic development with those countries possessing abundant reserves of mineral fuels having more of a chance to maintain or attain higher levels of prosperity.

Map 12b Resources: Critical Metals

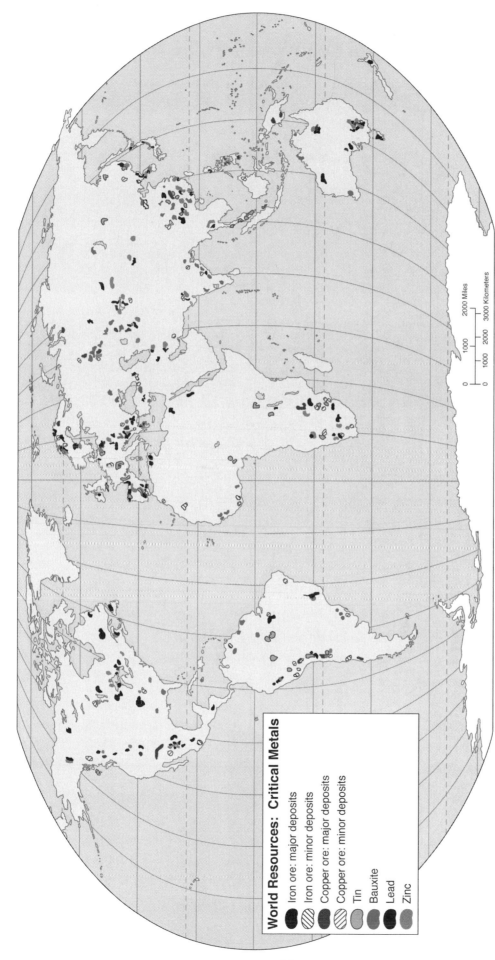

World Resources: Critical Metals

- Iron ore: major deposits
- Iron ore: minor deposits
- Copper ore: major deposits
- Copper ore: minor deposits
- Tin
- Bauxite
- Lead
- Zinc

0   1000   2000 Miles

0   1000   2000   3000 Kilometers

The location of deposits of critical metals such as iron, copper, tin, and others is an important determinant of the location of mining activities, like mineral fuel extraction, a "primary" economic activity. Also like mineral fuel extraction, mining for critical metallic ores makes significant environmental impact, particularly on vegetation, soils, and water resources. Some of the world's most dramatic examples of human modification of environments are located in areas of metallic ore extraction: the open pit copper mining areas of Arizona and Utah, for example. Environmental impact aside, those countries with significant critical metal deposits tend to stand a better chance of reaching higher levels of economic development, as long as they can extract and market the ores

themselves rather than having the extraction process controlled by outside concerns. The average Bolivian, for example, does not benefit greatly from the fact that his/her country is an important producer of tin and other metals. Bolivia is a "colonial dependency" country and the wealth generated by metallic ore production there tends to flow out of the country to Europe and North America. On the other hand, another South American country, Brazil, is paying for much of its own current economic development by utilizing its reserves of iron and other metals and more of the wealth from the extraction of those resources stays within the country.

Map 13 Natural Hazards

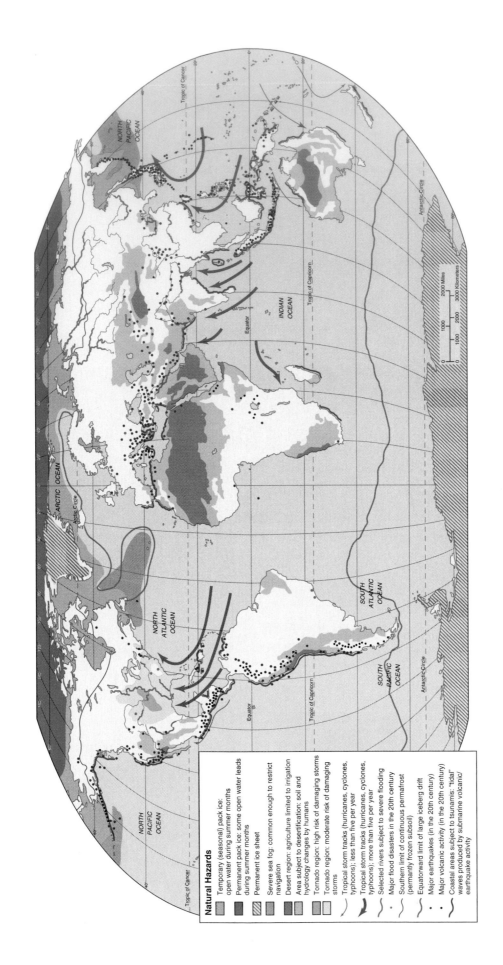

Natural Hazards

- Temporary (seasonal) pack ice: open water during summer months
- Permanent pack ice: some open water leads during summer months
- Permanent ice sheet
- Severe sea fog: common enough to restrict navigation
- Desert region: agriculture limited to irrigation
- Area subject to desertification: soil and hydrology changes by humans
- Tornado region: high risk of damaging storms
- Tornado region: moderate risk of damaging storms
- Tropical storm tracks (hurricanes, cyclones, typhoons): less than five per year
- Tropical storm tracks (hurricanes, cyclones, typhoons): more than five per year
- Selected rivers subject to severe flooding
- Major flood disasters in the 20th century
- Southern limit of continuous permafrost (permanently frozen subsoil)
- Equatorward limit of large iceberg drift
- Major earthquakes (in the 20th century)
- Major volcanic activity (in the 20th century)
- Coastal areas subject to tsunamis: "tidal" waves produced by submarine volcanic/ earthquake activity

Unlike other elements of physical geography, most natural hazards are unpredictable. However, there are certain regions where the probability of the occurrence of a particular natural hazard is high. This map shows regions affected by major natural hazards at rates that are higher than the global norm. The presence of persistent natural hazards may influence the types of modifications that people make in the environment and certainly influence the styles of housing and other elements of cultural geography. Natural hazards may also undermine the utility of an area for economic purposes and some scholars suggest that regions of environmental instability may be regions of political instability as well. The study of natural hazards has become an important activity for "resource geographers" whose areas of interest overlap both human and physical fields of geography.

Part II

Global Human Patterns

Map 14 Past Population Distributions and Densities

The map of the world at 100,000 B.P. (Before Present) shows the distributions of hominids who at that time had spread from their probable origin in Africa into parts of the Old World. At 30,000 B.P. few places in the Old World remained uninhabited. All the people then were hunters and gatherers who subsisted on wild foods. The environment probably was not at its carrying capacity for human populations until about 15, 000 B.P.

By 10,000 B.P. hunting and gathering people had spread throughout the world. Plant and animal domestication (farming and pastoralism) had begun in some parts of the Old World

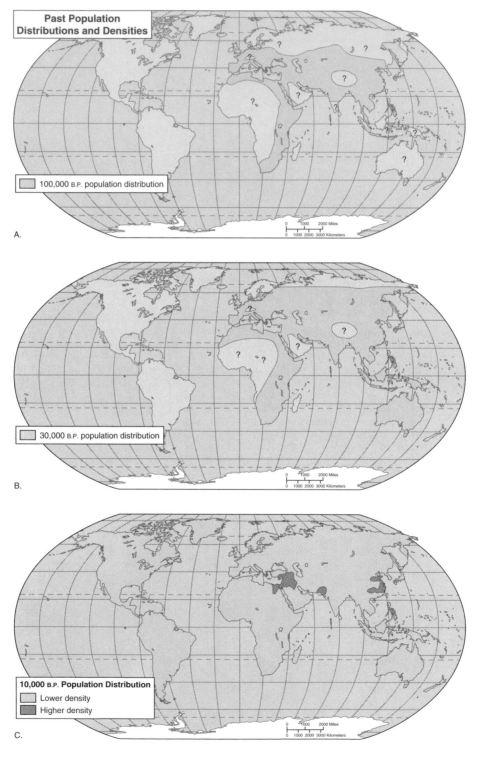

Past Population Distributions and Densities

A.

100,000 B.P. population distribution

B.

30,000 B.P. population distribution

C.

10,000 B.P. Population Distribution
Lower density
Higher density

perhaps as a response to behavioral changes necessitated by the population, which now exceeded the environmental carrying capacity. Farming supports higher population densities than hunting and gathering. Urban civilization with cities dependent on their hinterlands had developed by 5000 B.P. The maps of A.D. 1 and A.D. 1500 approximate actual population density on a scale of one dot to every million people. A chart provides total world population figures for different time periods. Compare these maps to the contemporary population density map on the following page.

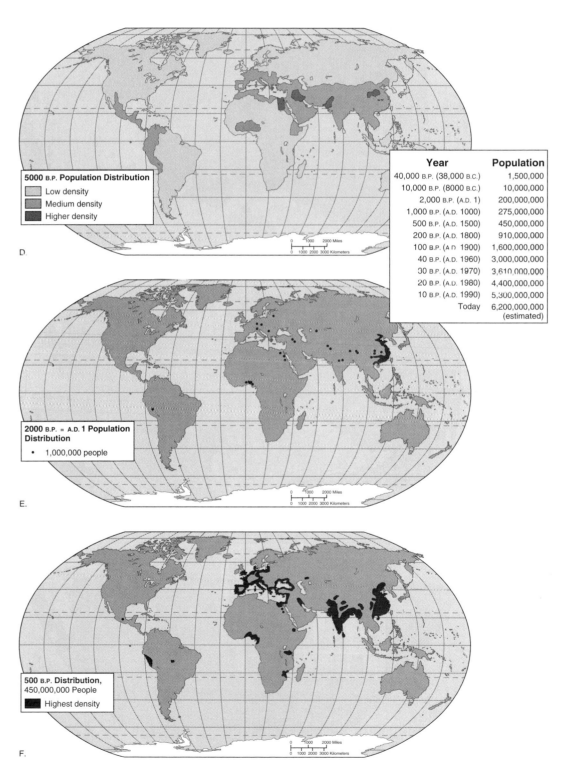

5000 B.P. Population Distribution
- Low density
- Medium density
- Higher density

D.

Year	Population
40,000 B.P. (38,000 B.C.)	1,500,000
10,000 B.P. (8000 B.C.)	10,000,000
2,000 B.P. (A.D. 1)	200,000,000
1,000 B.P. (A.D. 1000)	275,000,000
500 B.P. (A.D. 1500)	450,000,000
200 B.P. (A.D. 1800)	910,000,000
100 B.P. (A.D. 1900)	1,600,000,000
40 B.P. (A.D. 1960)	3,000,000,000
30 B.P. (A.D. 1970)	3,610,000,000
20 B.P. (A.D. 1980)	4,400,000,000
10 B.P. (A.D. 1990)	5,300,000,000
Today	6,200,000,000 (estimated)

2000 B.P. = A.D. 1 Population Distribution
- • 1,000,000 people

E.

500 B.P. Distribution, 450,000,000 People
- Highest density

F.

Map 15 Population Density

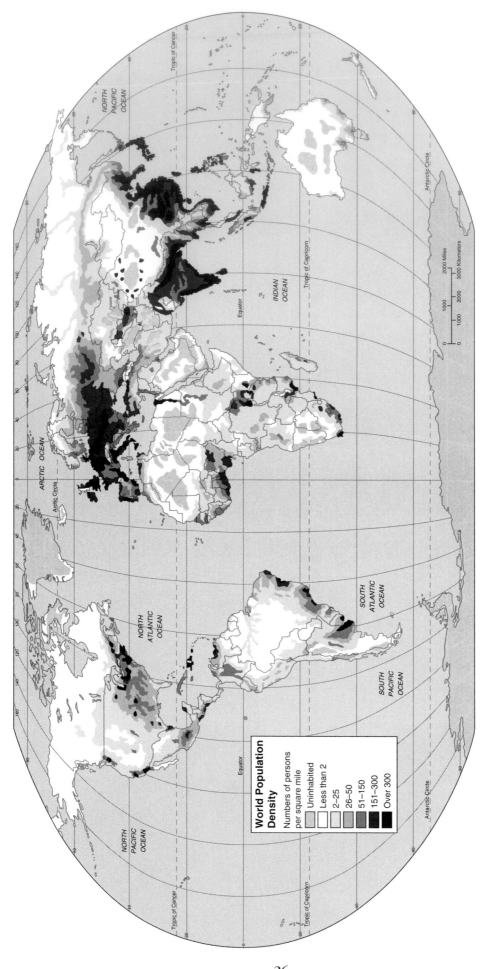

World Population Density

Numbers of persons per square mile

- Uninhabited
- Less than 2
- 2–25
- 26–50
- 51–150
- 151–300
- Over 300

No feature of human activity is more reflective of geographic relationships than where people live. In the areas of densest populations, a mixture of natural and human factors has combined to allow maximum food production, maximum urbanization, and maximum centralization of economic activities. Three great concentrations of human population appear on the map—East Asia, South Asia, and Europe—with a fourth, lesser concentration in eastern North America. While population growth is relatively slow in three of these population clusters, in the fourth—South Asia—growth is still rapid and South Asia is expected to become even more densely populated in the early years of the twenty-first century, while density of the other regions is expected to remain about as it now appears. In Europe and North America, the relatively stable population growth rates are the result of economic development that has caused population growth to level off within the last century. In East Asia, the growth rates have also begun to decline. In the case of Japan, Taiwan, the Koreas, and other more highly developed nations of the Pacific Rim, the reduced growth is the result of economic development. In China, at least until recently, lowered population growth rates have resulted from strict family planning. The areas of future high density of population, in addition to those already existing, are likely to be in Middle and South America and in Central Africa, where population growth rates are well above the world average.

-26-

Map 16 Land Use, 1500

Europeans began to explore the world in the late 1400s. They encountered many independent people with self-sustaining economies at that time. Foraging people practiced hunting and gathering, utilizing the wild forms of plants and animals in their environments. Horticultural people practiced a simple form of agriculture using hoes or digging sticks as their basic tools. They sometimes cleared their land by burning and then planted crops. Pastoralists herded animals as their basic subsistence pattern. Complex state-level societies, such as the Mongols, had pastoralism as their base. Intensive agriculturalists based their societies on complicated irrigation systems and/or the plow and draft animals. Wheat and rice were two kinds of crops that supported large populations. In many of the areas of intensive agriculture—particularly in MesoAmerica, Europe, Southwest Asia, South Asia, and East Asia-complex patterns of market economies had begun to develop well before the 15th century and the beginnings of European expansion.

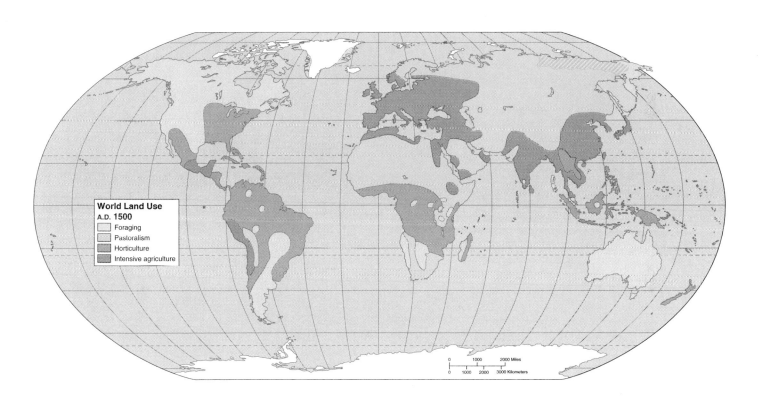

World Land Use
A.D. **1500**
- Foraging
- Pastoralism
- Horticulture
- Intensive agriculture

Map 17 Economic Activities

Land uses can be categorized as lying somewhere on a scale between extensive uses, in which human activities are dispersed over relatively large areas, and intensive uses, in which human activities are concentrated in relatively small areas. Many of the most important land use patterns of the world (such as urbanization, industry, mining, or transportation) are intensive and therefore relatively small in area and not easily seen on maps of this scale. Hence, even in the areas identified as "Manufacturing and Commerce" on the map there are many land uses that are not strictly industrial or commercial in nature, and, in fact, more extensive land uses (farming, residential, open space) may actually cover more ground than the intensive industrial or commercial activities. On the other hand, the more extensive land uses, like agriculture and forestry, tend to dominate the areas in which they are found. Thus, primary economic activities such as agriculture and forestry tend to dominate the world map of land use because of their extensive character. Much of this map is, therefore, a map that shows the global variations in agricultural patterns. Note, among other things, the differences between land use patterns in the more developed countries of the temperate zones and the less developed countries of the tropics.

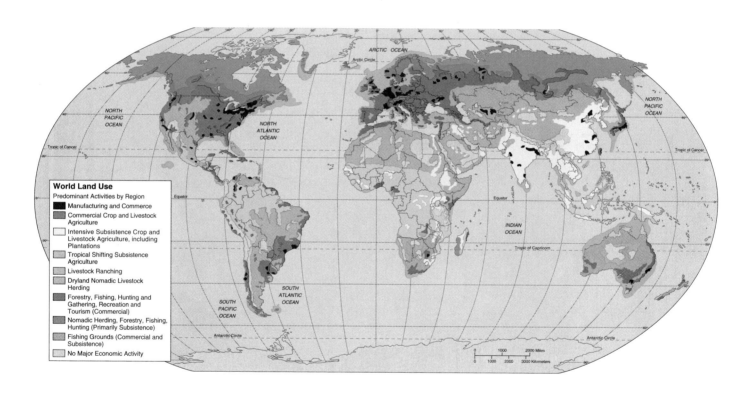

World Land Use

Predominant Activities by Region

- Manufacturing and Commerce
- Commercial Crop and Livestock Agriculture
- Intensive Subsistence Crop and Livestock Agriculture, including Plantations
- Tropical Shifting Subsistence Agriculture
- Livestock Ranching
- Dryland Nomadic Livestock Herding
- Forestry, Fishing, Hunting and Gathering, Recreation and Tourism (Commercial)
- Nomadic Herding, Forestry, Fishing, Hunting (Primarily Subsistence)
- Fishing Grounds (Commercial and Subsistence)
- No Major Economic Activity

Map 18 Urbanization

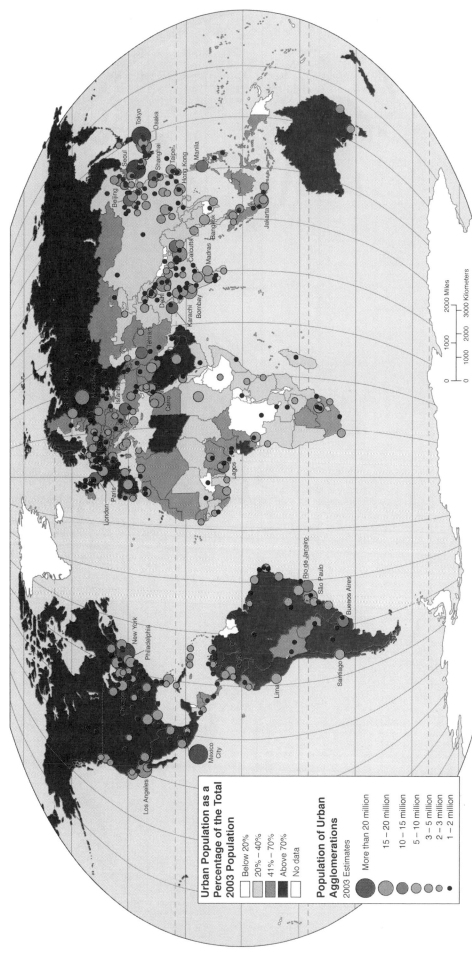

Urban Population as a Percentage of the Total 2003 Population

- Below 20%
- 20% – 40%
- 41% – 70%
- Above 70%
- No data

Population of Urban Agglomerations

2003 Estimates

- More than 20 million
- 15 – 20 million
- 10 – 15 million
- 5 – 10 million
- 3 – 5 million
- 2 – 3 million
- 1 – 2 million

Tokyo, Osaka, Seoul, Shanghai, Taipei, Hong Kong, Manila, Beijing, Jakarta, Calcutta, Madras, Bangkok, Delhi, Bombay, Karachi, Tehran, Istanbul, Cairo, Lagos, Paris, London, Rio de Janeiro, São Paulo, Buenos Aires, Bogotá, Santiago, Lima, New York, Philadelphia, Chicago, Los Angeles, Mexico City

0   1000   2000 Miles
0   1000   2000   3000 Kilometers

The degree to which a region's population is concentrated in urban areas is a major indicator of a number of things: the potential for environmental impact, the level of economic development, and the problems associated with human concentrations. Urban dwellers are rapidly becoming the norm among the world's people and rates of urbanization are increasing worldwide, with the greatest increases in urbanization taking place in developing or developing regions. Whether in developed or developing countries, those who live in cities exert an influence on the environment, politics, economics, and social systems that go far beyond the confines of the city itself. Acting as the focal

points for the flow of goods and ideas, cities draw resources and people not just from their immediate hinterland but from the entire world. This process creates far-reaching impacts as resources are extracted, converted through industrial processes, and transported over great distances to metropolitan regions, and as ideas spread or *diffuse* along with the movements of people to cities and the flow of communication from them. The significance of urbanization can be most clearly seen, perhaps, in North America where, in spite of vast areas of relatively unpopulated land, well over 90 percent of the population lives in urban areas.

Map 19 Religions

Religious adherence is one of the fundamental defining characteristics of human culture, the style of life adopted by a people and passed from one generation to the next. Because of the importance of religion for culture, a depiction of the spatial distribution of religions is as close as we can come to a map of cultural patterns. More than just a set of behavioral patterns having to do with worship and ceremony, religion is a vital conditioner of the ways that people deal with one another, with their institutions, and with the environments

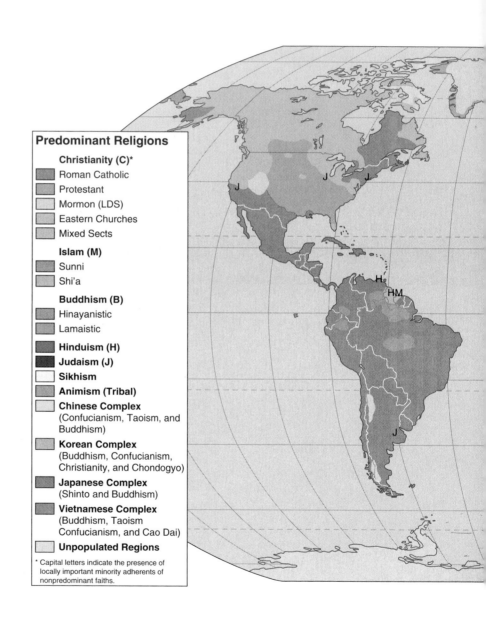

Predominant Religions

Christianity (C)*
- Roman Catholic
- Protestant
- Mormon (LDS)
- Eastern Churches
- Mixed Sects

Islam (M)
- Sunni
- Shi'a

Buddhism (B)
- Hinayanistic
- Lamaistic

Hinduism (H)

Judaism (J)

Sikhism

Animism (Tribal)

Chinese Complex
(Confucianism, Taoism, and Buddhism)

Korean Complex
(Buddhism, Confucianism, Christianity, and Chondogyo)

Japanese Complex
(Shinto and Buddhism)

Vietnamese Complex
(Buddhism, Taoism Confucianism, and Cao Dai)

Unpopulated Regions

* Capital letters indicate the presence of locally important minority adherents of nonpredominant faiths.

they occupy. In many areas of the world, the ways in which people make a living, the patterns of occupation that they create on the land, and the impacts that they make on ecosystems are the direct consequences of their adherence to a religious faith. An examination of the map in the context of international and intranational conflict will also show that tension between countries and the internal stability of states is also a function of the spatial distribution of religion.

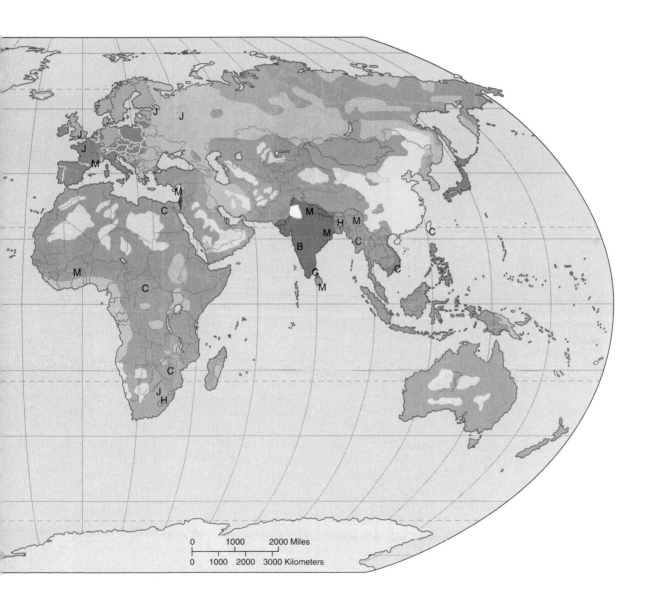

Map 20 Transportation Patterns

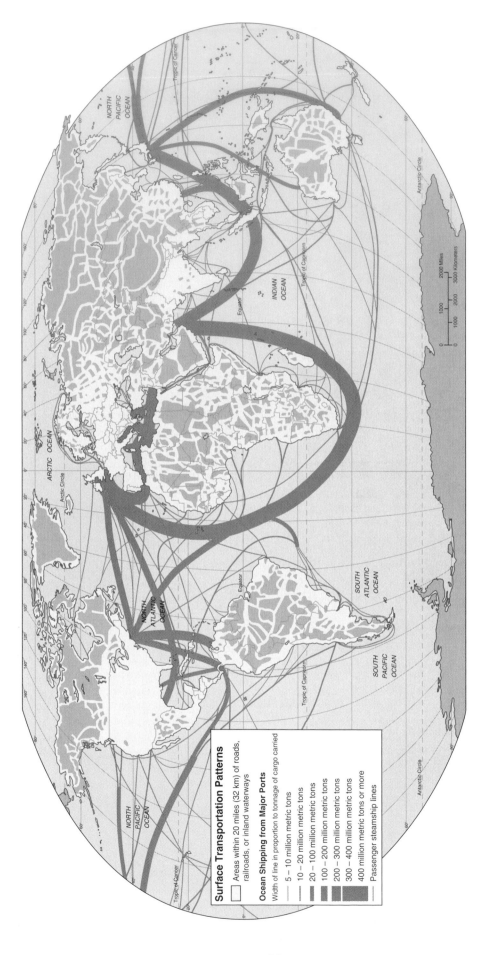

Surface Transportation Patterns

☐ Areas within 20 miles (32 km) of roads, railroads, or inland waterways

Ocean Shipping from Major Ports

Width of line in proportion to tonnage of cargo carried

- 5 – 10 million metric tons
- 10 – 20 million metric tons
- 20 – 100 million metric tons
- 100 – 200 million metric tons
- 200 – 300 million metric tons
- 300 – 400 million metric tons
- 400 million metric tons or more
- Passenger steamship lines

As a form of land use, transportation is second only to agriculture in its coverage of the earth's surface and is one of the clearest examples in the human world of a *network*, a linked system of lines allowing flows from one place to another. The global transportation network and its related communication web is responsible for most of the *spatial interaction*, or movement of goods, people, and ideas between places. As the chief mechanism of spatial interaction, transportation is linked firmly with the concept of a shrinking world and the development of a global community and economy. Because

transportation systems require significant modification of the earth's surface, transportation is also responsible for massive alterations in the quantity and quality of water, for major soil degradations and erosion, and (indirectly) for the air pollution that emanates from vehicles utilizing the transportation system. In addition, as improved transportation technology draws together places on the earth that were formerly remote, it allows people to impact environments a great distance away from where they live.

-32-

Map 21 Languages

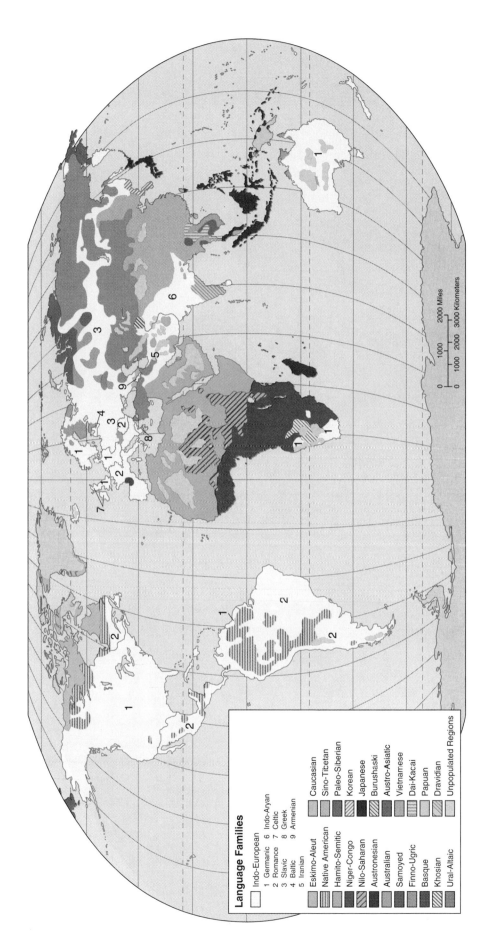

Language Families

☐ Indo-European	6 Indo-Aryan	
1 Germanic	7 Celtic	
2 Romance	8 Greek	
3 Slavic	9 Armenian	
4 Baltic		
5 Iranian		

Eskimo-Aleut — Caucasian
Native American — Sino-Tibetan
Hamito-Semitic — Paleo-Siberian
Niger-Congo — Korean
Nilo-Saharan — Japanese
Austronesian — Burushaski
Australian — Austro-Asiatic
Samoyed — Vietnamese
Finno-Ugric — Dai-Kacai
Basque — Papuan
Khosian — Dravidian
Ural-Altaic — Unpopulated Regions

Language, like religion, is an important identifying characteristic of culture. Indeed, it is perhaps the most durable of all those identifying characteristics or *cultural traits*: language, religion, institutions, material technologies, and ways of making a living. After centuries of exposure to other languages or even conquest by speakers of other languages, the speakers of a specific tongue will often retain their own linguistic identity. Language helps us to locate areas of potential conflict, particularly in regions where two or more languages overlap. Many, if not most, of the world's conflict zones are also areas of linguistic diversity. Knowing the distribution of languages helps us to understand some of the reasons behind important current events: for example, linguistic

identity differences played an important part in the disintegration of the Soviet Union in the early 1990s; and in areas emerging from recent colonial rule, such as Africa, the participants in conflicts over territory and power are often defined in terms of linguistic groups. Language distributions also help us to comprehend the nature of the human past by providing clues that enable us to chart the course of human migrations, as shown in the distribution of Indo-European, Austronesian, or Hamito-Semitic languages. Finally, because languages have a great deal to do with the way people perceive and understand the world around them, linguistic patterns help to explain the global variations in the ways that people interact.

-33-

Map 22 Linguistic Diversity

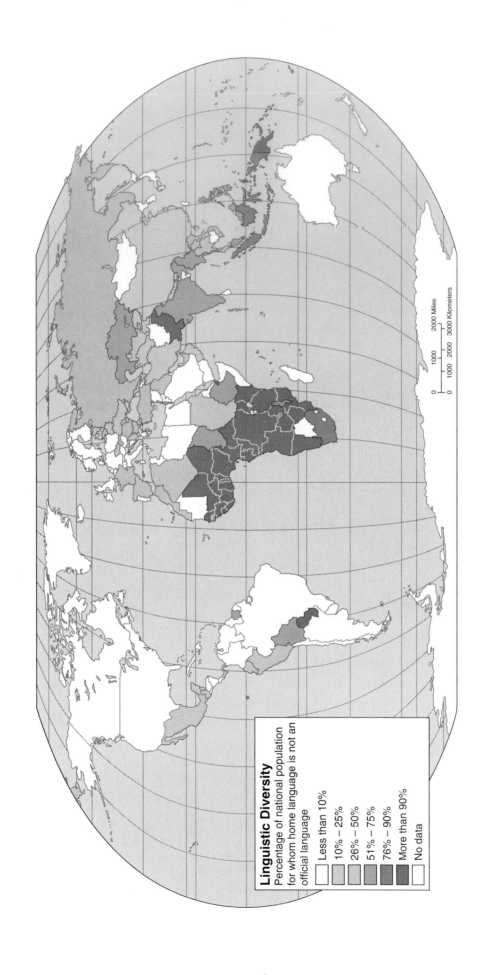

Linguistic Diversity
Percentage of national population
for whom home language is not an
official language

- Less than 10%
- 10% – 25%
- 26% – 50%
- 51% – 75%
- 76% – 90%
- More than 90%
- No data

0   1000   2000 Miles
0   1000   2000   3000 Kilometers

Of the world's approximately 6,000 languages, fewer than 100 are official languages, those designated by a country as the language of government, commerce, education, and information. This means that for much of the world's population, the language that is spoken in the home is different from the official language of the country of residence. The world's former colonial areas in Middle and South America, Africa, and South and Southeast Asia stand out on the map as regions in which there is significant disparity between home languages and official languages. To complicate matters further, for most of the world's population, the primary international languages of trade and tourism (French and English) are neither home nor official languages. China is a special case as the official language is the written form of Chinese while several spoken Chinese dialects such as Mandarin and Cantonese most of them mutually unintelligible are recognized as official languages. The formal language of government and business is Mandarin."

Map 23 External Migrations in Modern Times

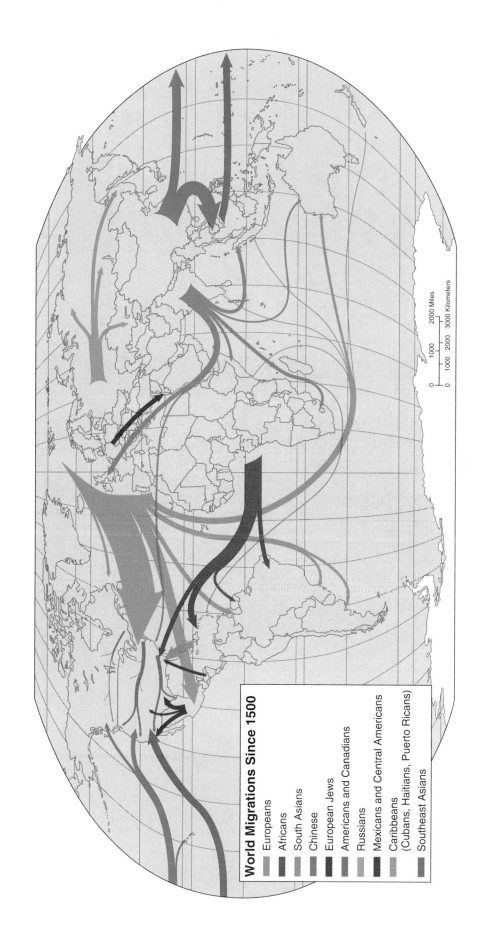

World Migrations Since 1500

- Europeans
- Africans
- South Asians
- Chinese
- European Jews
- Americans and Canadians
- Russians
- Mexicans and Central Americans
- Caribbeans (Cubans, Haitians, Puerto Ricans)
- Southeast Asians

0   1000   2000 Miles

0   1000   2000   3000 Kilometers

Migration has had a significant effect on world geography, contributing to cultural change and development, to the diffusion of ideas and innovations, and to the complex mixture of people and cultures found in the world today. *Internal migration* occurs within the boundaries of a country; *external migration* is movement from one country or region to another. Over the last 50 years, the most important migrations in the world have been internal, largely the rural-to-urban migration that has been responsible for the recent rise of global urbanization. Prior to the mid-twentieth century, three types of external migrations were most important: *voluntary*, most often in search of better eco-nomic conditions and opportunities; *involuntary* or *forced*, involving people who have been driven from their homelands by war, political unrest, or environmental disasters, or who have been transported as slaves or prisoners; and *imposed*, not entirely forced but which conditions make highly advisable. Human migrations in recorded history have been responsible for major changes in the patterns of languages, religions, ethnic composition, and economies. Particularly during the last 500 years, migrations of both the voluntary and involuntary or forced type have literally reshaped the human face of the earth.

Part III

Global Demographic Patterns

Map 24 Population Growth Rates

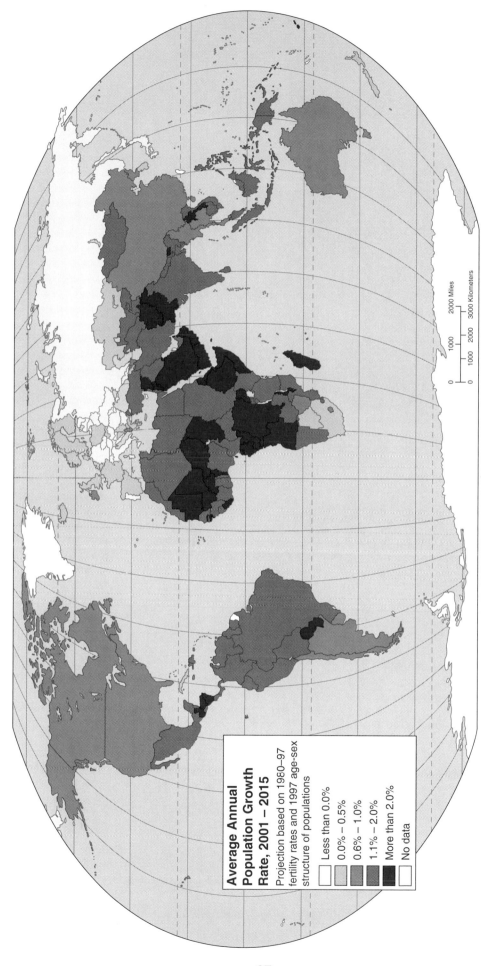

Average Annual Population Growth Rate, 2001 – 2015

Projection based on 1980–97 fertility rates and 1997 age-sex structure of populations

- Less than 0.0%
- 0.0% – 0.5%
- 0.6% – 1.0%
- 1.1% – 2.0%
- More than 2.0%
- No data

2000 Miles

3000 Kilometers

Of all the statistical measurements of human population, that of the rate of population growth is the most important. The growth rate of a population is a combination of natural change (births and deaths), in-migration, and out-migration; it is obtained by adding the number of births to the number of immigrants during a year and subtracting from that total the sum of deaths and emigrants for the same year. For a specific country, this figure will determine many things about the country's future ability to feed, house, educate, and provide medical services to its citizens. Some of the countries with the largest populations (such as India) also have high growth rates. Since these countries tend to be in developing regions, the combination of high population and high growth rates poses special problems for political stability and continuing economic development; the combination also carries heightened risks for environmental degradation. Many people believe that the rapidly expanding world population is a potential crisis that may cause environmental and human disaster by the middle of the twenty-first century.

Map 25 Total Fertility Rates

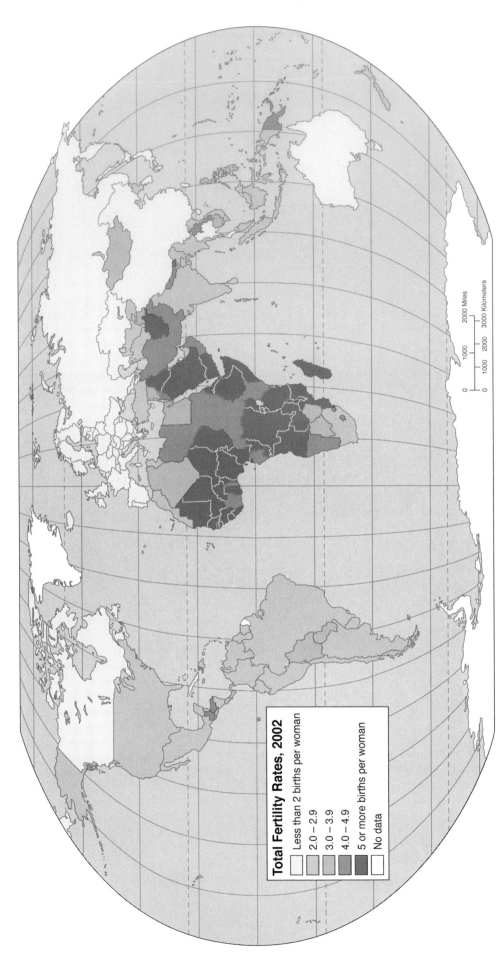

Total Fertility Rates, 2002

Less than 2 births per woman
2.0 – 2.9
3.0 – 3.9
4.0 – 4.9
5 or more births per woman
No data

0   1000   2000 Miles
0   1000   2000   3000 Kilometers

The fertility rate measures the number of children that a woman is expected to bear during her lifetime, based on the age-specific fertility figures of women between 15 and 40 (the normal childbearing years). While fertility rates tell us a great deal about present population growth, with high fertility rates indicating high population growth rates, they are also indicative of potential or projected growth. A country whose women can be expected to bear many children is a country with enormous potential for population growth in the future. Given present fertility rates, for example, the number of offspring from the average German woman over the next three generations (the total number of children, grandchildren, and great-grandchildren) will be 7. During the same three generations, the average American woman will have a total of 17 children, grandchildren, and great-grandchildren. But during this time, assuming that present fertility rates are maintained, the average woman in sub-Saharan Africa will have [&em]258[&stop] children, grandchildren, and great-grandchildren. You might be interested in working out some potential population growth rates over two or three generations, using the data as presented on the map.

Map 26 Infant Mortality Rates

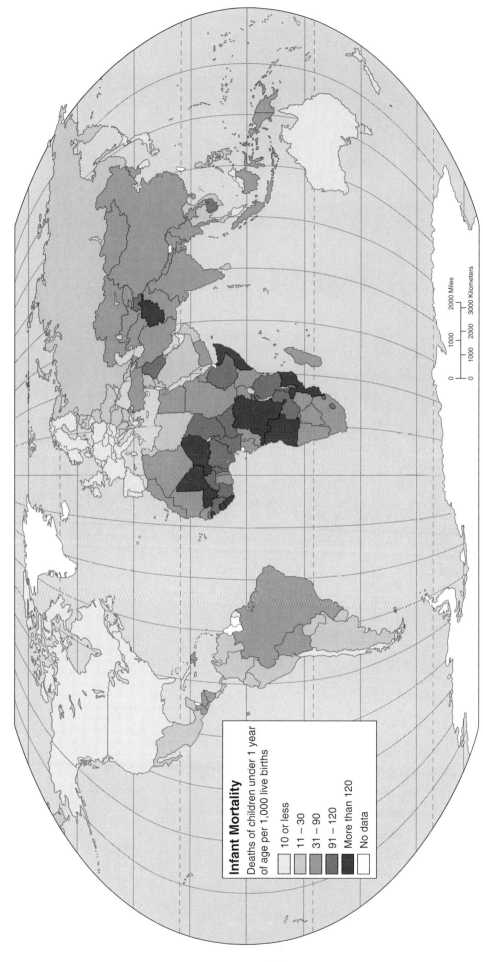

Infant Mortality

Deaths of children under 1 year
of age per 1,000 live births

- 10 or less
- 11 – 30
- 31 – 90
- 91 – 120
- More than 120
- No data

Infant mortality rates are calculated by dividing the number of children born in a given year who die before their first birthday by the total number of children born that year and then multiplying by 1,000; this shows how many infants have died for every 1,000 births. Infant mortality rates are prime indicators of economic development. In highly developed economies, with advanced medical technologies, sufficient diets, and adequate public sanitation, infant mortality rates tend to be quite low. By contrast, in less developed countries, with the disadvantages of poor diet, limited access to medical technology, and the other problems of poverty, infant mortality rates tend to be high.

Although worldwide infant mortality has decreased significantly during the last 2 decades, many regions of the world still experience infant mortality above the 10 percent level (100 deaths per 1,000 live births). Such infant mortality rates not only represent human tragedy at its most basic level, but also are powerful inhibiting factors for the future of human development. Comparing infant mortality rates in the midlatitudes and the tropics shows that children in most African countries are more than 10 times as likely to die within a year of birth as children in European countries.

-39-

Map **27** Child Mortality Rate

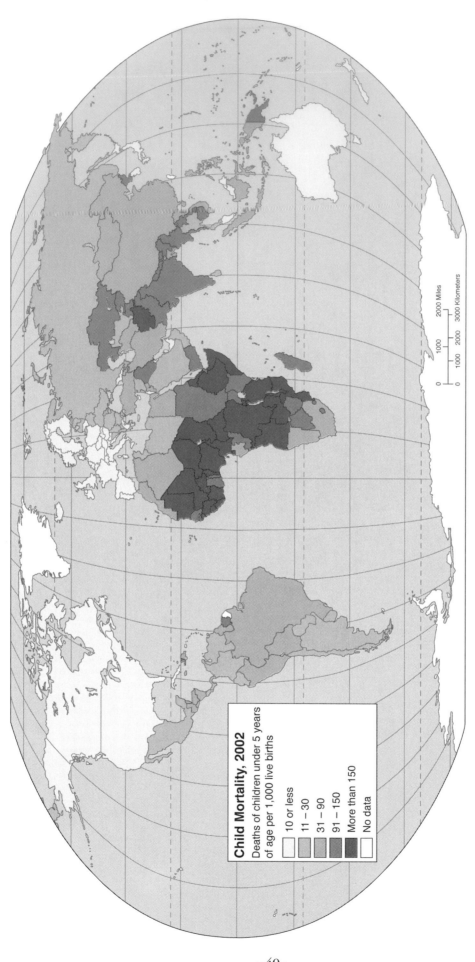

Child Mortality, 2002

Deaths of children under 5 years of age per 1,000 live births

- 10 or less
- 11 – 30
- 31 – 90
- 91 – 150
- More than 150
- No data

Child mortality rates are calculated by determining the probability that a child born in a specified year will die before reaching the age 5, using current age-specific mortality rates for a population. The major sources of mortality rates are vital registration systems and estimates made from surveys and/or census reports. Along with infant mortality and average life-expectancy rates, child mortality rates, according to the World Bank, "are probably the best general indicators of a community's current health status and are often cited as overall measures of a population's welfare or quality of life." Where infant mortality often reflects health care conditions, child mortality is usually a reflection of the inadequacy of nutrition, leading to early deaths from nutritionally related diseases. In some less developed countries in Africa and Asia, child mortality is also an indicator of the widespread presence of infectious diseases such as malaria, tuberculosis, and HIV/AIDS.

-40-

Map 28 Child Malnutrition

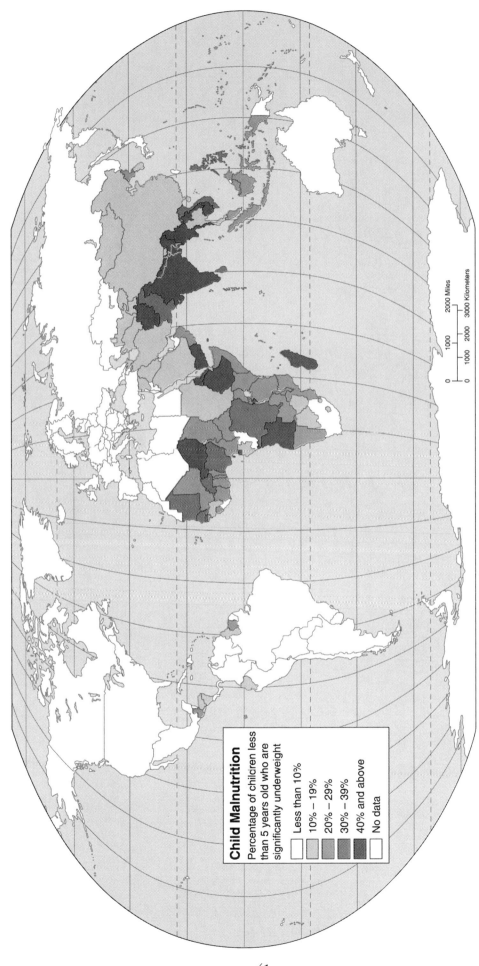

Child Malnutrition

Percentage of children less than 5 years old who are significantly underweight

- Less than 10%
- 10% – 19%
- 20% – 29%
- 30% – 39%
- 40% and above
- No data

The weight of poverty is not evenly spread among the members of a population, falling disproportionately upon the weakest and most disadvantaged members of society. In most societies, these individuals are children, particularly female children. Children simply do not compete as successfully as adults for their (meager) share of the daily food supply. Where food shortages prevail, children tend to have the quality of their future lives severely compromised by poor nutrition, which, in a downward spiral, robs them of the energy necessary to compete more effectively for food. Children who are inadequately fed are less likely to do well in school, are more prone to debilitating disease, and will more often become a drain on scarce societal resources than well-fed children. Recently, health care officials in the more developed world have become concerned over the trend to "overnutrition," leading to obesity and related health problems in the world's economically developed countries. Nevertheless, child malnourishment remains one of the primary distinguishing factors between the "haves" and "have-nots."

Map 29 Primary School Enrollment

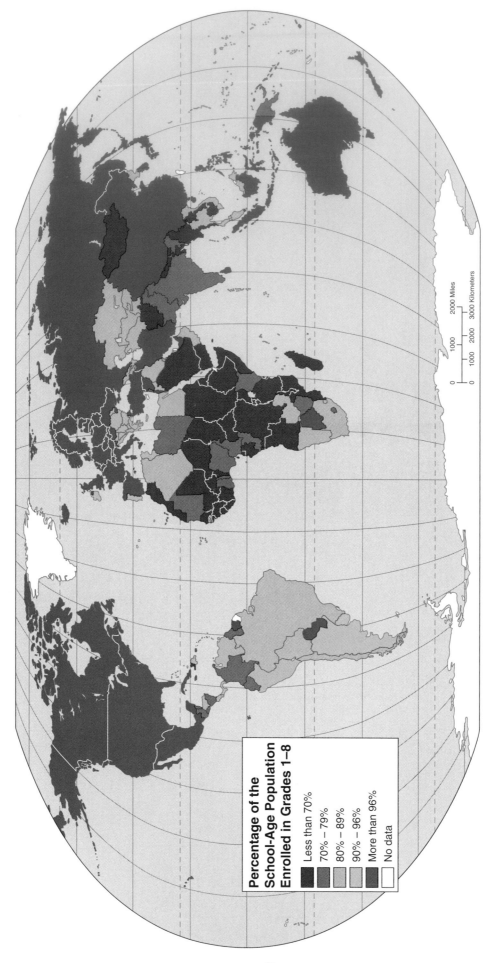

Percentage of the School-Age Population Enrolled in Grades 1–8

- Less than 70%
- 70% – 79%
- 80% – 89%
- 90% – 96%
- More than 96%
- No data

Like many of the other measures illustrated in this atlas, primary school enrollment is a clear reflection of the division of the world into "have" and "have-not" countries. It is also a measure that has changed more rapidly over the last decade than demographic and other indicators of development, as countries of even very modest means have made concerted attempts to attain relatively high percentages of primary school enrollment. That they have been able to do so is good evidence of the fact that reasonably respectable levels of human development are feasible at even modest income levels.

High primary school enrollment is also a reflection of the worldwide opinion that a major element in economic development is a well-educated, literate population. The links between human progress, as typified by higher levels of education, and economic growth are not automatic, however, and those countries without programs for maintaining the headway gained by improved education may be on the road to failure in terms of economic development.

Map 30 Average Life Expectancy at Birth

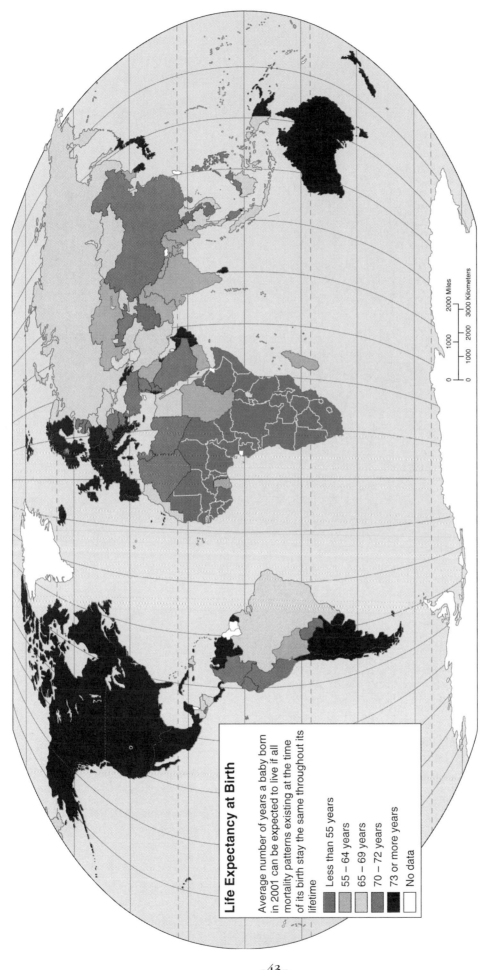

Life Expectancy at Birth

Average number of years a baby born in 2001 can be expected to live if all mortality patterns existing at the time of its birth stay the same throughout its lifetime

- Less than 55 years
- 55 – 64 years
- 65 – 69 years
- 70 – 72 years
- 73 or more years
- No data

Average life expectancy at birth is a measure of the average longevity of the population of a country. Like all average measures, it is distorted by extremes. For example, a country with a high mortality rate among children will have a low average life expectancy. Thus, an average life expectancy of 45 years does not mean that everyone can be expected to die at the age of 45. More normally, what the figure means is that a substantial number of children die between birth and 5 years of age, thus reducing the average life expectancy for the entire population. In spite of the dangers inherent in misinterpreting the data, average life expectancy (along with infant mortality and several other measures) is a valid way of judging the relative health of a population. It reflects the nature of the health care system, public sanitation and disease control, nutrition, and a number of other key human need indicators. As such, it is a measure of well-being that is significant in indicating economic development and predicting political stability.

Map 31 Population by Age Group

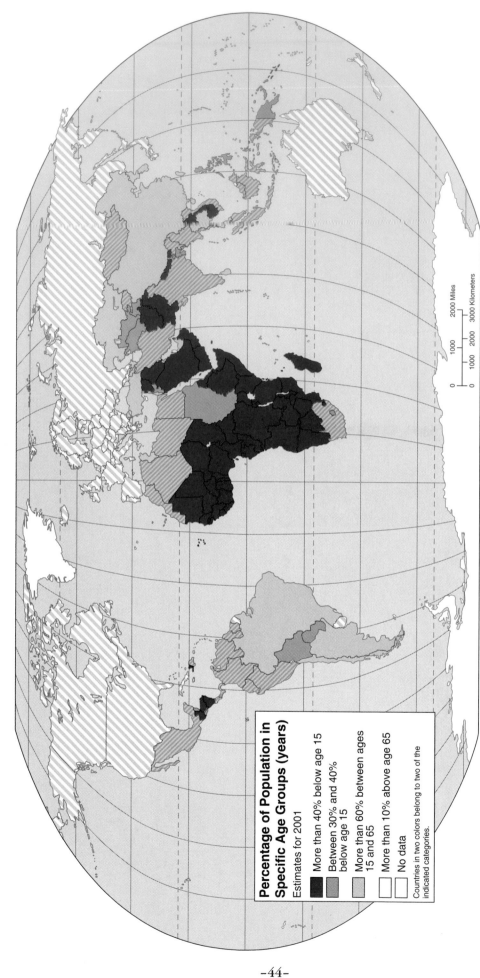

Percentage of Population in Specific Age Groups (years)

Estimates for 2001

- More than 40% below age 15
- Between 30% and 40% below age 15
- More than 60% between ages 15 and 65
- More than 10% above age 65
- No data

Countries in two colors belong to two of the indicated categories.

0 1000 2000 Miles
0 1000 2000 3000 Kilometers

Of all the measurements that illustrate the dynamics of a population, age distribution may be the most significant, particularly when viewed in combination with average growth rates. The particular relevance of age distribution is that it tells us what to expect from a population in terms of growth over the next generation. If, for example, approximately 40–50 percent of a population is below the age of 15, that suggests that in the next generation about one-quarter of the total population will be women of childbearing age. When age distribution is combined with fertility rates (the average number of children born per woman in a population), an especially valid measurement of future growth potential may be derived. A simple example: Nigeria, with a 2002 population of 130 million, has 43.6 percent of its population below the age of 15 and a fertility rate of 5.5; the United States, with a 2002 population of 280 million, has 21 percent of its population below the age of 15 and a fertility rate of 2.07. During the period in which those women presently under the age of 15 are in their childbearing years, Nigeria can be expected to add a total of approximately 155 million persons to its total population. Over the same period, the United States can be expected to add only 61 million.

Map 32 Average Daily Per Capita Supply of Calories (Kilocalories)

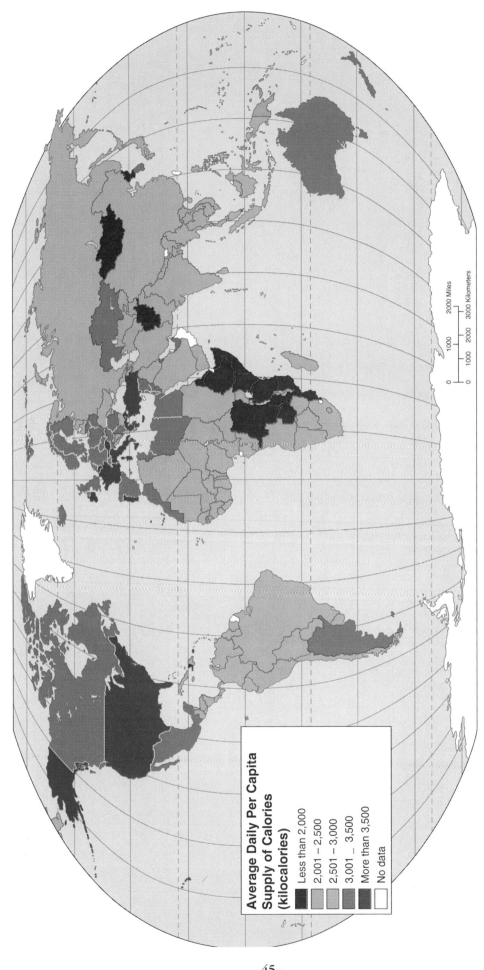

Average Daily Per Capita Supply of Calories (kilocalories)

- Less than 2,000
- 2,001 – 2,500
- 2,501 – 3,000
- 3,001 – 3,500
- More than 3,500
- No data

0   1000   2000 Miles

0   1000   2000   3000 Kilometers

The data shown on this map, which indicate the presence or absence of critical food shortages, do not necessarily indicate the presence of starvation or famine. But they certainly do indicate potential problem areas for the next decade. The measurements are in calories from *all* food sources: domestic production, international trade, drawdown on stocks or food reserves, and direct foreign contributions or aid. The quantity of calories available is that amount, estimated by the UN's Food and Agriculture Organization (FAO), that reaches consumers. The calories actually consumed may be lower than the figures shown, depending on how much is lost in a variety of ways: in home storage (to pests such as rats and mice), in preparation and cooking, through consumption by pets and domestic animals, and as discarded foods, for example. The estimate of need is not a global uniform value but is calculated for each country on the basis of the age and sex distribution of the population, and the estimated level of activity of the population. Compare this map with Map 59 for a good measure of potential problem areas for food shortages within the next decade.

-45-

Map 33 Illiteracy Rates

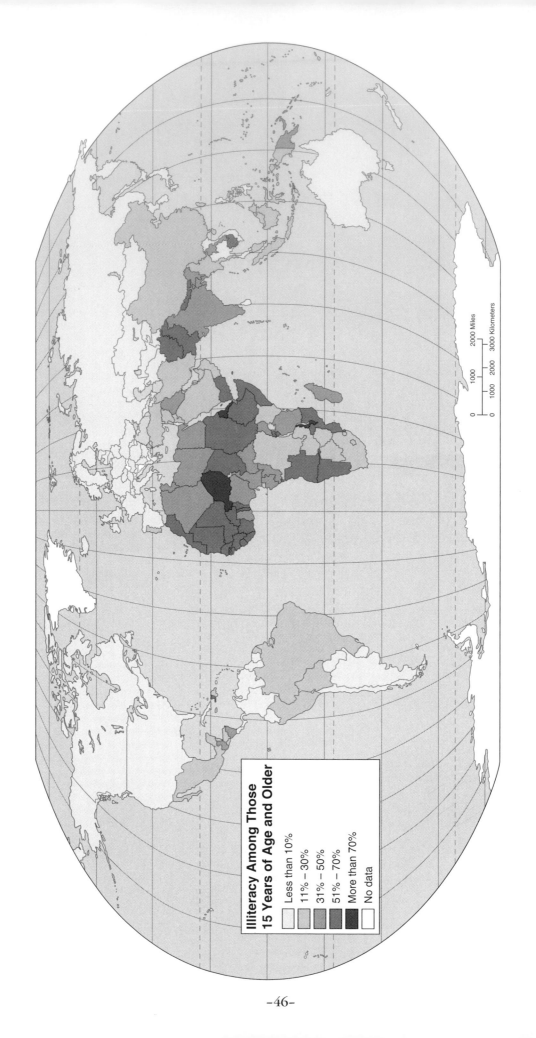

Illiteracy Among Those 15 Years of Age and Older

- Less than 10%
- 11% – 30%
- 31% – 50%
- 51% – 70%
- More than 70%
- No data

0   1000   2000 Miles

0   1000   2000   3000 Kilometers

Illiteracy rates are based on the percentages of people age 15 or above (classed as adults in most countries) who are not able to write and read, with understanding, a brief, simple statement about everyday life written in their home- or official language. As might be expected, illiteracy rates tend to be higher in the less developed states, where educational systems are a low government priority. Rates of literacy or illiteracy also tend to be gender-differentiated, with women in many countries experiencing educational neglect or discrimination that makes it more likely they will be illiterate. In many developing countries, between five and ten times as many women will be illiterate as men, and the illiteracy rate for women may even exceed 90 percent. Both male and female illiteracy severely compromise economic development.

Map 34 Female/Male Inequality in Education and Employment

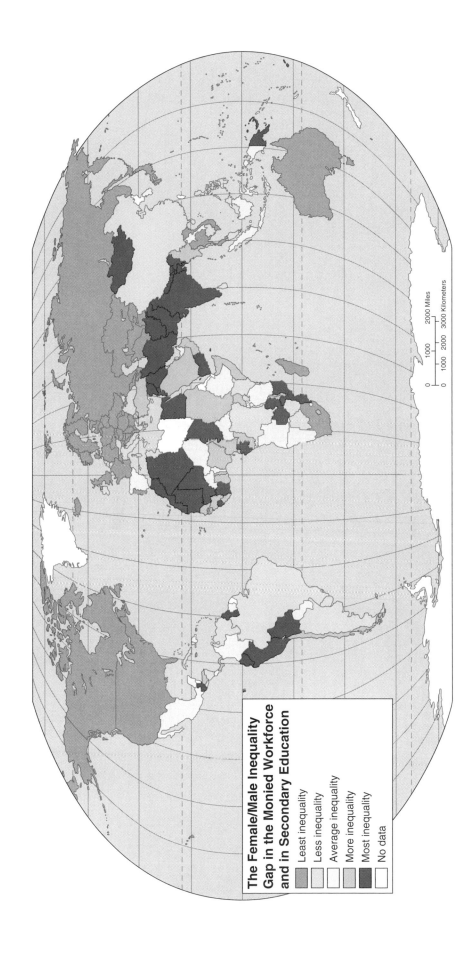

The Female/Male Inequality Gap in the Monied Workforce and in Secondary Education

- Least inequality
- Less inequality
- Average inequality
- More inequality
- Most inequality
- No data

0   1000   2000 Miles
0 1000 2000 3000 Kilometers

While women in developed countries, particularly in North America and Europe, have made significant advances in socioeconomic status in recent years, in most of the world they suffer from significant inequality when compared with their male counterparts. Although women have received the right to vote in most of the world's countries, in over 90 percent of these countries that right has only been granted in the last 50 years. In most regions, literacy rates for women still fall far short of those for men; In Africa and Asia, for example, only about half as many women are literate as are men. Women marry considerably younger than men and attend school for shorter periods of time. Inequalities in education and employment are perhaps the most telling indicators of the unequal status of women in most of the world. Lack of secondary education in comparison with men prevents women from entering the workforce with equally high-paying jobs. Even where women are employed in positions similar to those held by men, they still tend to receive less compensation. The gap between rich and poor involves not only a clear geographic differentiation, but a clear gender differentiation as well.

Map 35 Global Scourges: Major Infectious Diseases

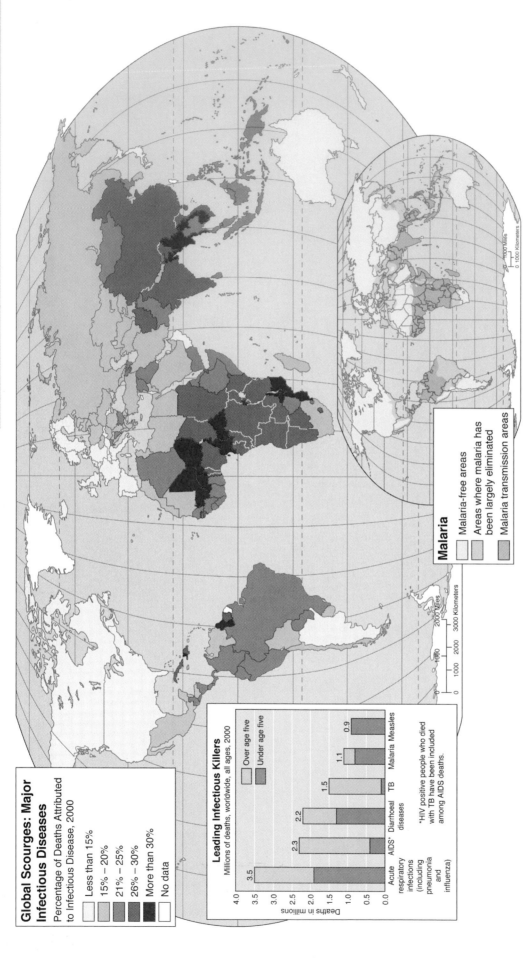

Global Scourges: Major Infectious Diseases

Percentage of Deaths Attributed to Infectious Disease, 2000

- Less than 15%
- 15% – 20%
- 21% – 25%
- 26% – 30%
- More than 30%
- No data

Malaria

- Malaria-free areas
- Areas where malaria has been largely eliminated
- Malaria transmission areas

Leading Infectious Killers

Millions of deaths, worldwide, all ages, 2000

- Over age five
- Under age five

Deaths in millions

- Acute respiratory infections (including pneumonia and influenza) — 3.5
- AIDS* — 2.3
- Diarrhoeal diseases — 2.2
- TB — 1.5
- Malaria — 1.1
- Measles — 0.9

*HIV positive people who died with TB have been included among AIDS deaths.

Infectious diseases are the world's leading cause of premature death and at least half of the world's population is, at any time, at risk of contracting an infectious disease. Although we often think of infectious diseases as being restricted to the tropical world (malaria, dengue fever), many if not most of them have attained global proportions. A major case in point is HIV/AIDS, which quite probably originated in Africa but has, over the last two decades, spread throughout the entire world. Major diseases of the nineteenth century, such as cholera and tuberculosis, are making a major comeback in many parts of the world, in spite of being preventable or treatable. Part of the problem with infectious diseases is that they tend to be associated with poverty (poor nutrition, poor sanitation, substandard housing, and so on) and, therefore, are seen as a problem

of undeveloped countries, with the consequent lack of funding for prevention and treatment. Infectious diseases are also tending to increase because lifesaving drugs, such as antibiotics and others used in the fight against diseases, are losing their effectiveness as bacteria develop genetic resistance to them. The problem of global warming is also associated with a spread of infectious diseases as many disease vectors (certain species of mosquito, for example) are spreading into higher latitudes with increasingly warm temperatures and are spreading disease into areas where populations have no resistance to them. Infectious diseases have become something greater than simply a health issue of poor countries. They are now major social problems with potentially enormous consequences for the entire world.

-48-

Map 36 The Index of Human Development

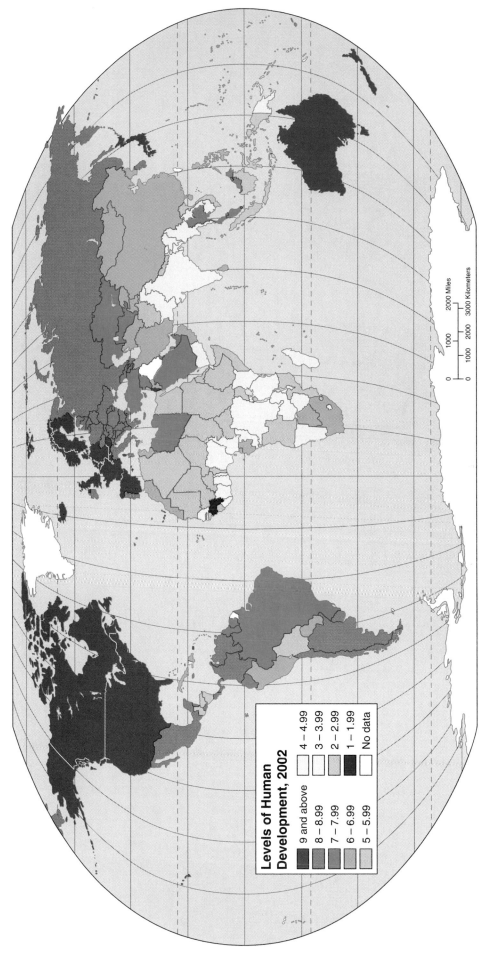

Levels of Human Development, 2002

- 9 and above
- 8 – 8.99
- 7 – 7.99
- 6 – 6.99
- 5 – 5.99
- 4 – 4.99
- 3 – 3.99
- 2 – 2.99
- 1 – 1.99
- No data

The development index upon which this map is based takes into account a wide variety of demographic, health, and educational data, including population growth, per capita gross domestic income, longevity, literacy, and years of schooling. The map reveals significant improvement in the quality of life in Middle and South America, although it is questionable whether the gains made in those regions can be maintained in the face of the dramatic population increases expected over the next 30 years. More clearly than anything else, the map illustrates the near-desperate situation in Africa and South Asia. In those regions, the unparalleled growth in population threatens to overwhelm all efforts to improve the quality of life. In Africa, for example, the population is increasing by 20 million persons per year. With nearly 45 percent of the continent's population aged 15 years or younger, this growth rate will accelerate as the women reach child-bearing age. Africa, along with South Asia, faces the very difficult challenge of providing basic access to health care, education, and jobs for a rapidly increasing population. The map also illustrates the striking difference in quality of life between those who inhabit the world's equatorial and tropical regions and those fortunate enough to live in the temperate zones, where the quality of life is significantly higher.

Part IV

Global Economic Patterns

Map 37 Rich and Poor Countries: Gross National Income

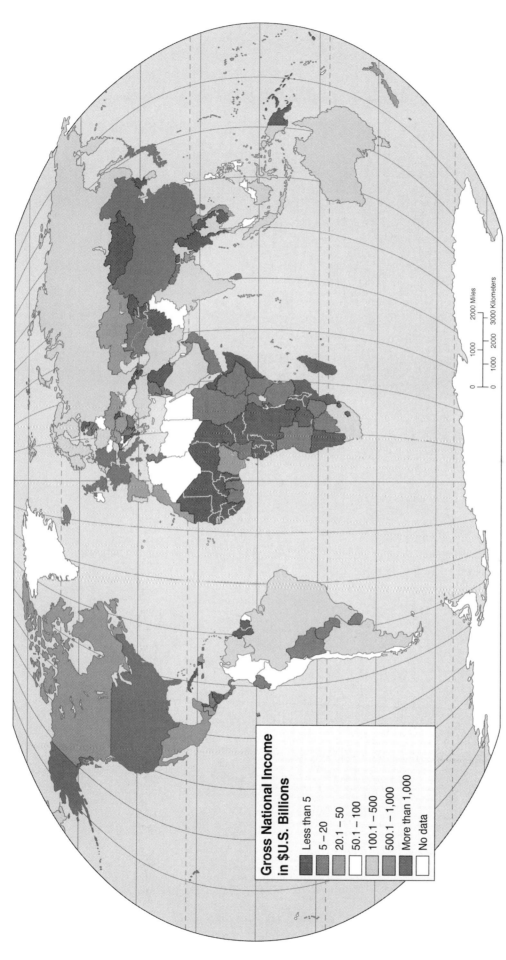

Gross National Income in $U.S. Billions

- Less than 5
- 5 – 20
- 20.1 – 50
- 50.1 – 100
- 100.1 – 500
- 500.1 – 1,000
- More than 1,000
- No data

Gross National Income (GNI) is the broadest measure of national income and measures the total claims of a country's residents to all income from domestic and foreign products during a year. Although GNI is often misleading and commonly incomplete, it is often used by economists, geographers, political scientists, policy makers, development experts, and others not only as a measure of relative well-being but also as an instrument of assessing the effectiveness of economic and political policies. What is wrong with GNI? First of all, it does not take into account a number of real economic factors,

such as, environmental deterioration, the accumulation or degradation of human and social capital, or the value of household work. Yet in spite of these deficiencies, GNI is still a reasonable way to assess the relative wealth of nations: the vast differences in wealth that separate the poorest countries from the richest. One of the more striking features of the map is the evidence it presents that such a small number of countries possess so many of the world's riches (keeping in mind that GNI provides no measure of the distribution of wealth within a country).

Map 38 Gross National Income Per Capita

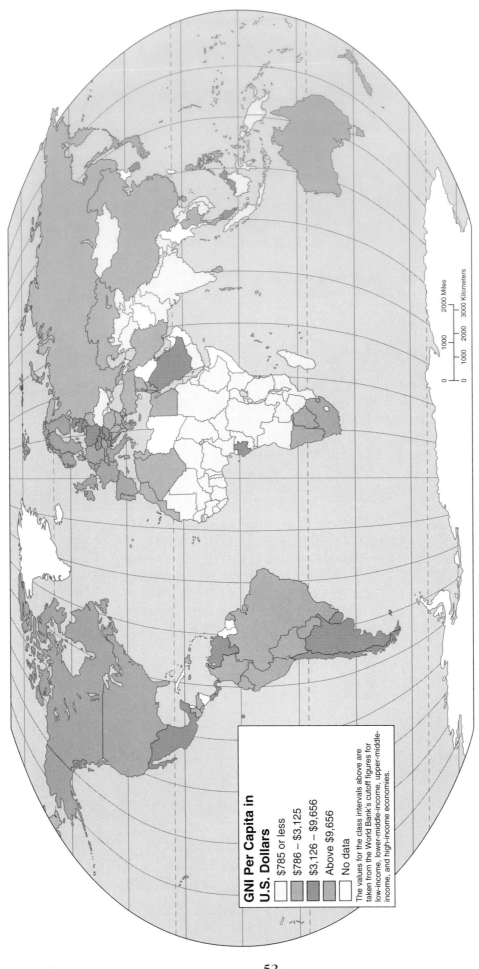

GNI Per Capita in U.S. Dollars

- $785 or less
- $786 – $3,125
- $3,126 – $9,656
- Above $9,656
- No data

The values for the class intervals above are taken from the World Bank's cutoff figures for low-income, lower-middle-income, upper-middle-income, and high-income economies.

0   1000   2000 Miles

0 1000 2000 3000 Kilometers

Gross National Income (GNI) in either absolute or per capita form should be used cautiously as a yardstick of economic strength because it does not measure the distribution of wealth among a population. There are countries (most notably, the oil-rich countries of the Middle East) where per capita GNI is high but where the bulk of the wealth is concentrated in the hands of a few individuals, leaving the remainder in poverty. Even within countries in which wealth is more evenly distributed (such as those in North America or Western Europe), there is a tendency for dollars or pounds sterling or euros to concentrate in the bank accounts of a relatively small percentage of the population. Yet the mal-distribution of wealth tends to be greatest in the less developed countries, where the per capita GNI is far lower than in North America and Western Europe, and poverty is widespread. In fact, a map of GNI per capita offers a reasonably good picture of comparative economic well-being. It should be noted that a low per capita GNI does not automatically condemn a country to low levels of basic human needs and services. There are a few countries, such as Costa Rica and Sri Lanka, that have relatively low per capita GNI figures but rank comparatively high in other measures of human well-being, such as average life expectancy, access to medical care, and literacy.

Map 39 Economic Growth: GDP Change Per Capita

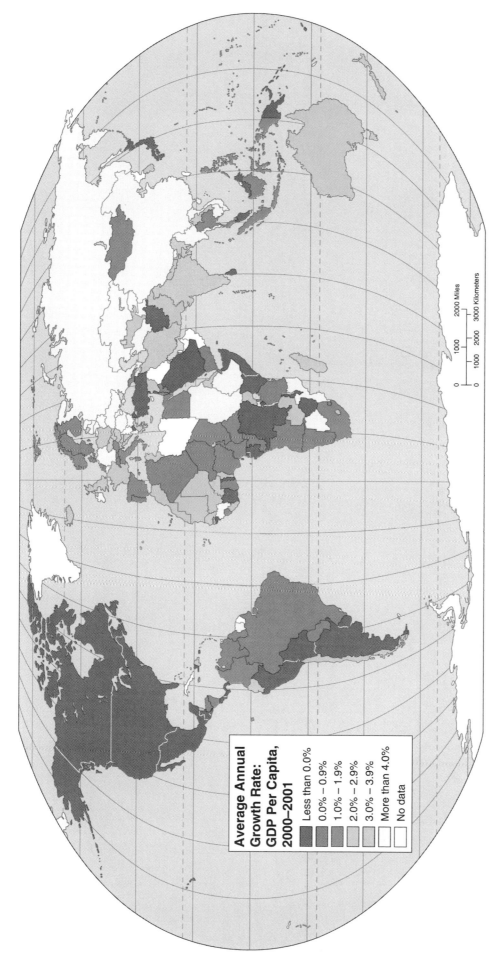

Average Annual Growth Rate: GDP Per Capita, 2000–2001

- Less than 0.0%
- 0.0% – 0.9%
- 1.0% – 1.9%
- 2.0% – 2.9%
- 3.0% – 3.9%
- More than 4.0%
- No data

Gross Domestic Product or GDP is Gross National Income (GNI) less receipts of primary income from foreign sources. While the calculations of GDP growth per capita are complex, the growth rate is considered by the World Bank and international economists to be a particularly good measure of economic growth. One of the worldwide tendencies measured by GDP growth per capita is for continued economic development in Africa, and in South, Southeast, and East Asia where GDP grew at rates higher than the growth rates of "richer" countries in Europe and North and South America. This should not necessarily be viewed as a case of the poor catching up with the rich; in fact, it shows the huge impact that even relatively small production increases will have in countries with small GNIs and GDPs. Nevertheless, in spite of the continuing low economic growth through most of sub-Saharan Africa, the GDP growth rate of some of the world's poorer countries is an encouraging trend.

Map 40 Total Labor Force

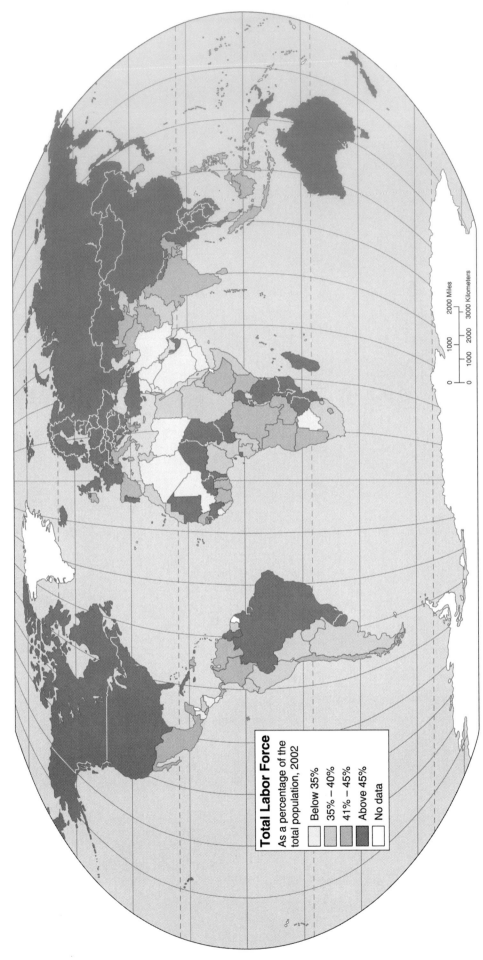

Total Labor Force

As a percentage of the total population, 2002

- Below 35%
- 35% – 40%
- 41% – 45%
- Above 45%
- No data

0	1000	2000 Miles
0	1000	2000	3000 Kilometers

The term *labor force* refers to the economically active portion of a population, that is, all people who work or are without work but are available for and are seeking work to produce economic goods and services. The total labor force thus includes both the employed and the unemployed (as long as they are actively seeking employment). Labor force is considered a better indicator of economic potential than employment/unemployment figures, since unemployment figures will include experienced workers with considerable potential who are temporarily out of work. Unemployment figures will also incorporate persons seeking employment for the first time (many recent college graduates, for example). Generally, countries with higher percentages of total population within the labor force will be countries with higher levels of economic development. This is partly a function of levels of education and training and partly a function of the age distribution of populations. In developing countries, substantial percentages of the total population are too young to be part of the labor force. Also in developing countries a significant percentage of the population consist of women engaged in household activities or subsistence cultivation. These people seldom appear on lists of either employed or unemployed seeking employment and are the world's forgotten workers.

Map 41 Employment by Economic Activity

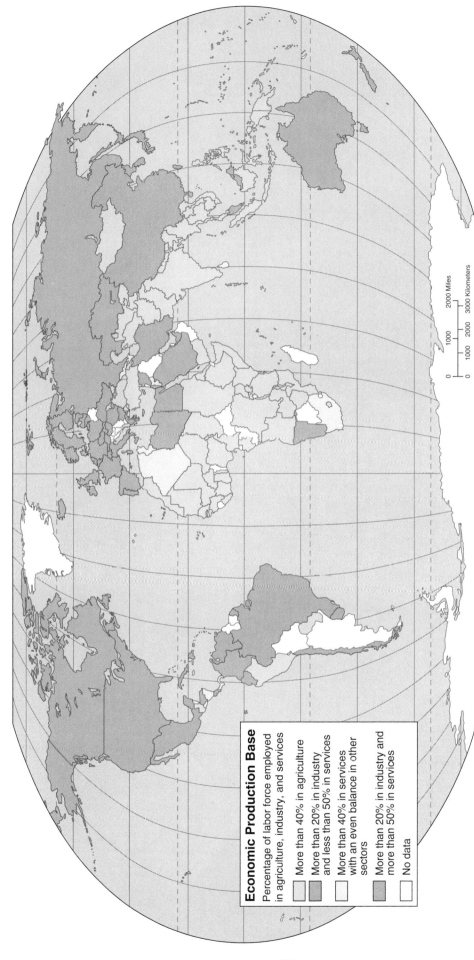

Economic Production Base

Percentage of labor force employed in agriculture, industry, and services

- More than 40% in agriculture
- More than 20% in industry and less than 50% in services
- More than 40% in services with an even balance in other sectors
- More than 20% in industry and more than 50% in services
- No data

The employment structure of a country's population is one of the best indicators of the country's position on the scale of economic development. At one end of the scale are those countries with more than 40 percent of their labor force employed in agriculture. These are almost invariably the least developed, with high population growth rates, poor human services, significant environmental problems, and so on. In the middle of the scale are two types of countries: those with more than 20 percent of their labor force employed in industry and those with a fairly even balance among agricultural, industrial, and service employment but with at least 40 percent of their labor force employed in service activities. Generally, these countries have undergone the industrial revolution fairly recently and are still developing an industrial base while building up their service activities. This category also includes countries with a disproportionate share of their economies in service activities primarily related to resource extraction. On the other end of the scale from the agricultural economies are countries with more than 20 percent of their labor force employed in industry and more than 50 percent in service activities. These countries are, for the most part, those with a highly automated industrial base and a highly mechanized agricultural system (the "postindustrial," developed countries). They also include, particularly in Middle and South America and Africa, industrializing countries that are also heavily engaged in resource extraction as a service activity.

-55-

Map 42 Economic Output per Sector

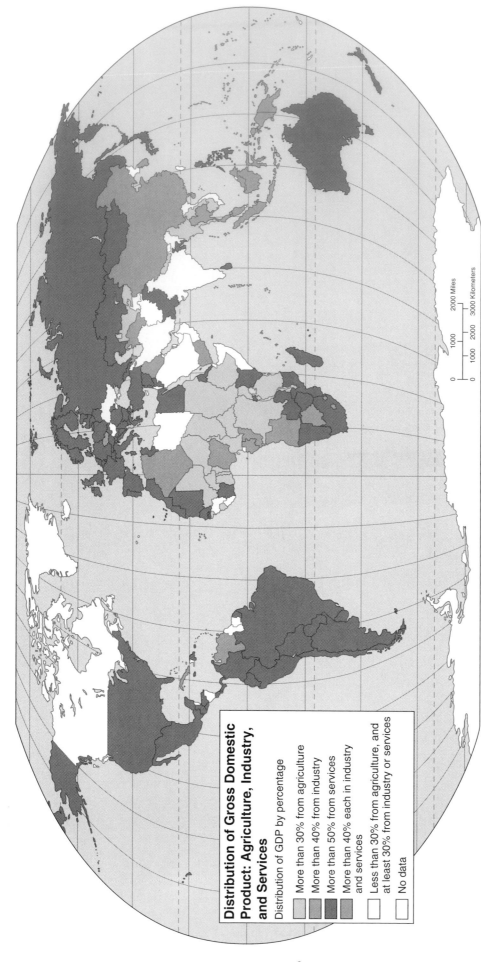

Distribution of Gross Domestic Product: Agriculture, Industry, and Services

Distribution of GDP by percentage

- More than 30% from agriculture
- More than 40% from industry
- More than 50% from services
- More than 40% each in industry and services
- Less than 30% from agriculture, and at least 30% from industry or services
- No data

0   1000   2000 Miles

0   1000   2000   3000 Kilometers

The percentage of the gross domestic product (the final output of goods and services produced by the domestic economy, including net exports of goods and nonfactor—nonlabor, noncapital—services) that is devoted to agricultural, industrial, and service activities is considered a good measure of the level of economic development. In general, countries with more than 40 percent of their GDP derived from agriculture are still in a *colonial dependency economy*—that is, raising agricultural goods primarily for the export market and dependent upon that market (usually the richer countries). Similarly, countries with more than 40 percent of GDP devoted to both agriculture and services often emphasize resource extractive (primarily mining and forestry) activities. These also tend to be *colonial dependency* countries, providing raw materials for foreign mar-

kets. Countries with more than 40 percent of their GDP obtained from industry are normally well along the path to economic development. Countries with more than half of their GDP based on service activities fall into two ends of the development spectrum. On the one hand are countries heavily dependent upon both extractive activities and tourism and other low-level service functions. On the other hand are countries that can properly be termed *postindustrial*: they have already passed through the industrial stage of their economic development and now rely less on the manufacture of products than on finance, research, communications, education, and other service-oriented activities.

Map 43 Agricultural Production Per Capita

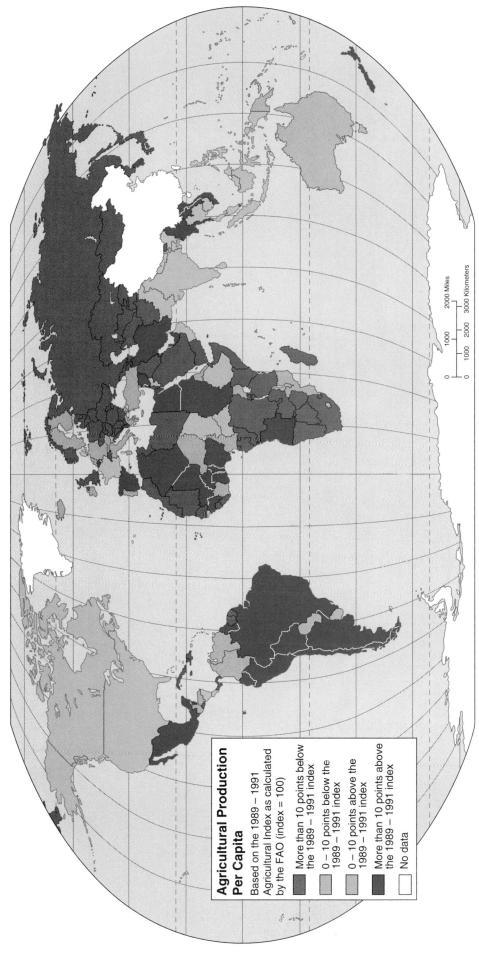

Agricultural Production Per Capita

Based on the 1989 – 1991
Agricultural Index as calculated
by the FAO (index = 100)

- More than 10 points below the 1989 – 1991 index
- 0 – 10 points below the 1989 – 1991 index
- 0 – 10 points above the 1989 – 1991 index
- More than 10 points above the 1989 – 1991 index
- No data

Agricultural production includes the value of all crop and livestock products originating within a country for the base period of 1999–2001. The index value portrays the disposable output (after deductions for livestock feed and seed for planting) of a country's agriculture in comparison with the base period 1989–1991. Thus, the production values show not only the relative ability of countries to produce food but also show whether or not that ability has increased or decreased over a 10-year period. In general, global food production has kept up with or very slightly exceeded population growth. However, there are significant regional variations in the trend of food production keeping up with or surpassing population growth. For example, agricultural production in Africa and in Middle America has fallen, while production in South America, Asia, and Europe has risen. In the case of Africa, the drop in production reflects a population growing more rapidly than

agricultural productivity. Where rapid increases in food production per capita exist (as in certain countries in South America, Asia, and Europe), most often the reason is the development of new agricultural technologies that have allowed food production to grow faster than population. In much of Asia, for example, the so-called Green Revolution of new, highly productive strains of wheat and rice made positive index values possible. Also in Asia, the cessation of major warfare allowed some countries (Cambodia, Laos, and Vietnam) to show substantial increases over the 1989–1991 index. In some cases, a drop in production per capita reflects government decisions to limit production in order to maintain higher prices for agricultural products. The United States and Japan fall into this category.

Map 44 Exports of Primary Products

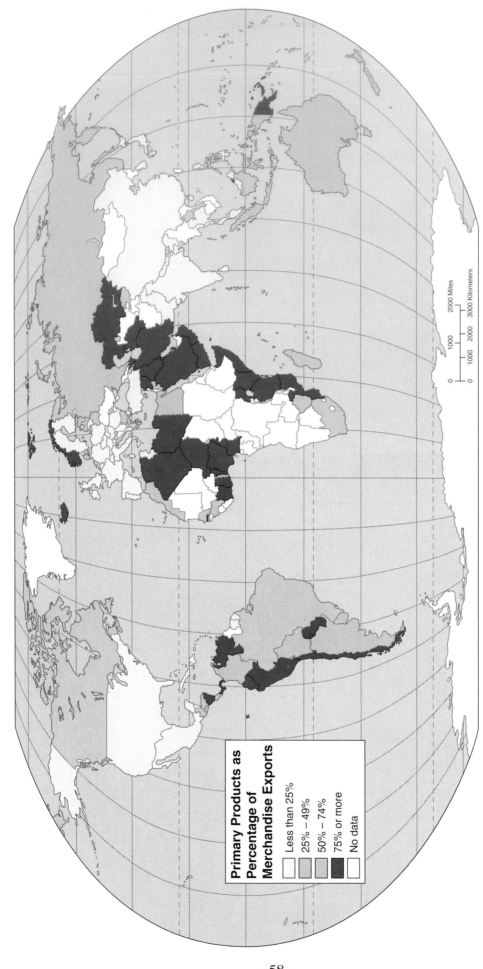

Primary Products as Percentage of Merchandise Exports

- Less than 25%
- 25% – 49%
- 50% – 74%
- 75% or more
- No data

0   1000   2000 Miles

0   1000   2000   3000 Kilometers

Primary products are those that require additional processing before they enter the consumer market: metallic ores that must be converted into metals and then into metal products such as automobiles or refrigerators; forest products such as timber that must be converted to lumber before they become suitable for construction purposes; and agricultural products that require further processing before being ready for human consumption. It is an axiom in international economics that the more a country relies on primary products for its export commodities, the more vulnerable its economy is to market fluctuations. Those countries with only primary products to export are hampered in their economic growth. A country dependent on only one or two products for export revenues is unprotected from economic shifts, particularly a changing market demand for its products. Imagine what would happen to the thriving economic status of the oil-exporting states of the Persian Gulf, for example, if an alternate source of cheap energy were found. A glance at this map, together with Map 57, shows that those countries with the lowest levels of economic development tend to be concentrated on primary products and, therefore, have economies that are especially vulnerable to economic instability.

Map 45 Dependence on Trade

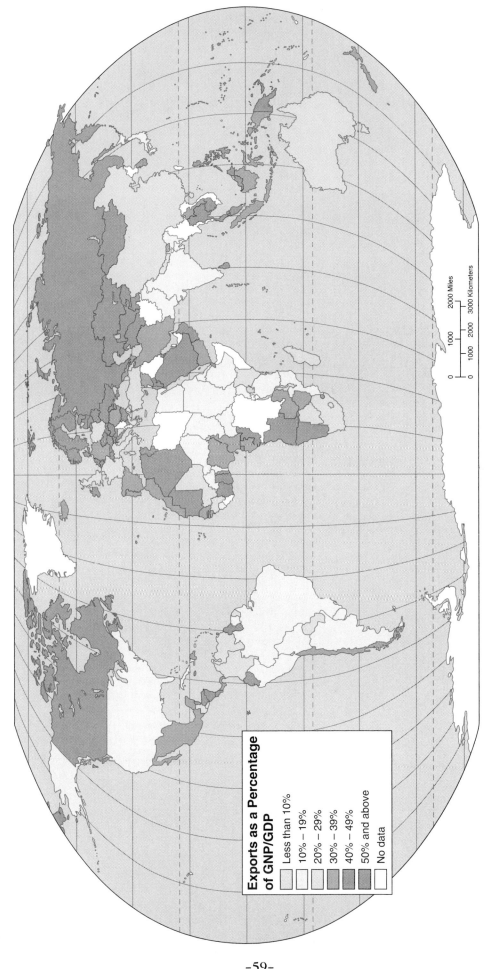

Exports as a Percentage of GNP/GDP

- Less than 10%
- 10% – 19%
- 20% – 29%
- 30% – 39%
- 40% – 49%
- 50% and above
- No data

0   1000   2000 Miles

0 1000 2000 3000 Kilometers

As the global economy becomes more and more a reality, the economic strength of virtually all countries is increasingly dependent upon trade. For many developing nations, with relatively abundant resources and limited industrial capacity, exports provide the primary base upon which their economies rest. Even countries like the United States, Japan, and Germany, with huge and diverse economies, depend on exports to generate a significant percentage of their employment and wealth. Without imports, many products that consumers want would be unavailable or more expensive; without exports, many jobs would be eliminated.

Map 46 The Indebtedness of States

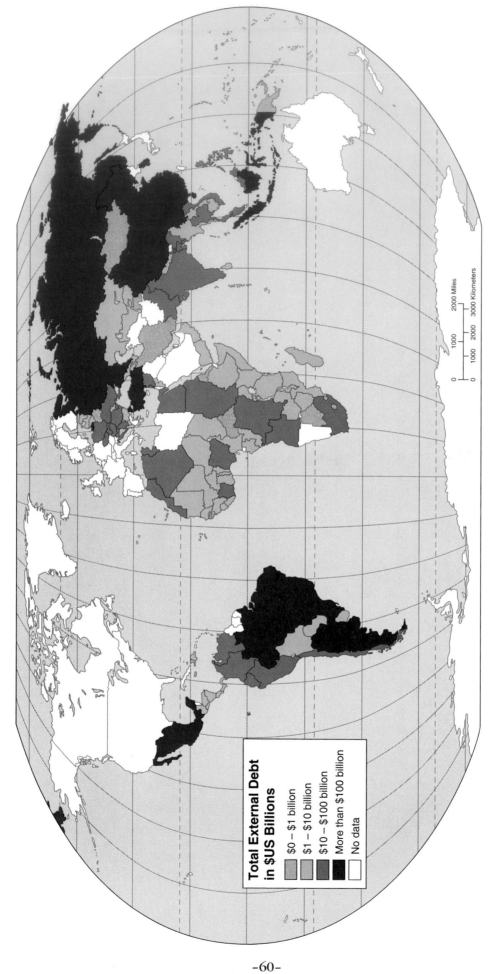

Total External Debt in $US Billions

- $0 – $1 billion
- $1 – $10 billion
- $10 – $100 billion
- More than $100 billion
- No data

Many governments spend more on a wide variety of services and activities than they collect in taxes and other revenues. In order to finance this deficit spending, governments borrow money—often from banks or other investors outside their country. Repayment of these debts, or even meeting interest payments on them, often means expending a country's export income—in other words, exchanging a country's wealth in production or, more often, resources, for debt service. Where the debt is external, as it is in most developing countries, governments become more open to outside influence in political as well as economic terms. Even internal debt service or repayment of monies owed to investors within a country gives financial establishments a measure of influence over government decisions. The amounts of debt shown on the map indicate the total external indebtedness of states.

-60-

Map 47 Global Flows of Investment Capital

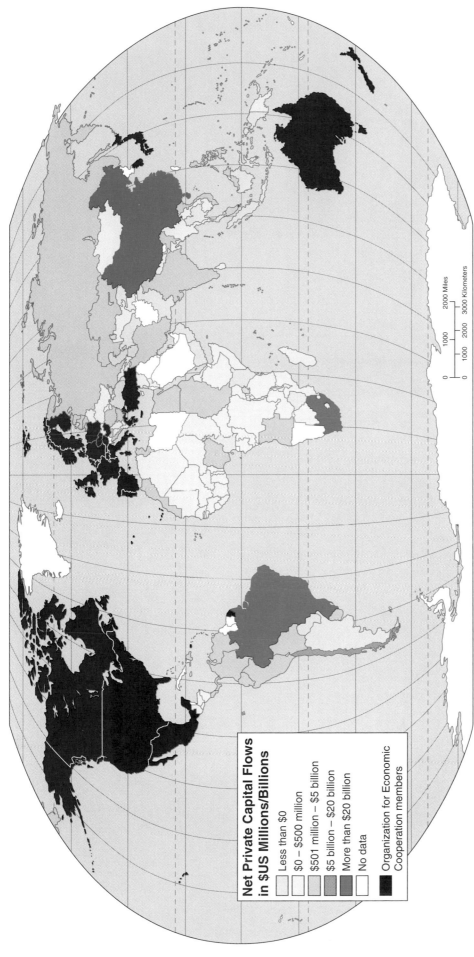

Net Private Capital Flows in $US Millions/Billions

- Less than $0
- $0 – $500 million
- $501 million – $5 billion
- $5 billion – $20 billion
- More than $20 billion
- No data
- Organization for Economic Cooperation members

0   1000   2000 Miles
0   1000   2000   3000 Kilometers

International capital flows include private debt and nondebt flows from one country to another, shown on the map as flows into a country. Nearly all of the capital comes from those countries that are members of the Organization for Economic Cooperation and Development (OECD), shown in black on the map. Capital flows include commercial bank lending, bonds, other private credits, foreign direct investment, and portfolio investment. Most of these flows are indicators of the increasing influence developed countries exert over the developing economies. Foreign direct investment or FDI, for example, is a measure of the net inflow of investment monies used to acquire long-term management interest in businesses located somewhere other than in the economy of the investor. Usually this means the acquisition of at least 10 percent of the stock of a company by a foreign investor and is, then, a measure of what might be termed "economic colonialism": control of a region's economy by foreign investors that could, in the world of the future, be as significant as colonial political control was in the past. International capital flows have increased greatly in the last decade as the result of the increasing liberalization of developing countries, the strong economic growth exhibited by many developing countries, and the falling costs and increased efficiency of communication and transportation services.

Map 48 Central Government Expenditures Per Capita

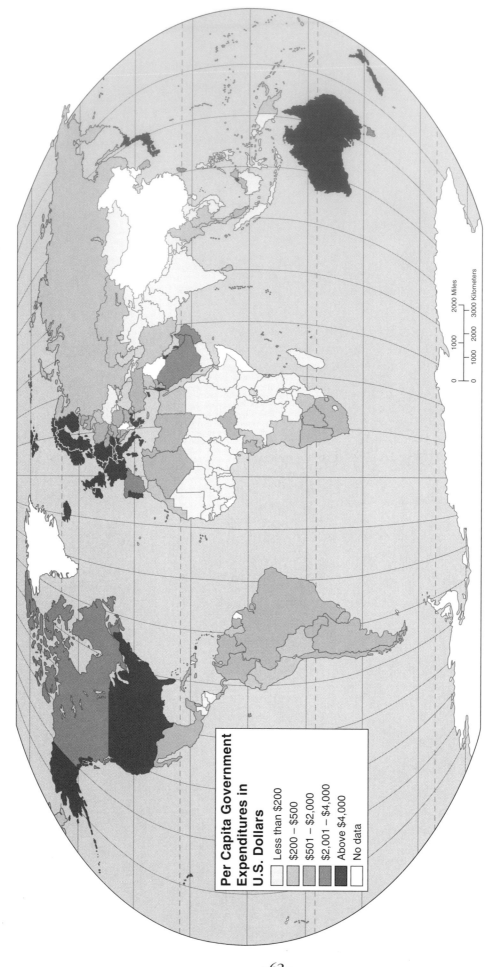

Per Capita Government Expenditures in U.S. Dollars

- Less than $200
- $200 – $500
- $501 – $2,000
- $2,001 – $4,000
- Above $4,000
- No data

The amount of money that the central government of a country spends upon a variety of essential governmental functions is a measure of relative economic development, particularly when it is viewed on a per-person basis. These functions include such governmental responsibilities as agriculture, communications, culture, defense, education, fishing and hunting, health, housing, recreation, religion, social security, transportation, and welfare. Generally, the higher the level of economic development, the greater the per capita expenditures on these services. However, the data do mask some internal variations. For example, countries that spend 20 percent or more of their central government expenditures on defense will often show up in the more developed category when, in fact, all that the figures really show is that a disproportionate amount of the money available to the government is devoted to purchasing armaments and maintaining a large standing military force. Thus, the fact that Libya spends more than the average for Africa does not suggest that the average Libyan is much better off than the average Tanzanian. Nevertheless, this map—particularly when compared with Map 51, Energy Consumption Per Capita—does provide a reasonable approximation of economic development levels.

Map 49 The Relative Wealth of Nations: Purchasing Power Parity

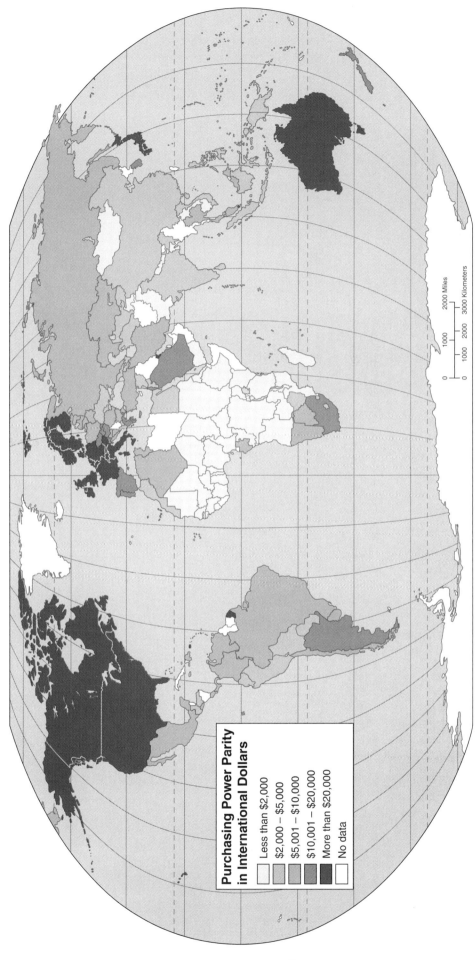

Purchasing Power Parity in International Dollars

- Less than $2,000
- $2,000 – $5,000
- $5,001 – $10,000
- $10,001 – $20,000
- More than $20,000
- No data

0   1000   2000 Miles
0   1000   2000   3000 Kilometers

Of all the economic measures that separate the "haves" from the "have-nots," perhaps per capita Purchasing Power Parity (PPP) is the most meaningful. While per capita figures can mask significant uneven distributions within a country, they are generally useful for demonstrating important differences between countries. Per capita GNP and GDP (Gross Domestic Product) figures, and even per capita income, have the limitation of seldom reflecting the true purchasing power of a country's currency at home. In order to get around this limitation, international economists seeking to compare national currencies developed the PPP measure, which shows the level of goods and services that holders of a country's money can acquire locally. By converting all currencies to the "international dollar," the World Bank and other organizations using PPP can now show more truly comparative values, since the new currency value shows the

number of units of a country's currency required to buy the same quantity of goods and services in the local market as one U.S. dollar would buy in an average country. The use of PPP currency values can alter the perceptions about a country's true comparative position in the world economy. More than per capita income figures, PPP provides a valid measurement of the ability of a country's population to provide for itself the things that people in the developed world take for granted: adequate food, shelter, clothing, education, and access to medical care. A glance at the map shows a clear-cut demarcation between temperate and tropical zones, with most of the countries with a PPP above $5,000 in the midlatitude zones and most of those with lower PPPs in the tropical and equatorial regions. Where exceptions to this pattern occur, they usually stem from a tremendous maldistribution of wealth among a country's population.

Map 50 Energy Production Per Capita

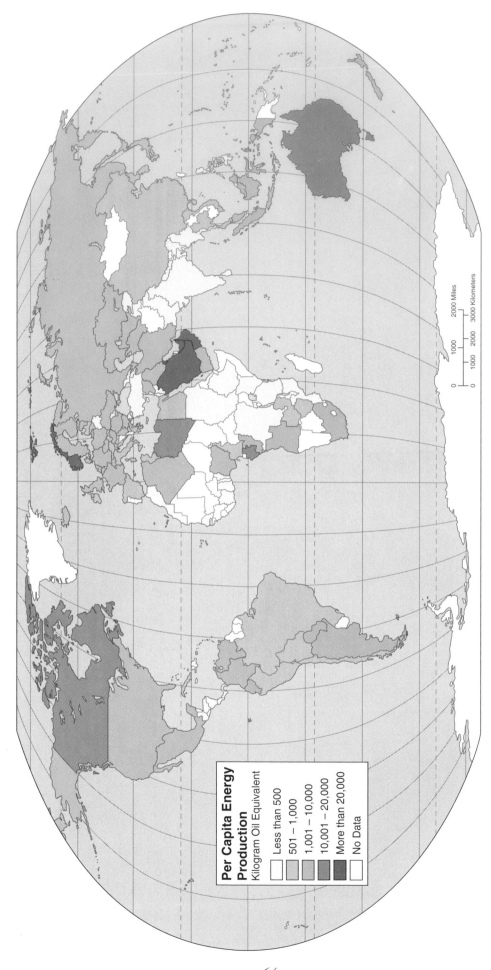

Per Capita Energy Production
Kilogram Oil Equivalent

Less than 500
501 – 1,000
1,001 – 10,000
10,001 – 20,000
More than 20,000
No Data

Energy production per capita is a measure of the availability of mechanical energy to assist people in their work. This map shows the amount of all kinds of energy—solid fuel (primarily coal), liquid fuel (primarily petroleum), natural gas, geothermal, wind, solar, hydroelectric, nuclear, waste recycling, and indigenous heat pumps—produced per person in each country. With some exceptions, wealthier countries produce more energy per capita than poor ones. Countries such as Japan and many European states rank among the world's wealthiest, but are energy-poor and produce relatively little of their own energy. They have the ability, however, to pay for imports. On the other hand, countries such as those of the Persian Gulf or the oil-producing states of Central and South America may rank relatively low on the scale of economic development but rank high as producers of energy. In many poor countries, especially in Central and South America, Africa, South Asia, and East Asia, large proportions of energy come from traditional fuels such as firewood and animal dung. Indeed for many in the developing world, the real energy crisis is a shortage of wood for cooking and heating.

-64-

Map 51 Energy Consumption Per Capita

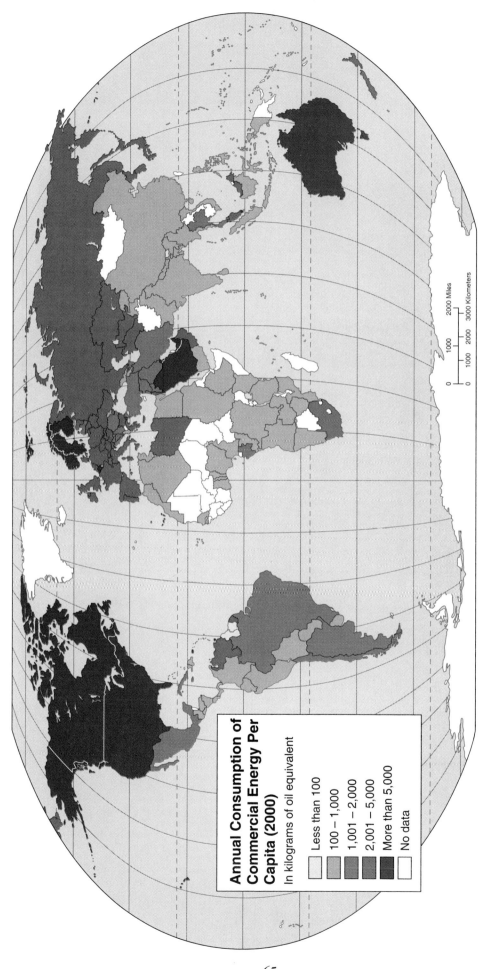

Annual Consumption of Commercial Energy Per Capita (2000)

In kilograms of oil equivalent

- Less than 100
- 100 – 1,000
- 1,001 – 2,000
- 2,001 – 5,000
- More than 5,000
- No data

Of all the quantitative measures of economic well-being, energy consumption per capita may be the most expressive. All of the countries defined by the World Bank as having high incomes consume at least 100 gigajoules of commercial energy (the equivalent of about 3.5 metric tons of coal) per person per year, with some, such as the United States and Canada, having consumption rates in the 300 gigajoule range (the equivalent of more than 10 metric tons of coal per person per year). With the exception of the oil-rich Persian Gulf states, where consumption figures include the costly "burning off" of excess energy in the form of natural gas flares at wellheads, most of the highest-consuming countries are in the Northern Hemisphere, concentrated in North America and Western Europe. At the other end of the scale are low-income countries, whose consumption rates are often less than 1 percent of those of the United States and other high consumers. These figures do not, of course, include the consumption of noncommercial energy—the traditional fuels of firewood, animal dung, and other organic matter—widely used in the less developed parts of the world.

Map **52** Energy Dependency

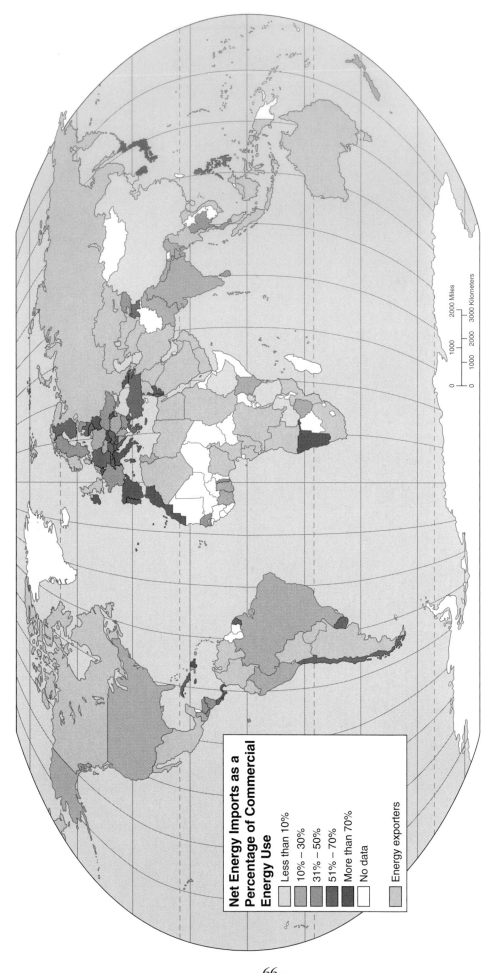

Net Energy Imports as a Percentage of Commercial Energy Use

- Less than 10%
- 10% – 30%
- 31% – 50%
- 51% – 70%
- More than 70%
- No data
- Energy exporters

The patterns on the map show dependence on commercial energy before transformation to other end-use fuels such as electricity or refined petroleum products; energy from traditional sources such as fuelwood or dried animal dung is not included. Energy dependency is the difference between domestic consumption and domestic production of commercial energy and is most often expressed as a net energy import or export. A few of the world's countries are net exporters of energy; most are importers. The growth in global commercial energy use over the last decade indicates growth in the modern sectors of the economy—industry, transportation, and urbanization—particularly in the lesser developed countries. Still, the primary consumers of energy—and those having the greatest dependence on foreign sources of energy—are the more highly developed countries of Europe, North America, and Japan.

Map **53** Flows of Oil

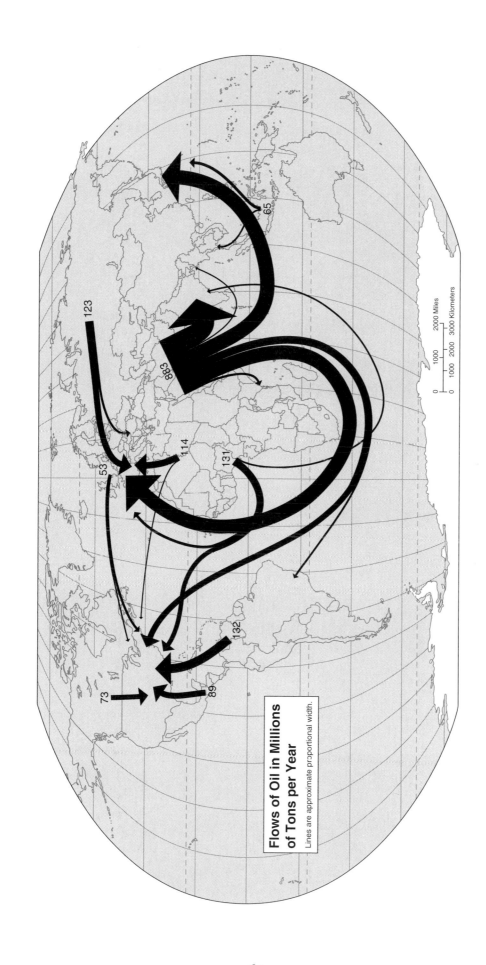

Flows of Oil in Millions of Tons per Year

Lines are approximate proportional width.

The pattern of oil movements from producing region to consuming region is one of the dominant facts of contemporary international maritime trade. Supertankers carry a million tons of crude oil and charge rates in excess of $0.10 per ton per mile, making the transportation of oil not only a necessity for the world's energy-hungry countries, but also an enormously profitable proposition. One of the major negatives of these massive oil flows is the damage done to the oceanic ecosystems—not just from the well-publicized and dramatic events like the wrecking of the *Exxon Valdez* but from the ircalculable amounts of oil from leakage, scrubbings, purgings, and so on, which are a part of the oil transport technology. It is clear from the map that the primary recipients of these oil flows are the world's most highly developed economies.

Part V

Global Patterns of Environmental Disturbance

Map 54 Global Air Pollution: Sources and Wind Currents

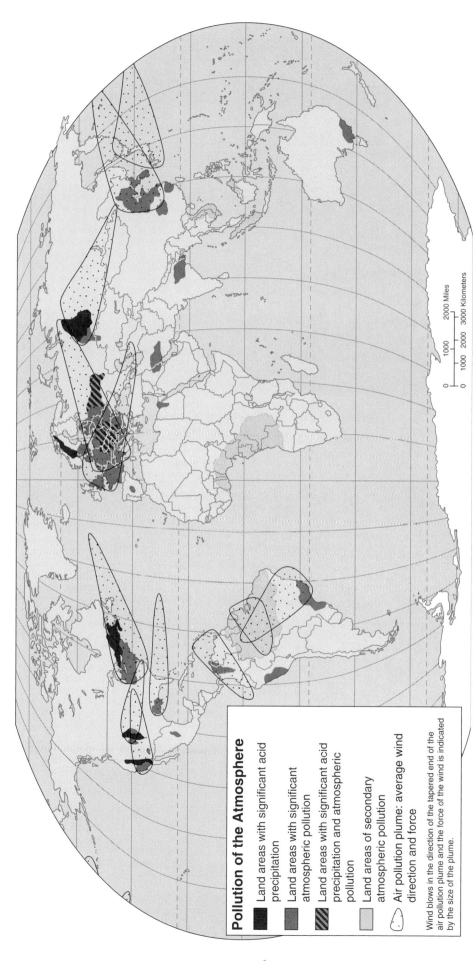

Pollution of the Atmosphere

- Land areas with significant acid precipitation
- Land areas with significant atmospheric pollution
- Land areas with significant acid precipitation and atmospheric pollution
- Land areas of secondary atmospheric pollution
- Air pollution plume: average wind direction and force

Wind blows in the direction of the tapered end of the air pollution plume and the force of the wind is indicated by the size of the plume.

0   1000   2000 Miles
0   1000   2000   3000 Kilometers

Almost all processes of physical geography begin and end with the flows of energy and matter among land, sea, and air. Because of the primacy of the atmosphere in this exchange system, air pollution is potentially one of the most dangerous human modifications in environmental systems. Pollutants such as various oxides of nitrogen or sulfur cause the development of acid precipitation, which damages soil, vegetation, and wildlife and fish. Air pollution in the form of smog is often dangerous for human health. And most atmospheric scientists believe that the efficiency of the atmosphere in retaining heat—the so-called greenhouse effect—is being enhanced by increased carbon dioxide, methane, and other gases produced by agricultural and industrial activities. The result, they fear, will be a period of global warming that will dramatically alter climates in all parts of the world.

Map 55 The Acid Deposition Problem: Air, Water, Soil

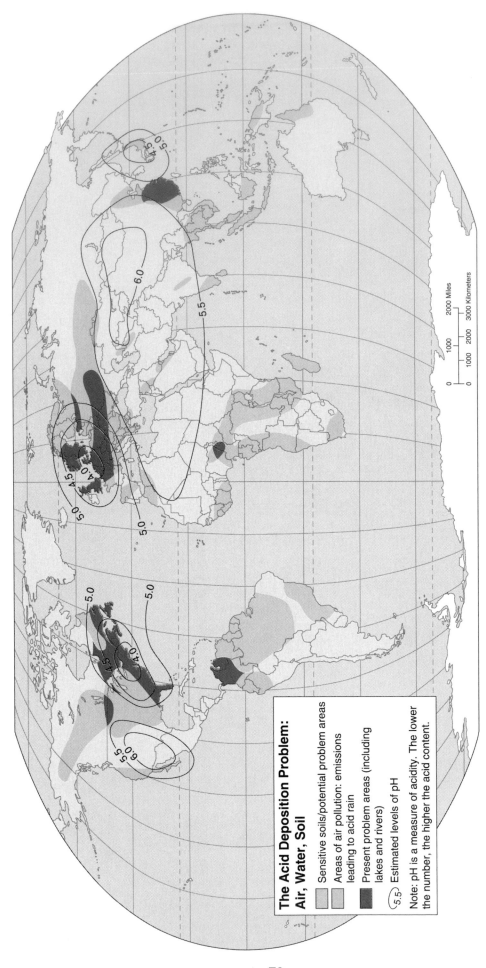

The Acid Deposition Problem: Air, Water, Soil

- Sensitive soils/potential problem areas
- Areas of air pollution: emissions leading to acid rain
- Present problem areas (including lakes and rivers)
- 5.5 Estimated levels of pH

Note: pH is a measure of acidity. The lower the number, the higher the acid content.

0   1000   2000 Miles
0   1000   2000   3000 Kilometers

The term "acid precipitation" refers to increasing levels of acidity in snowfall and rainfall caused by atmospheric pollution. Oxides of nitrogen and sulfur resulting from incomplete combustion of fossil fuels (coal, oil, and natural gas) combine with water vapor in the atmosphere to produce weak acids that then "precipitate" or fall along with water or ice crystals. Some atmospheric acids formed by this process are known as "dry-acid" precipitates and they too will fall to earth, although not necessarily along with rain or snow. In some areas of the world, the increased acidity of streams and lakes stemming from high levels of acid precipitation or dry acid fallout has damaged or destroyed aquatic life. Acid precipitation and dry acid fallout also harms soil systems and vegetation, producing a characteristic burned appearance in forests that lends the same quality to landscapes that forest fires would. The region most dramatically impacted by acid precipitation is Central Europe where decades of destructive environmental practices, including the burning of high sulfur coal for commercial, industrial, and residential purposes, has produced the destruction of hundreds of thousands of acres of woodlands—a phenomenon described by the German foresters who began their study of the area following the lifting of the Iron Curtain as "Waldsterben": Forest Death.

Map 56 Major Polluters and Common Pollutants

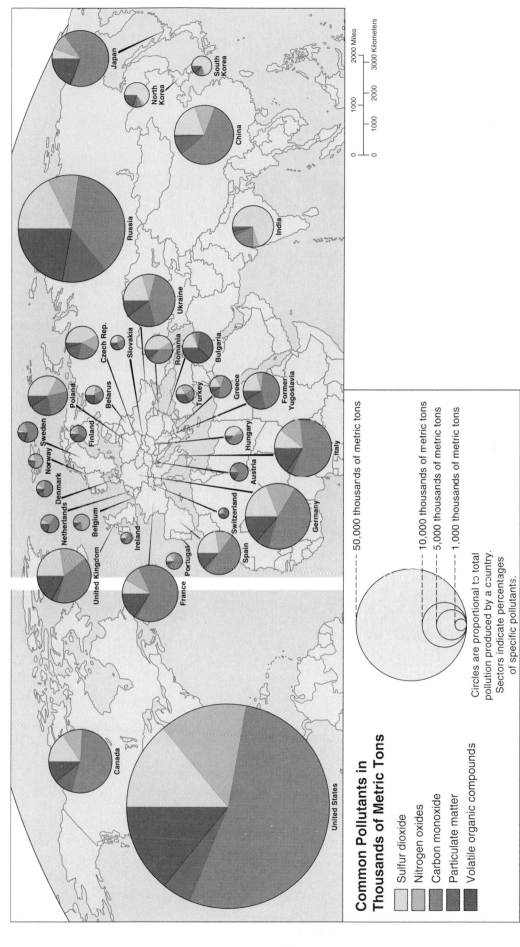

Common Pollutants in Thousands of Metric Tons

- Sulfur dioxide
- Nitrogen oxides
- Carbon monoxide
- Particulate matter
- Volatile organic compounds

---- 50,000 thousands of metric tons

---- 10,000 thousands of metric tons
---- 5,000 thousands of metric tons
---- 1,000 thousands of metric tons

Circles are proportional to total
pollution produced by a country.
Sectors indicate percentages
of specific pollutants.

More than 90 percent of the world's total of anthropogenic (human-generated) air pollutants come from the heavily populated industrial regions of North America, Europe, South Asia (primarily in India), and East Asia (mainly in China, Japan, and the two Koreas). This map shows the origins of the five most common pollutants: sulfur dioxide, nitrogen oxide, carbon monoxide, particulate matter, and volatile organic compounds. These substances are produced both by industry and by the combustion of fossil fuels that generate electricity and power trains, planes, automobiles, buses, and trucks. In addition to combining with other components of the atmosphere and with one another to produce smog, they are the chief ingredients in acid accumulations in the atmosphere, which ultimately result in acid deposition, either as acid precipitation or dry acid fallout. Like other forms of pollutants, these air pollutants do not recognize political boundaries, and regions downwind of major polluters receive large quantities of pollutants from areas over which they often have no control.

Map 57 Global Carbon Dioxide Emissions

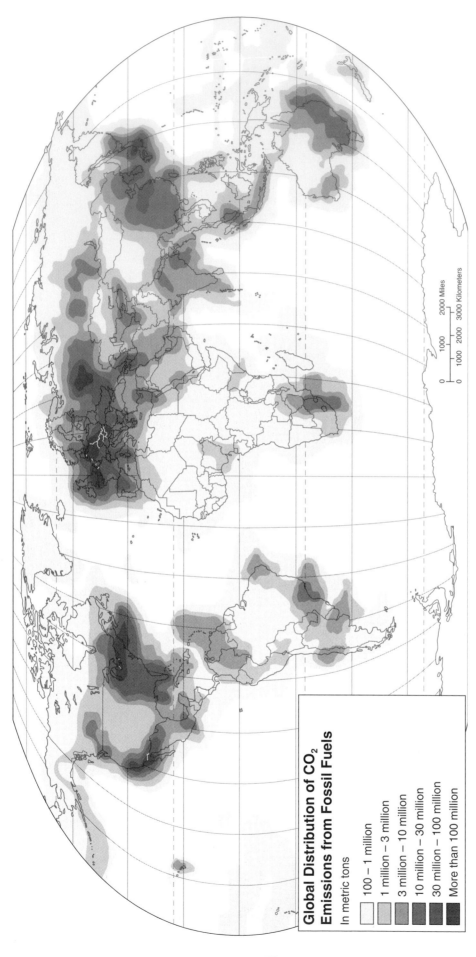

Global Distribution of CO₂ Emissions from Fossil Fuels

In metric tons

- 100 – 1 million
- 1 million – 3 million
- 3 million – 10 million
- 10 million – 30 million
- 30 million – 100 million
- More than 100 million

One of the most important components of the atmosphere is the gas carbon dioxide (CO₂), the byproduct of animal respiration, decomposition, and combustion. During the past 200 years, atmospheric CO₂ has risen dramatically, largely as the result of the tremendous increase in fossil fuel combustion brought on by the industrialization of the world's economy and the burning and clearing of forests by the expansion of farming. While CO₂ by itself is relatively harmless, it is an important "greenhouse gas." The gases in the atmosphere act like the panes of glass in a greenhouse roof, allowing light in but preventing heat from escaping. The greenhouse capacity of the atmosphere is

crucial for organic life and is a purely natural component of the global energy cycle. But too much CO₂ and other greenhouse gases such as methane could cause the earth's atmosphere to warm up too much, producing the global warming that atmospheric scientists are concerned about. Researchers estimate that if greenhouse gases such as CO₂ continue to increase at their present rates, the earth's mean temperature could rise between 1.5 and 4.5 degrees Celsius by the middle of the next century. Such a rise in global temperatures would produce massive alterations in the world's climate patterns.

Map 58 Potential Global Temperature Change

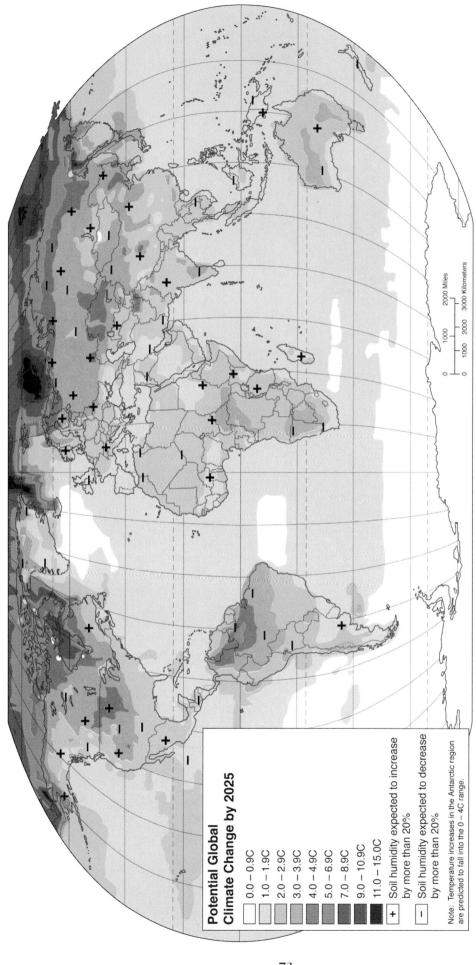

Potential Global Climate Change by 2025

- 0.0 – 0.9C
- 1.0 – 1.9C
- 2.0 – 2.9C
- 3.0 – 3.9C
- 4.0 – 4.9C
- 5.0 – 6.9C
- 7.0 – 8.9C
- 9.0 – 10.9C
- 11.0 – 15.0C

+ Soil humidity expected to increase by more than 20%

| Soil humidity expected to decrease by more than 20%

Note: Temperature increases in the Antarctic region are predicted to fall into the 0 – 4C range.

0   1000   2000 Miles
0   1000   2000   3000 Kilometers

According to atmospheric scientists, one of the major problems of the twenty-first century will be "global warming," produced as the atmosphere's natural ability to trap and retain heat is enhanced by increased percentages of carbon dioxide, methane, chlorinated fluorocarbons or "CFCs," and other "greenhouse gases" in the earth's atmosphere. Computer models based on atmospheric percentages of carbon dioxide resulting from present use of fossil fuels show that warming is not just a possibility but a probability. Increased temperatures would cause precipitation patterns to alter significantly as well and would produce a number of other harmful effects, including a rise in the level of the world's oceans that could flood most coastal cities. International conferences on the topic of the enhanced greenhouse effect have resulted in several international agreements to reduce the emission of carbon dioxide or to maintain it at present levels. Unfortunately, the solution is not that simple since reduction of carbon dioxide emissions is, in the short run, expensive—particularly as long as the world's energy systems continue to be based on fossil fuels. Chief among the countries that could be hit by serious international mandates to reduce emissions are those highest on the development scale who use the highest levels of fossil fuels and, therefore, produce the highest emissions, and those on the lowest end of the development scale whose efforts to industrialize could be severely impeded by the more expensive energy systems that would replace fossil fuels.

Map **59** Water Resources: Availability of Renewable Water Per Capita

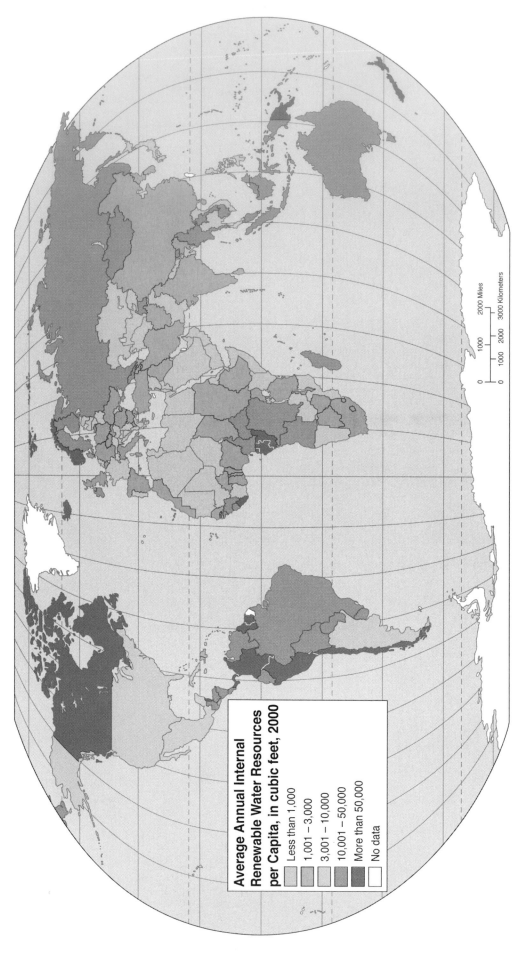

**Average Annual Internal
Renewable Water Resources
per Capita, in cubic feet, 2000**

Less than 1,000
1,001 – 3,000
3,001 – 10,000
10,001 – 50,000
More than 50,000
No data

0   1000   2000 Miles

0   1000   2000   3000 Kilometers

Renewable water resources are usually defined as the total water available from streams and rivers (including flows from other countries), ponds and lakes, and groundwater storage or aquifers. Not included in the total of renewable water would be water that comes from such nonrenewable sources as desalinization plants or melted icebergs. While the concept of renewable or flow resources is a traditional one in resource management, in fact, few resources, including water, are truly renewable when their use is excessive. The water resources shown here are indications of that principle. A country like the United States possesses truly enormous quantities of water. But the United States also uses enormous quantities of water. The result is that, largely because of excessive use, the availability of renewable water is much less than in many other parts of the world where the total supply of water is significantly less.

Map 60 Water Resources: Annual Withdrawal Per Capita

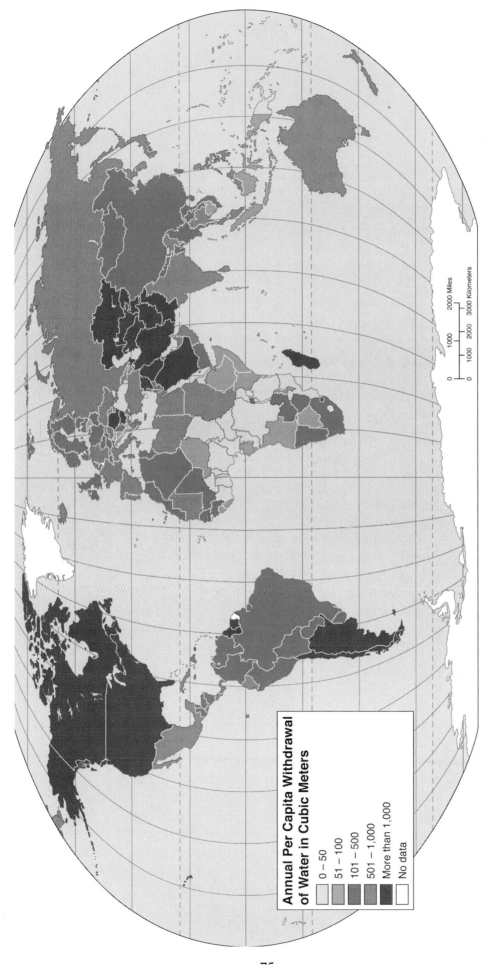

Annual Per Capita Withdrawal of Water in Cubic Meters

- 0 – 50
- 51 – 100
- 101 – 500
- 501 – 1,000
- More than 1,000
- No data

Water resources must be viewed like a bank account in which deposits and withdrawals are made. As long as the deposits are greater than the withdrawals, a positive balance remains. But when the withdrawals begin to exceed the deposits, sooner or later (depending on the relative sizes of the deposits and withdrawals) the account becomes overdrawn. For many of the world's countries, annual availability of water is insufficient to cover the demand. In these countries, reserves stored in groundwater are being tapped, resulting in depletion of the water supply (think of this as shifting money from a savings account to a checking account). The water supply can maintain its status as a renewable resource only if deposits continue to be greater than withdrawals, and that seldom happens. In general, countries with high levels of economic development and countries that rely on irrigation agriculture are the most spendthrift when it comes to their water supplies.

Map 61 Pollution of the Oceans

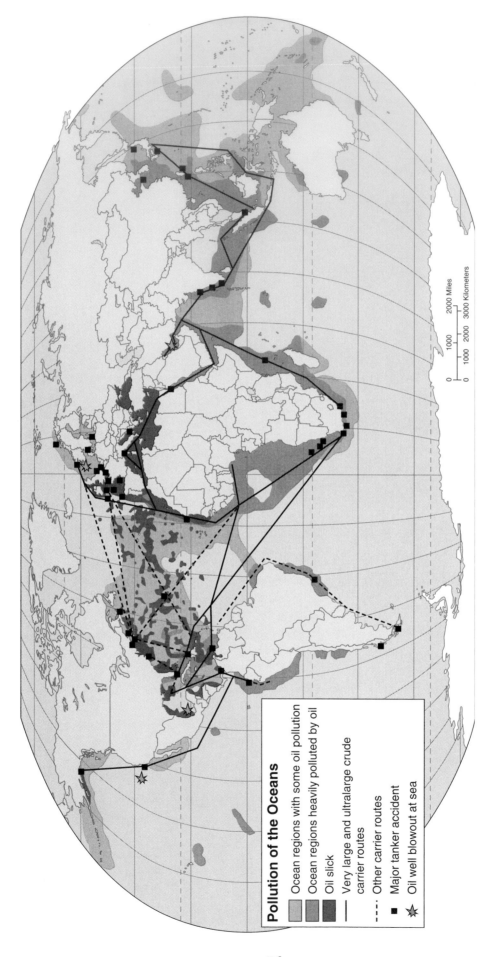

Pollution of the Oceans

- Ocean regions with some oil pollution
- Ocean regions heavily polluted by oil
- Oil slick
- —— Very large and ultralarge crude carrier routes
- - - - Other carrier routes
- ■ Major tanker accident
- ☆ Oil well blowout at sea

| 0 | 1000 | 2000 Miles |
| 0 | 1000 | 2000 | 3000 Kilometers |

The pollution of the world's oceans has long been a matter of concern to physical geographers, oceanographers, and other environmental scientists. The great circulation systems of the ocean are one of the controlling factors of the earth's natural environment, and modifications to those systems have unknown consequences. This map is based on what we can measure: (1) areas of oceans where oil pollution has been proven to have inflicted significant damage to ocean ecosystems and life-forms (including phytoplankton, the oceans' primary food producers, equivalent to land-based vegetation) and (2) areas of oceans where unusually high concentrations of hydrocarbons from oil spills may have inflicted some damage to the oceans' biota. A glance at the map shows that there are few areas of the world's oceans where some form of pollution is not a part of the environmental system. What the map does not show in detail, because of the scale, are the dramatic consequences of large individual pollution events: the wreck of the *Exxon Valdez* and the polluting of Prince William Sound, or the environmental devastation produced by the 1991 Gulf War in the Persian Gulf.

Map 62 Food Supply From Marine and Freshwater Systems

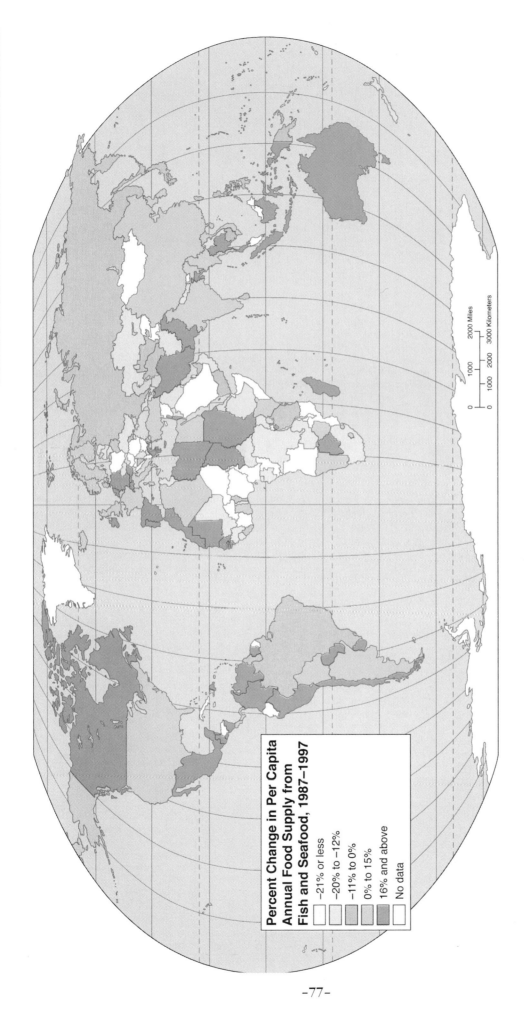

Percent Change in Per Capita Annual Food Supply from Fish and Seafood, 1987–1997

- −21% or less
- −20% to −12%
- −11% to 0%
- 0% to 15%
- 16% and above
- No data

0   1000   2000   2000 Miles
0   1000   2000   3000 Kilometers

Not that many years ago, food supply experts were confidently predicting that the "starving millions" of the world of the future could be fed from the unending bounty of the world's oceans. While the annual catch from the sea helped to keep hunger at bay for a time, by the late 1980s it had become apparent that without serious human intervention in the form of aquaculture, the supply of fish would not be sufficient to offset the population/food imbalance that was beginning to affect so many of the world's regions. The development of factory-fishing with advanced equipment to locate fish and process them before they went to market increased the supply of food from the ocean, but in that increase was sown the seeds of future problems. The factory-fishing system, efficient in terms of economics, was costly in terms of fish populations. In some

well-fished areas, the stock of fish that was viewed as near infinite just a few decades ago has dwindled nearly to the point of disappearance. This map shows both increases and decreases in the amount of individual countries' food supplies from the ocean. The increases are often the result of more technologically advanced fishing operations. The decreases are usually the result of the same thing: increased technology has brought increased harvests, which has reduced the supply of fish and shellfish and that, in turn, has increased prices. Most of the countries that have experienced sharp decreases in their supply of food from the world's oceans are simply no longer able to pay for an increasingly scarce commodity.

Map 63 Cropland Per Capita: Changes, 1987–1997

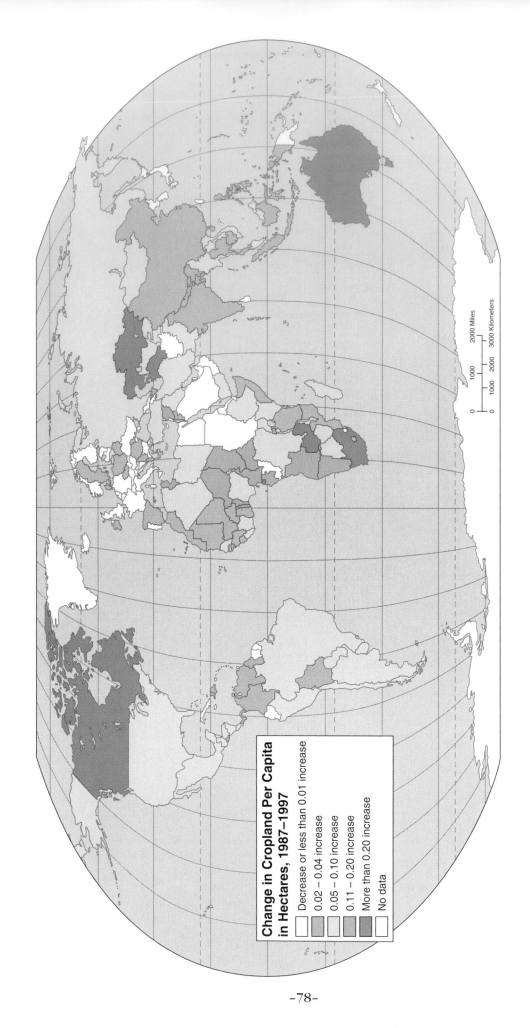

Change in Cropland Per Capita in Hectares, 1987–1997

- Decrease or less than 0.01 increase
- 0.02 – 0.04 increase
- 0.05 – 0.10 increase
- 0.11 – 0.20 increase
- More than 0.20 increase
- No data

0   1000   2000 Miles

0 1000 2000 3000 Kilometers

As population has increased rapidly throughout the world, the area of cultivated land has increased at the same time; in fact, the amount of farmland per person has gone up slightly. Unfortunately, the figures that show this also tell us that since most of the best (or even good) agricultural land in 1985 was already under cultivation, most of the agricultural area added since the early 1980s involves land that would have been viewed as marginal by the fathers and grandfathers of present farmers—marginal in that it was too dry, too wet, too steep to cultivate, too far from a market, and so on. The continued expansion of agricultural area is one reason that serious famine and starvation have struck only a few regions of the globe. But land, more than any other resource we deal with, is finite, and the expansion cannot continue indefinitely. Future gains in agricultural production are most probably going to come through more intensive use of existing cropland, heavier applications of fertilizers and other agricultural chemicals, and genetically engineered crops requiring heavier applications of energy and water, than from an increase in the amount of the world's cropland.

-78-

Map 64 Annual Change in Forest Cover, 1990–1995

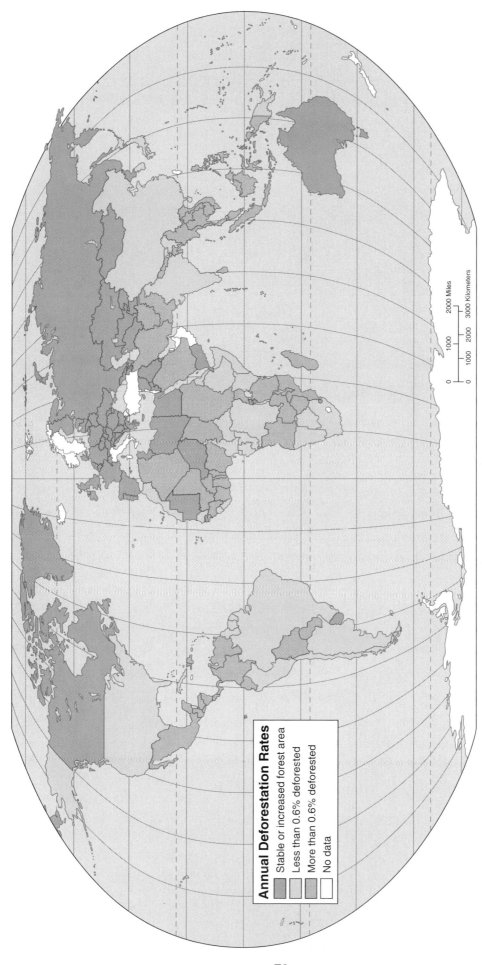

Annual Deforestation Rates
- Stable or increased forest area
- Less than 0.6% deforested
- More than 0.6% deforested
- No data

One of the most discussed environmental problems is that of deforestation. For most people, deforestation means clearing of tropical rain forests for agricultural purposes. Yet nearly as much forest land per year—much of it in North America, Europe, and Russia—is impacted by commercial lumbering as is cleared by tropical farmers and ranchers. Even in the tropics, much of the forest clearance is undertaken by large corporations producing high-value tropical hardwoods for the global market in furniture, ornaments, and other fine wood products. Still, it is the agriculturally driven clear-ing of the great rain forests of the Amazon Basin, west and central Africa, Middle America, and Southeast Asia that draws public attention. Although much concern over forest clearance focuses on the relationship between forest clearance and the reduction in the capacity of the world's vegetation system to absorb carbon dioxide (and thus delay global warming), of just as great concern are issues having to do with the loss of biodiversity (large numbers of plants and animals), the near-total destruction of soil systems, and disruptions in water supply that accompany clearing.

Map 65

Map 65 The Loss of Biodiversity: Globally Threatened Animal Species

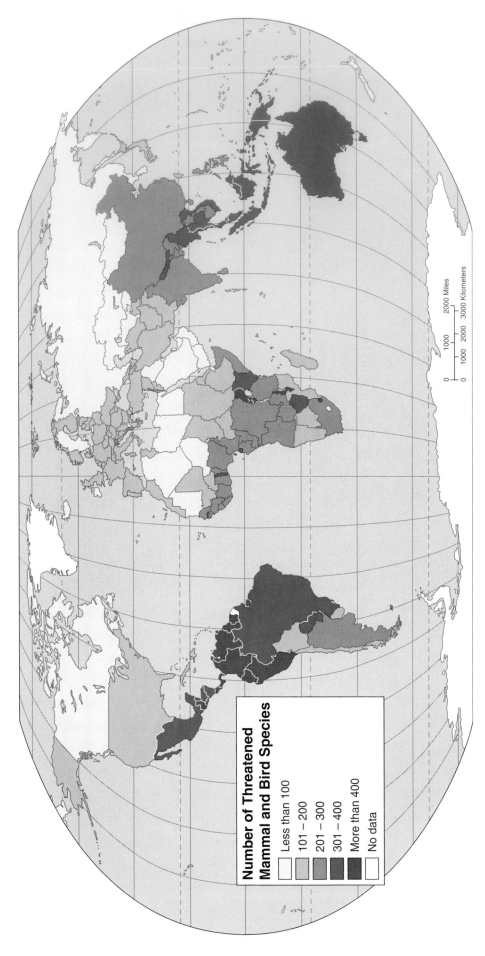

Number of Threatened Mammal and Bird Species

- Less than 100
- 101 – 200
- 201 – 300
- 301 – 400
- More than 400
- No data

0 1000 2000 Miles
0 1000 2000 3000 Kilometers

Threatened species are those in grave danger of going extinct. Their populations are becoming restricted in range, and the size of the populations required for sustained breeding is nearing a critical minimum. *Endangered species* are in immediate danger of becoming extinct. Their range is already so reduced that the animals may no longer be able to move freely within an ecozone, and their populations are at the level where the species may no longer be able to sustain breeding. Most species become threatened first and then endangered as their range and numbers continue to decrease. When people think of animal extinction, they think of large herbivorous species like the rhinoc-

eros or fierce carnivores like lions, tigers, or grizzly bears. Certainly these animals make almost any list of endangered or threatened species. But there are literally hundreds of less conspicuous animals that are equally threatened. Extinction is normally nature's way of informing a species that it is inefficient. But conditions in the late twentieth century are controlled more by human activities than by natural evolutionary processes. Species that are endangered or threatened fall into that category because, somehow, they are competing with us or with our domesticated livestock for space and food. And in that competition the animals are always going to lose.

Map 66 The Loss of Biodiversity: Globally Threatened Plant Species

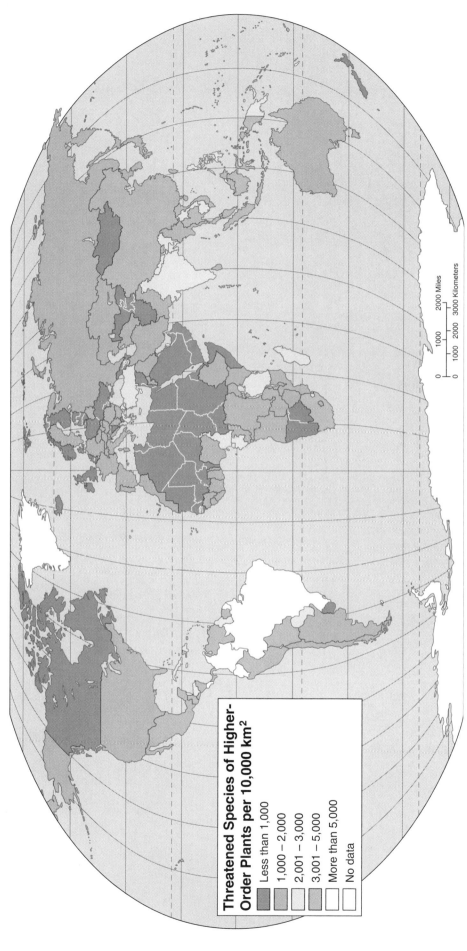

Threatened Species of Higher-Order Plants per 10,000 km²

- Less than 1,000
- 1,000 – 2,000
- 2,001 – 3,000
- 3,001 – 5,000
- More than 5,000
- No data

While most people tend to be more concerned about the animals on threatened and endangered species lists, the fact is that many more plants are in jeopardy, and the loss of plant life is, in all ecological regions, a more critical occurrence than the loss of animal populations. Plants are the primary producers in the ecosystem; that is, plants produce the food upon which all other species in the food web, including human beings, depend for sustenance. It is plants from which many of our critical medicines come, and it is plants that maintain the delicate balance between soil and water in most of the world's regions. When environmental scientists speak of a loss of biodiversity, what they are most often describing is a loss of the richness and complexity of plant life that lends stability to ecosystems. Systems with more plant life tend to be more stable than those with less. For these and other reasons, the scientific concern over extinction is greater when applied to plants than to animals. It is difficult for people to become as emotional over a teak tree as they would over an elephant. But as great a tragedy as the loss of the elephant would be, the loss of the teak would be greater.

Map 67 Hotspots of Biodiversity

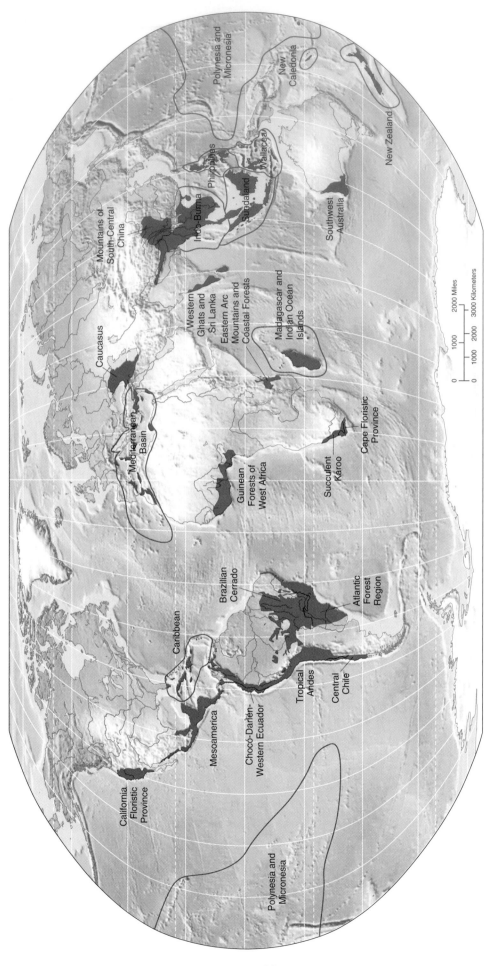

Where we have normally thought of tropical forest basins such as Amazonia as the worlds most biologically diverse ecosystems, recent research has discovered the surprising fact that a number of hotspots of biological diversity exist outside the major tropical forest regions. These hotspot regions contain slightly less than 2 percent of the world's total land area but may contain up to 60 percent of the total world's terrestrial species of plants and animals. Geographically, the hotspot areas are characterized by vertical zonation (that is, they tend to be hilly to mountainous regions), long known to

be a factor in biological complexity. They are also in coastal locations or near large bodies of water, locations that stimulate climatic variability and, hence, biological complexity. Although some of the hotspots are sparsely populated, others, such as Sundaland, are occupied by some of the world's densest populations. Protection of the rich biodiversity of these hotspots is, most biologists feel, of crucial importance to the preservation of the world's biological heritage.

-82-

Map 68 The Risks of Desertification

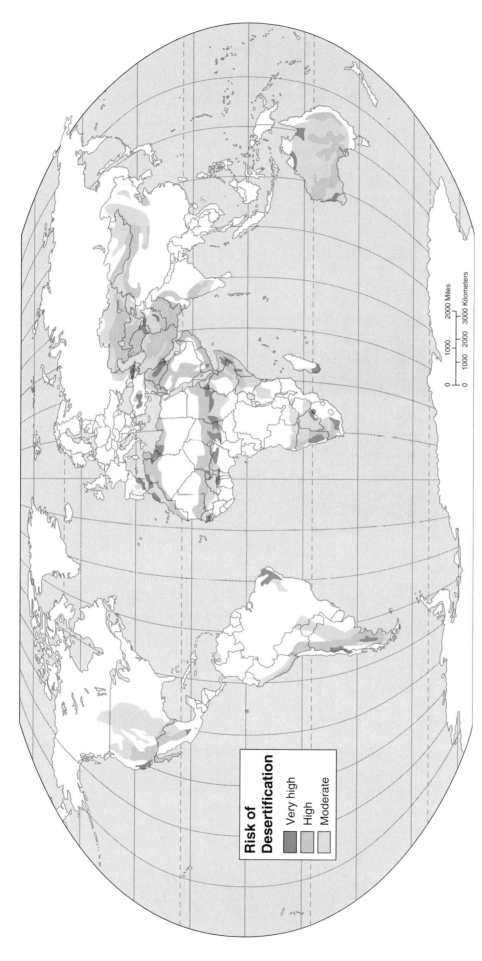

Risk of Desertification

- Very high
- High
- Moderate

0	1000	2000 Miles
0	1000	2000 3000 Kilometers

The awkward-sounding term *desertification* refers to a reduction in the food-producing capacity of drylands through vegetation, soil, and water changes that culminate in either a drier climate or in soil and plant systems that are less efficient in their use of water. Most of the world's existing drylands—the shortgrass steppes, the tropical savannas, the bunchgrass regions of the desert fringe—are fairly intensively used for agriculture and are, therefore, subject to the kinds of pressures that culminate in desertification. Most desertification is a natural process that occurs near the margins of desert regions. It is caused by dehydration of the soil's surface layers during periods of drought and by high water loss through evaporation in an environment of high temperature and high winds. This natural process is greatly enhanced by human agricultural activities that expose topsoil to wind and water erosion. Among the most important practices that cause desertification are (1) overgrazing of rangelands, resulting from too many livestock on too small an area of land; (2) improper management of soil and water resources in irrigation agriculture, leading to accelerated erosion and to salt buildup in the soil; (3) cultivation of marginal terrain with soils and slopes that are unsuitable for farming; (4) surface disturbances of vegetation (clearing of thorn scrub, mesquite, chaparral, and similar vegetation) without soil protection efforts being made or replanting being done; and (5) soil compaction by agricultural implements, domesticated livestock, and rain falling on an exposed surface.

-83-

Map 69 Soil Degradation

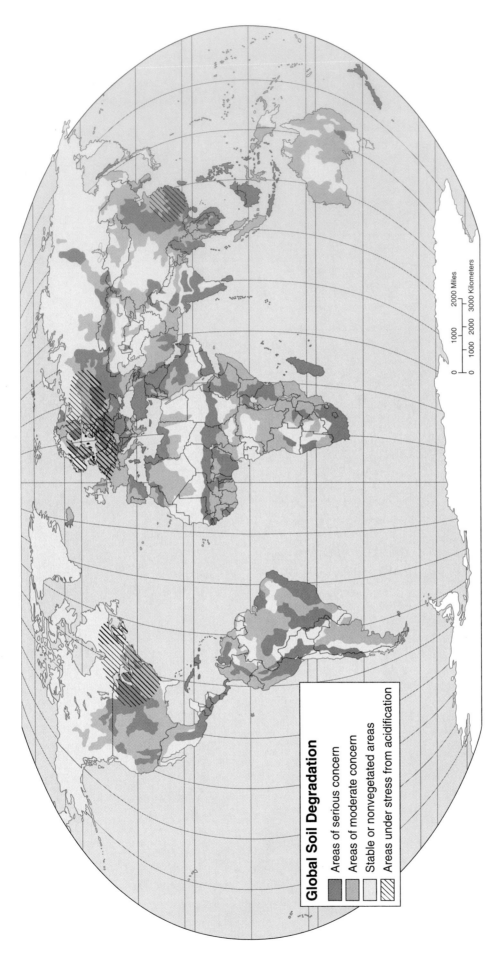

Global Soil Degradation

- Areas of serious concern
- Areas of moderate concern
- Stable or nonvegetated areas
- Areas under stress from acidification

0   1000   2000 Miles
0   1000   2000   3000 Kilometers

Recent research has shown that more than 3 billion acres of the world's surface suffer from serious soil degradation, with more than 22 million acres so severely eroded or poisoned with chemicals that they can no longer support productive crop agriculture. Most of this soil damage has been caused by poor farming practices, overgrazing of domestic livestock, and deforestation. These activities strip away the protective cover of natural vegetation forests and grasslands, allowing wind and water erosion to remove the topsoil that contains necessary nutrients and soil microbes for plant growth. But millions of acres of topsoil have been degraded by chemicals as well. In some instances these chemicals are the result of overapplication of fertilizers, herbicides, pesticides, and other agricultural chemicals. In other instances, chemical deposition from industrial and urban wastes and from acid precipitation has poisoned millions of acres of soil. As the map shows, soil erosion and pollution are not problems just in developing countries with high population densities and increasing use of marginal lands. They also afflict the more highly developed regions of mechanized, industrial agriculture. While many methods for preventing or reducing soil degradation exist, they are seldom used because of ignorance, cost, or perceived economic inefficiency.

Map 70 The Degree of Human Disturbance

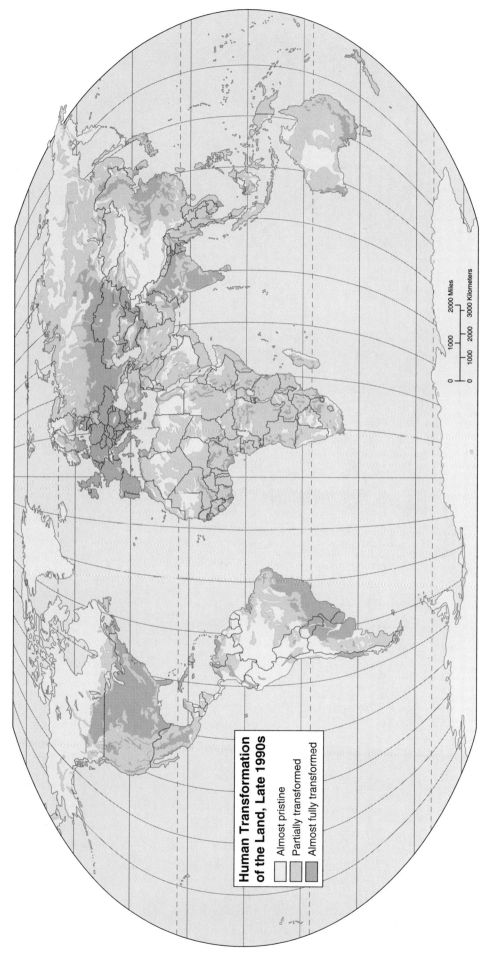

Human Transformation of the Land, Late 1990s

- Almost pristine
- Partially transformed
- Almost fully transformed

0   1000   2000 Miles
0   1000   2000   3000 Kilometers

The data on human disturbance have been gathered from a wide variety of sources, some of them conflicting and not all of them reliable. Nevertheless, at a global scale this map fairly depicts the state of the world in terms of the degree to which humans have modified its surface. The almost pristine areas, covered with natural vegetation, generally have population densities under 10 persons per square mile. These areas are, for the most part, in the most inhospitable parts of the world: too high, too dry, too cold for permanent human habitation in large numbers. The partially transformed areas are normally agricultural areas, either subsistence (such as shifting cultivation) or extensive (such as livestock grazing). They often contain areas of secondary vegetation, regrown after removal of original vegetation by humans. They are also sometimes marked by a density of livestock in excess of carrying capacity, leading to overgrazing, which further alters the condition of the vegetation. The almost fully transformed areas are those of permanent and intensive agriculture and urban settlement. The primary vegetation of these regions has been removed, with no evidence of regrowth or with current vegetation that is quite different from natural (potential) vegetation. Soils are in a state of depletion and degradation, and, in drier lands, desertification is a factor of human occupation. The disturbed areas match closely those areas of the world with the densest human populations.

Part VI

Global Political Patterns

Map 71 Political Systems

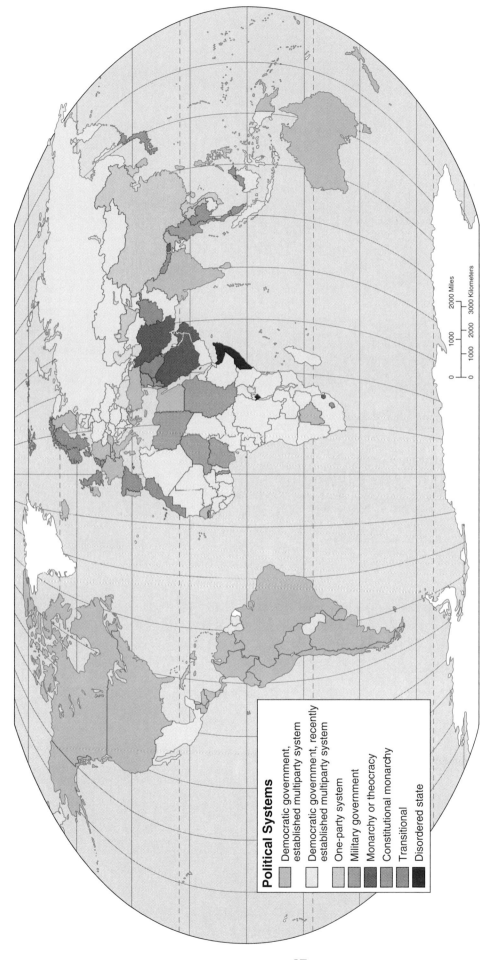

Political Systems

- Democratic government, established multiparty system
- Democratic government, recently established multiparty system
- One-party system
- Military government
- Monarchy or theocracy
- Constitutional monarchy
- Transitional
- Disordered state

World political systems have changed dramatically during the last decade and may change even more in the future. The categories of political systems shown on the map are subject to some interpretation: established multiparty democracies are those in which elections by secret ballot with adult suffrage are and have been long-term features of the political landscape; recently established multiparty democracies are those in which the characteristic features of multiparty democracies have only recently emerged. The former Soviet satellites of eastern Europe and the republics that formerly constituted the USSR are in this category; so are states in emerging regions that are beginning to throw off the single-party rule that often followed the violent upheavals of the immediate postcolonial governmental transitions. The other categories are more or less obvious. One-party systems are states where single-party rule is constitutionally guaranteed or where a one-party regime is a fact of political life. Monarchies are countries with heads of state who are members of a royal family. In a constitutional monarchy, such as the U.K. and the Netherlands, the monarchs are titular heads of state only. Theocracies are countries in which rule is within the hands of a priestly or clerical class; today, this means primarily fundamentalist Islamic countries such as Iran. Military governments are frequently organized around a junta that has seized control of the government from civil authority; such states are often technically transitional, that is, the military claims that it will return the reins of government to civil authority when order is restored. Finally, disordered states are countries so beset by civil war or widespread ethnic conflict that no organized government can be said to exist within them.

Map 72 Sovereign States: Duration of Independence

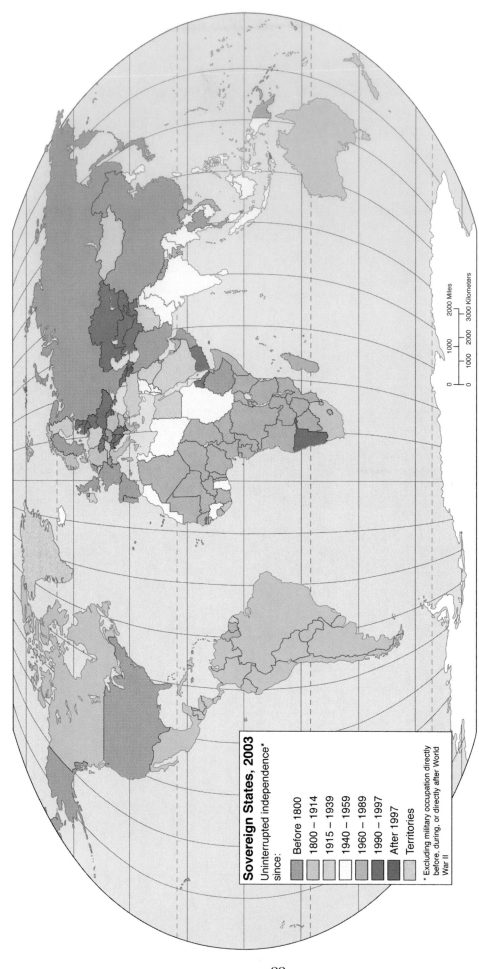

Sovereign States, 2003

Uninterrupted independence* since:

- Before 1800
- 1800 – 1914
- 1915 – 1939
- 1940 – 1959
- 1960 – 1989
- 1990 – 1997
- After 1997
- Territories

* Excluding military occupation directly before, during, or directly after World War II

0   1000   2000 Miles
0   1000   2000   3000 Kilometers

Most countries of the modern world, including such major states as Germany and Italy, became independent after the beginning of the nineteenth century. Of the world's current countries, only 27 were independent in 1800. (Ten of the 27 were in Europe; the others were Afghanistan, China, Colombia, Ethiopia, Haiti, Iran, Japan, Mexico, Nepal, Oman, Paraguay, Russia, Taiwan, Thailand, Turkey, the United States, and Venezuela). Following 1800, there have been five great periods of national independence. During the first of these (1800–1914), most of the mainland countries of the Americas achieved independence. During the second period (1915–1939), the countries of Eastern Europe emerged as independent entities. The third period (1940–1959) includes World War II and the years that followed, when independence for African and Asian nations that had been under control of colonial powers first began to occur. During the fourth period (1960–1989), independence came to the remainder of the colonial African and Asian nations, as well as to former colonies in the Caribbean and the South Pacific. More than half of the world's countries came into being as independent political entities during this period. Finally, in the last few years (1990–1997), the breakup of the existing states of the Soviet Union, Yugoslavia, and Czechoslovakia created 22 countries where only 3 had existed before.

Map 73   European Colonialism 1500–2000

European nations have controlled many parts of the world during the last 500 years. The period of European expansion began when European explorers sailed the oceans in search of new trading routes and ended after World War II when many colonies in Africa and Asia gained independence. The process of colonization was very complex but normally involved the acquisition, extraction, or production of raw materials (including minerals, forest products, products from the sea, agricultural products, and animal furs/pelts) from the areas being controlled by the European colonial power in exchange for items of European manufacture. The concept of colonial dependency implied an economic structure in which the European country obtained raw materials from the colonial country in exchange for those manufactured items upon which populations in the colonial areas quickly came to depend. The colors on this map represent colonial control at its maximum extent and do not take into account shifting colonial control. In North America, for example, "New France" became British territory and "Louisiana" became Spanish territory after the Seven Years' (French and Indian) War.

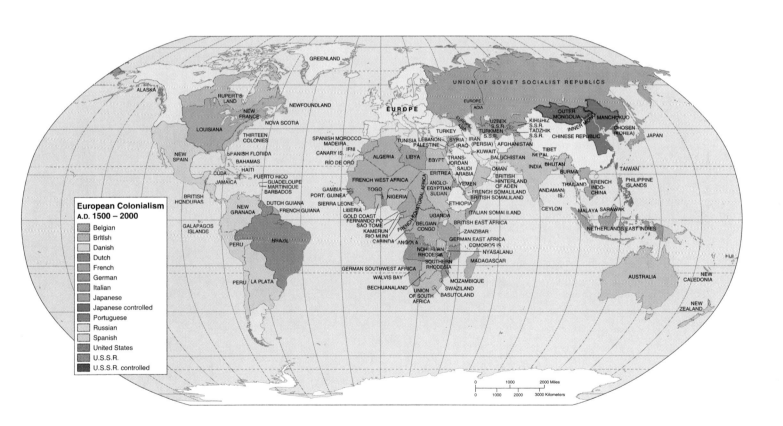

European Colonialism
A.D. 1500 – 2000

- Belgian
- British
- Danish
- Dutch
- French
- German
- Italian
- Japanese
- Japanese controlled
- Portuguese
- Russian
- Spanish
- United States
- U.S.S.R.
- U.S.S.R. controlled

Map 74 International Conflicts in the Post–World War II World

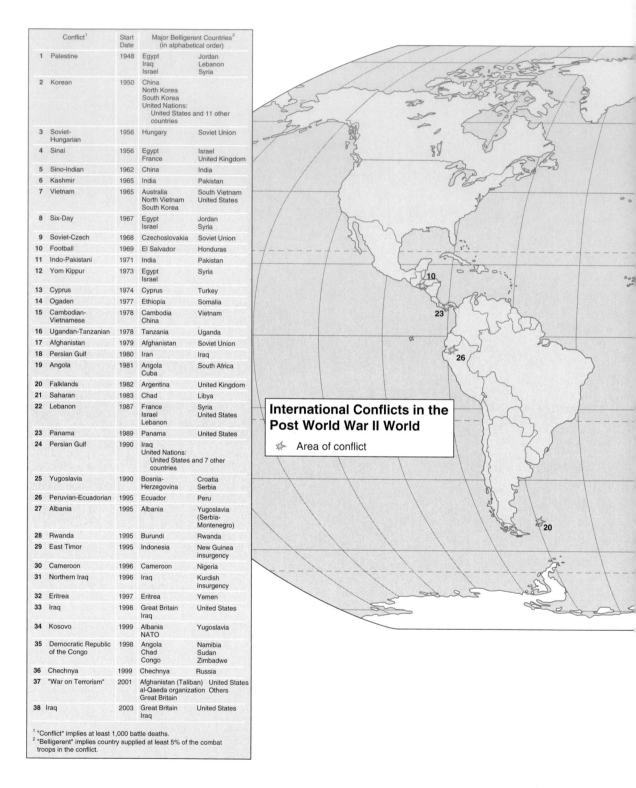

	Conflict[1]	Start Date	Major Belligerent Countries[2] (in alphabetical order)	
1	Palestine	1948	Egypt Iraq Israel	Jordan Lebanon Syria

2	Korean	1950	China
			North Korea
			South Korea
			United Nations:
			    United States and 11 other
			    countries	

3	Soviet-Hungarian	1956	Hungary	Soviet Union
4	Sinai	1956	Egypt France	Israel United Kingdom
5	Sino-Indian	1962	China	India
6	Kashmir	1965	India	Pakistan
7	Vietnam	1965	Australia North Vietnam South Korea	South Vietnam United States
8	Six-Day	1967	Egypt Israel	Jordan Syria
9	Soviet-Czech	1968	Czechoslovakia	Soviet Union
10	Football	1969	El Salvador	Honduras
11	Indo-Pakistani	1971	India	Pakistan
12	Yom Kippur	1973	Egypt Israel	Syria
13	Cyprus	1974	Cyprus	Turkey
14	Ogaden	1977	Ethiopia	Somalia
15	Cambodian-Vietnamese	1978	Cambodia China	Vietnam
16	Ugandan-Tanzanian	1978	Tanzania	Uganda
17	Afghanistan	1979	Afghanistan	Soviet Union
18	Persian Gulf	1980	Iran	Iraq
19	Angola	1981	Angola Cuba	South Africa
20	Falklands	1982	Argentina	United Kingdom
21	Saharan	1983	Chad	Libya
22	Lebanon	1987	France Israel Lebanon	Syria United States
23	Panama	1989	Panama	United States

24	Persian Gulf	1990	Iraq
			United Nations:
			    United States and 7 other
			    countries	

25	Yugoslavia	1990	Bosnia-Herzegovina	Croatia Serbia
26	Peruvian-Ecuadorian	1995	Ecuador	Peru
27	Albania	1995	Albania	Yugoslavia (Serbia-Montenegro)
28	Rwanda	1995	Burundi	Rwanda
29	East Timor	1995	Indonesia	New Guinea insurgency
30	Cameroon	1996	Cameroon	Nigeria
31	Northern Iraq	1996	Iraq	Kurdish insurgency
32	Eritrea	1997	Eritrea	Yemen
33	Iraq	1998	Great Britain Iraq	United States
34	Kosovo	1999	Albania NATO	Yugoslavia
35	Democratic Republic of the Congo	1998	Angola Chad Congo	Namibia Sudan Zimbadwe
36	Chechnya	1999	Chechnya	Russia
37	"War on Terrorism"	2001	Afghanistan (Taliban) al-Qaeda organization Great Britain	United States Others
38	Iraq	2003	Great Britain Iraq	United States

[1] "Conflict" implies at least 1,000 battle deaths.
[2] "Belligerent" implies country supplied at least 5% of the combat troops in the conflict.

International Conflicts in the Post World War II World

✻ Area of conflict

The Korean War and the Vietnam War dominated the post–World War II period in terms of international military conflict. But numerous smaller conflicts have taken place, with fewer numbers of belligerents and with fewer battle and related casualties. These smaller international conflicts have been mostly territorial conflicts, reflecting the continual readjustment of political boundaries and loyalties brought about by the end of colonial empires, and the dissolution of the Soviet Union. Many of these conflicts were not wars in the more traditional sense, in which two or more countries formally declare war on one another, sever-

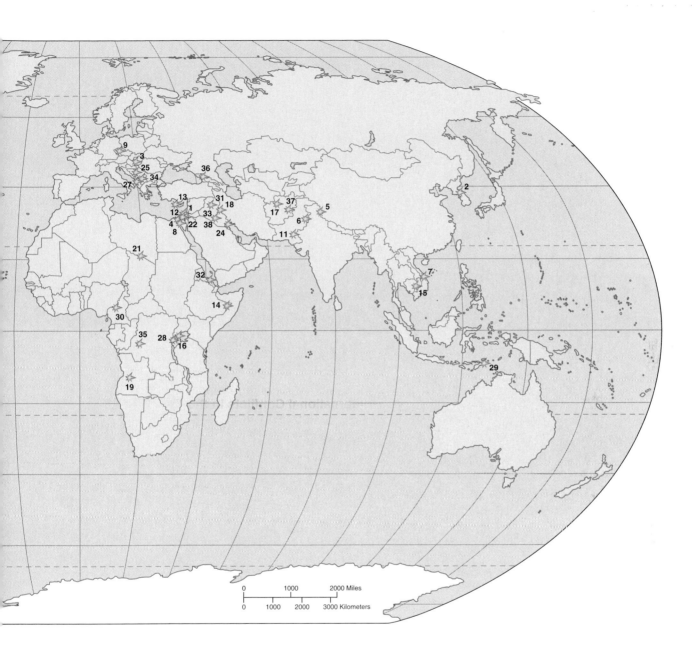

ing diplomatic ties and devoting their entire national energies to the war effort. Rather, many of these conflicts were and are undeclared wars, sometimes fought between rival groups within the same country with outside support from other countries. The aftermath of the September 11, 2001, terrorist attacks on the United States indicate the dawn of yet another type of international conflict, namely a "war" fought between traditional nation-states and non-state actors.

Map 75 Post–Cold War International Alliances

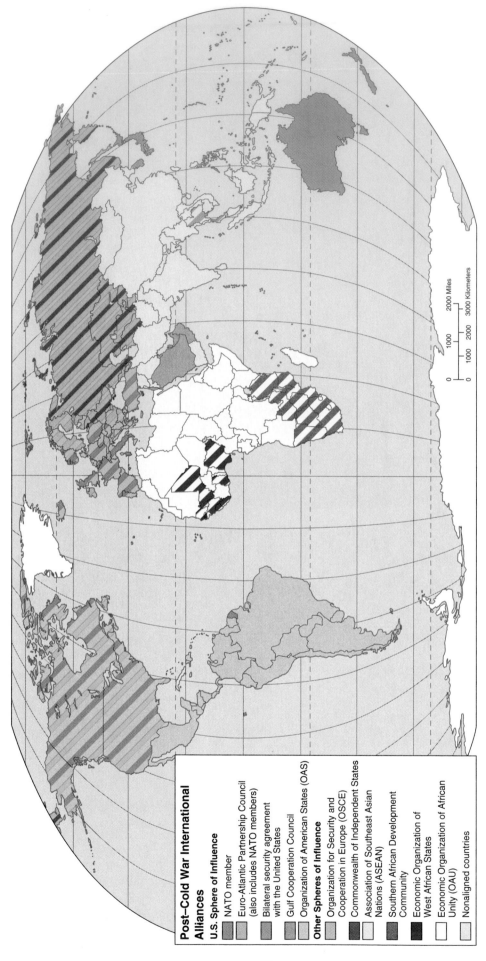

Post–Cold War International Alliances

U.S. Sphere of Influence
- NATO member
- Euro-Atlantic Partnership Council (also includes NATO members)
- Bilateral security agreement with the United States
- Gulf Cooperation Council
- Organization of American States (OAS)

Other Spheres of Influence
- Organization for Security and Cooperation in Europe (OSCE)
- Commonwealth of Independent States
- Association of Southeast Asian Nations (ASEAN)
- Southern African Development Community
- Economic Organization of West African States
- Economic Organization of African Unity (OAU)
- Nonaligned countries

When the Warsaw Pact dissolved in 1992, the North Atlantic Treaty Organization (NATO) was left as the only major military alliance in the world. Some former Warsaw Pact members (Czech Republic, Hungary, and Poland) have joined NATO and others are petitioning for entry. The bipolar division of the world into two major military alliances is over, at least temporarily, leaving the United States alone as the world's dominant political and military power. But other international alliances, such as the Commonwealth of Independent States (including most of the former republics of the Soviet Union), will continue to be important. It may well be that during the first few decades of the twenty-first century economic alliances will begin to overshadow military ones in their relevance for the world's peoples.

-92-

Map 76 Flashpoints, 2004

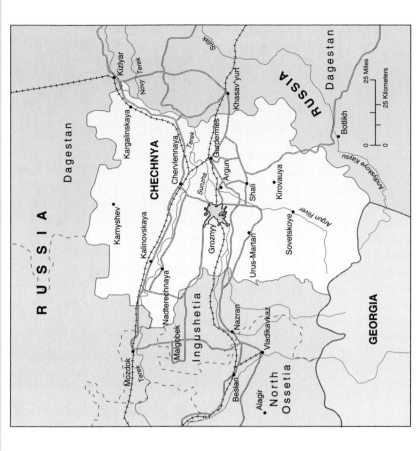

Chechnya: The area in southern Russia known as the Caucasus Region is home to a large variety of non-Russian ethnic groups; many are Muslim and resent centuries of Russian domination and Soviet-era totalitarianism. After the Soviet Union disintegrated in 1991, several of these ethnic groups began agitating for more autonomy from Moscow or for outright independence. One of the more vocal groups with a history of opposition to Moscow's rule were the Chechens. The Chechens declared themselves a sovereign nation in 1991 and by 1994 relations between the Macedonian government in Chechnya and the Russian government had drastically deteriorated. In December of that year, Russian forces attacked Chechnya, beginning the first of two (1994–96 and 1999–present) full-scale military conflicts that have also crept into the neighboring Russian autonomous area of Dagestan, itself largely Muslim. In the mid- and late 1990s Russia experienced several terrorist attacks in cities throughout the nation, which the Russian government attributed to Islamic extremists supporting Chechen independence. As a result, a second round of the conflict began in August 1999 with a full-scale Russian military assault on Dagestan and Chechnya. This assault is ongoing and continues to face intense resistance, with heavy casualties on both sides. In 2003 and 2004 Chechen rebels increased their presure with urban terrorist activities in Russian cities, including Moscow.

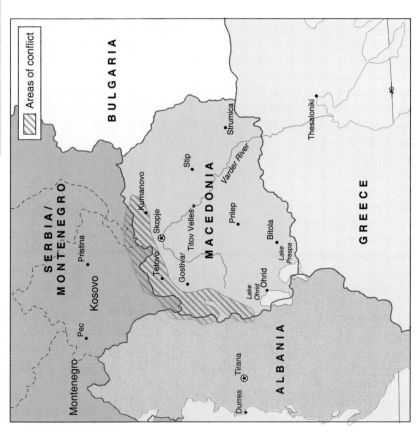

Macedonia: Since the fifteenth-century conquest of much of southeastern Europe by the Turkish Ottoman Empire, conflict in the Balkan region has been precipitated by ethnic and religious enmity between Orthodox Christians and Muslims. An area of particular concern has been the interface between predominantly Orthodox Macedonia and predominantly Muslim Albania, particularly in the border region the two share with the former Yugoslav republic of Kosovo. Tensions between the Macedonian government and its Albanian (Muslim) minority were heightened by the fallout from the conflict in Kosovo in the late 1990s between the Serbs and the Kosovar Albanians, backed by Albania. The resolution of this conflict by a NATO-led force in favor of the Kosovars led to emboldened feelings of Albanian patriotism in the region. Several sporadic incidents occurred along the Macedonian-Albanian border in late 2000, with more protracted and heavier combat between rebels and the Macedonian government forces occurring in the northern part of Macedonia (near the capital of Skopje) throughout 2001. The rebel forces assert they are only seeking to revise the Macedonian constitution and attain better rights for the Albanian minority in Macedonia. The Macedonian government is concerned that the Albanian minority centered in northern and western Macedonia wishes to secede and merge (along with Kosovo) into a Greater Albania, and suspects that Albania itself has encouraged this objective.

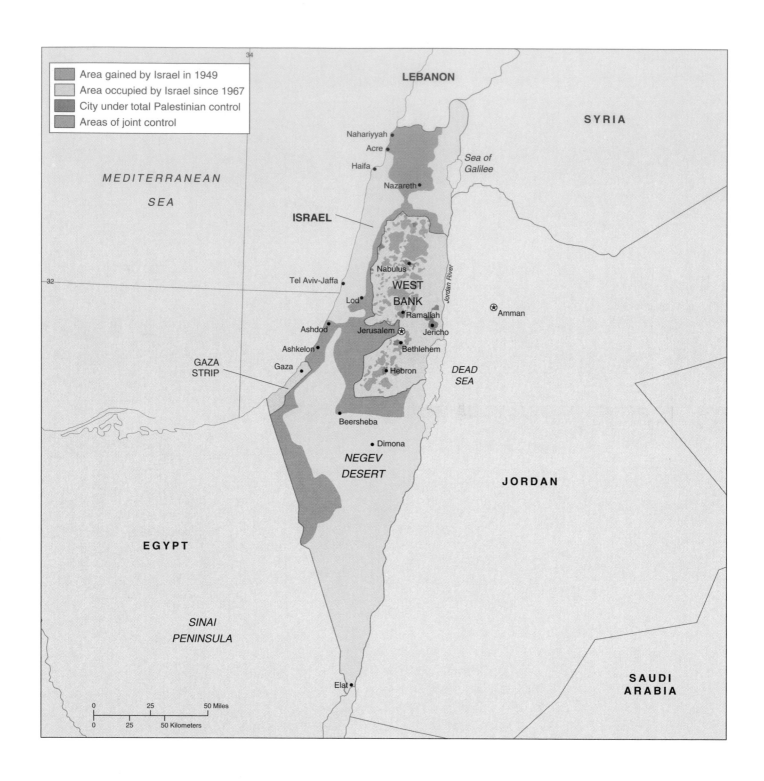

Legend:
- Area gained by Israel in 1949
- Area occupied by Israel since 1967
- City under total Palestinian control
- Areas of joint control

LEBANON

SYRIA

MEDITERRANEAN
SEA

Nahariyyah
Acre
Haifa
Sea of Galilee
Nazareth

ISRAEL

Nabulus
Tel Aviv-Jaffa
WEST BANK
Lod
Ramallah
Jordan River
Amman

Ashdod
Jerusalem
Jericho
Ashkelon
Bethlehem
GAZA STRIP
Gaza
Hebron
DEAD SEA

JORDAN

Beersheba

Dimona

NEGEV
DESERT

EGYPT

SINAI
PENINSULA

SAUDI
ARABIA

Elat

0   25   50 Miles
0   25   50 Kilometers

Israel and Its Neighbors: The modern state of Israel was created out of the former British Protectorate of Palestine, inhabited primarily by Muslim Arabs, after World War II. Conflict between Arabs and Israeli Jews has been a constant ever since. Much of the present tension revolves around the West Bank area, not part of the original Israeli state but taken from Jordan, an Arab country, in the Six-Day War of 1967. Many Palestinians had settled this part of Jordan after the creation of Israel and remain as a majority population in the West Bank region today. Israel has established many agricultural settlements within the region since 1967, angering Palestinian Arabs. For Israel, the West Bank is the region of ancient Judea and this region, won in battle, will not be ceded back to Palestinian Arabs without protracted or severe military action. The West Bank, inhabited by nearly 400,000 Israeli settlers and 4 million Palestinians, is also the location of most of the suicide bombings carried out by Islamic militant groups from 2001 to 2004. The U.S. government has recently agreed that Israel should retain most of its rights to the Israeli-settled areas of the West Bank.

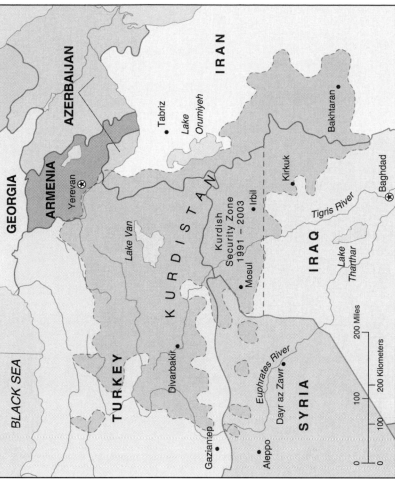

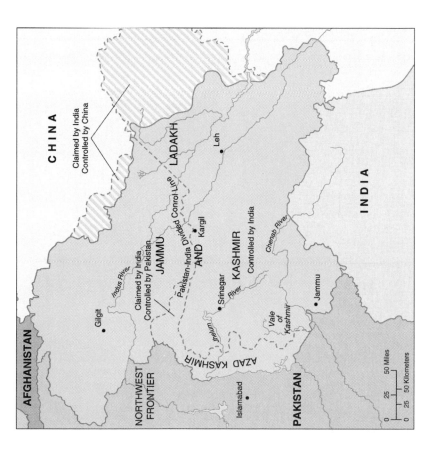

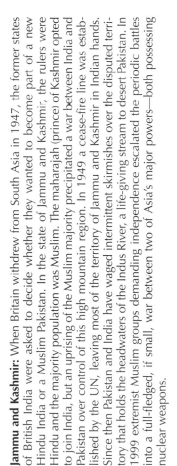

Kurdistan: Where Turkey, Iran, and Iraq meet in the high mountain region of the Tauros and Zagros mountains, a nation of 25 million people exists. This nation is "Kurdistan," but the Kurds, the occupants of this area for over 3,000 years, have no state, and receive much less attention than other stateless nations like the Palestinians. Following the 1991 Gulf War between Iraq and a U.S.-led coalition of European and Arabic states, the United Nations demarcated a Kurdish "security zone" in northern Iraq. From 1991 to 2003 the Security Zone was anything but secure as Iraqi militants from the south and Turks from the north infringed on Kurdish territory, and the internal militant extremist groups, such as the Kurdish Workers' Party, staged periodic attacks on rival villages. During the 2003 U.S.-led invasion of Iraq that eliminated the Baathist regime of Saddam Hussein, the Kurds played an important role in securing the northern portions of Iraq for the U.S.-British coalition and fought alongside American troops in expelling elements of the Iraqi army from cities like Mosul and Kirkuk. Rich in oil and history, Kurdistan will probably remain as a nation without a state, shared by Iraq, Turkey, and Iran—none of which is likely to give up substantial portions of territory for the establishment of a Kurdish state.

Jammu and Kashmir: When Britain withdrew from South Asia in 1947, the former states of British India were asked to decide whether they wanted to become part of a new Hindu India or a Muslim Pakistan. In the state of Jammu and Kashmir, the rulers were Hindu and the majority population was Muslim. The maharajah (prince) of Kashmir opted to join India, but an uprising of the Muslim majority population precipitated a war between India and Pakistan over control of this high mountain region. In 1949 a cease-fire line was established by the UN, leaving most of the territory of Jammu and Kashmir in Indian hands. Since then Pakistan and India have waged intermittent skirmishes over the disputed territory that holds the headwaters of the Indus River, a life-giving stream to desert Pakistan. In 1999 extremist Muslim groups demanding independence escalated the periodic battles into a full-fledged, if small, war between two of Asia's major powers—both possessing nuclear weapons.

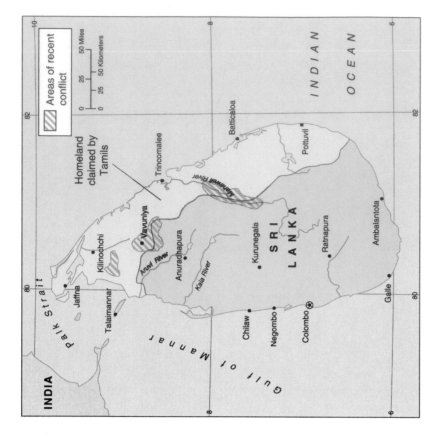

Sri Lanka: The island state of Sri Lanka, historically known as Ceylon, is potentially one of the most agriculturally productive regions of Asia. Unfortunately for plans related to agricultural development, two quite different peoples have occupied the island country: The Buddhist Sinhalese originally from northern India and long the dominant population in Sri Lanka, and the minority Hindu Tamil, a Dravidian people from south India. Since independence from Britain, Sri Lankan governments have sought to "resettle" the Tamil population in south India, actions that finally precipitated an armed rebellion by Tamils against the Sinhalese-dominated government. The Tamils at present are demanding a complete separation of the state into two parts, with a Tamil homeland in the north and along the east coast. At one time viewed as an island paradise, Sri Lanka is now a troubled country with an uncertain future. A cease fire between Sinhalese and Tamil fighters was brokered in 2001 by Norway but fell apart in late 2003 with the resumption of violence. In May 2004, Norwegian diplomats again made an attempt to negotiate a cease fire but with questionable results.

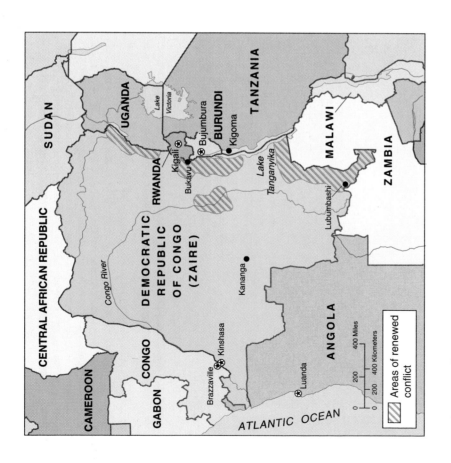

Congo: The war in the Democratic Republic of the Congo (formerly Zaire) has preoccupied the United Nations and African diplomats since 1999. Troops from Zimbabwe, Angola, Sudan, Chad and Namibia are now joined with the Congo's President Laurent Kabila against his former allies Rwanda, Burundi, and Uganda, who each back several separate Congolese rebel groups. The origins of the conflict lie in the overthrow of longtime dictator Mobutu Sese Seko by Kabila's army in May 1997 after a year of civil war. Since Kabila's failure to call elections or stabilize the country's economy led to further rounds of rebellion in the huge but fractious nation, rebellion supported by the economic and military assistance of neighboring Rwanda, Burundi, and Uganda. In October 2002, accord seemed to have been reached, and the various conflicting parties had agreed to withdraw troops. But in early 2003, new fighting flared along the country's eastern border, threatening a new and broadened war and the addition of more deaths to the 3.3 million since 1998. Diplomats have called the conflict "Africa's first world war," and fear that it may destabilize the southern half of the continent, leading to massive refugee flows and abject poverty.

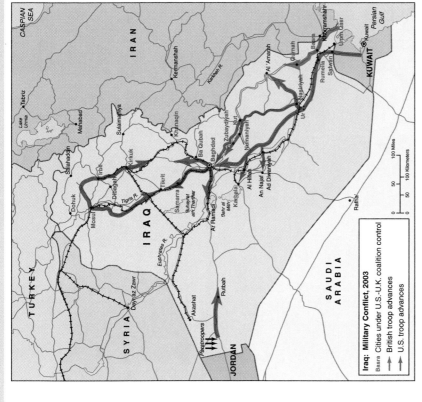

Iraq: Military Conflict, 2003–2004: Following the failure of the United States and allies Great Britain and Spain to secure approval from the United Nations to begin a UN-sponsored military conflict to "disarm" Iraq and effect "regime change" by removing the Baathist party dictator Saddam Hussein from power, the United States and its allies launched an independent military attack on Iraq in April 2003. The military campaign began with massive air and sea bombardments on government and military targets then ground troops moved from Kuwait along highway routes to secure the oil fields and major urban areas. By early May 2003, virtually all of the country was under the control of the U.S.-led coalition of forces. The Iraqi military, for the most part, melted away into the general civilian population and there were no major pitched battles for territory in the short-lived war. Conflicts between British forces and paramilitary/political forces in the important southern city of Al Basrah produced casualties on both sides before the city was secured. And the U.S. troops met guerilla resistance from political affiliates of Hussein's regime in cities such as An Najaf and Al Nasiriyah. In the northern parts of the country, U.S. troops inserted by parachute joined forces with Kurdish paramilitary groups to secure major cities like Mosul, Arbit, and Kirkuk. Although the U.S. and coalition forces controlled the country by June 2003, civil authority and infrastructure continued to be unavailable throughout large areas of Iraq by mid-2004. The spring of 2004 brought significant insurgency action against U.S. and coalition forces, who were increasingly viewed by the Iraqis as occupiers rather than liberators. A June 30, 2004, date was set to return Iraq to civilian Iraqi control.

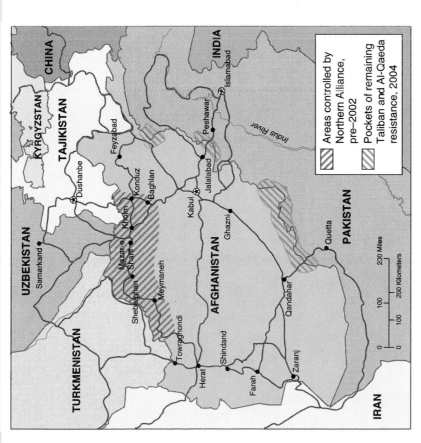

Afghanistan: In the aftermath of the tragic September 11, 2001, terrorist attacks on the World Trade Center and the Pentagon, the United States (backed to varying degrees by its allies) has declared a massive and global "war on terrorism" and any states that may provide "safe harbor" to terrorists. To date, the most prominent target of this U.S. declaration of war has been the Taliban regime of Islamic extremists who controlled about 95 percent of the territory of the beleaguered nation of Afghanistan. International observers believe that the Taliban regime has welcomed and provided a base for the al-Qaeda terrorist network dominated by Saudi expatriate and millionaire Osama bin Laden since the late 1990s. As a result of this intelligence, the U.S. and Britain pursued a daily bombardment of key al-Qaeda and Taliban installations inside Afghanistan for several months, with key logistical support in the form of air bases and supply depots provided by the government of Pakistan (in exchange for financial considerations and political support of the non-elected Pakistani government). U.S. and allied ground troops, aided by members of the Northern Alliance of Afghan rebels opposed to the Taliban regime, and expelled the Taliban government in 2002. While now under home rule and with a duly elected government, Afghanistan still is plagued by warlords in remote areas of the country who refuse to recognize the legally constituted government. In addition, along the Afghanistan-Pakistan border, significant pockets of resistance from remnants of the former Taliban regime and from al-Qaeda forces are engaged in ongoing military conflict with American and Pakistani troops.

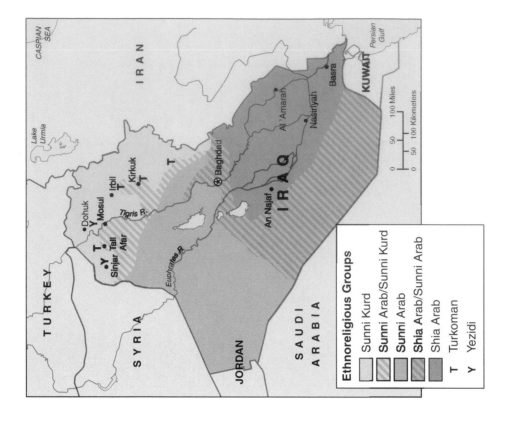

Ethnoreligious Groups

Sunni Kurd
Sunni Arab/Sunni Kurd
Sunni Arab
Shia Arab/Sunni Arab
Shia Arab
T Turkoman
Y Yezidi

IRAQ

CASPIAN SEA

IRAN

TURKEY

SYRIA

JORDAN

SAUDI ARABIA

KUWAIT

Persian Gulf

Lake Urmia

Dohuk
Mosul
Tall Afar
Sinjar
Irbil
Kirkuk
Baghdad
An Najaf
Al 'Amarah
Nasiriyah
Basra

Tigris R.
Euphrates R.

0   50   100 Miles
0   50   100 Kilometers

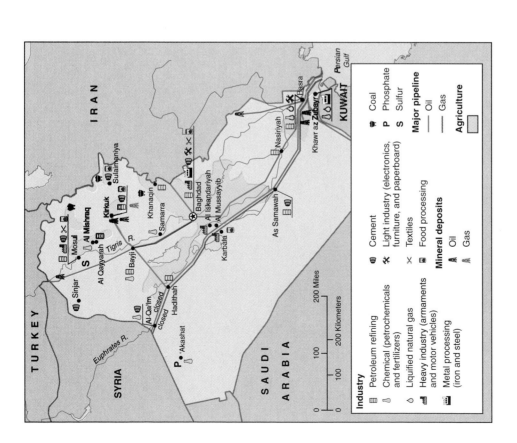

Industry

- Petroleum refining
- Chemical (petrochemicals and fertilizers)
- Liquified natural gas
- Heavy industry (armaments and motor vehicles)
- Metal processing (iron and steel)
- Cement
- Light industry (electronics, furniture, and paperboard)
- Textiles
- Food processing

Mineral deposits

- Oil
- Gas
- Coal
- P Phosphate
- S Sulfur

Major pipeline

— Oil
— Gas

Agriculture

IRAQ

TURKEY

SYRIA

SAUDI ARABIA

IRAN

KUWAIT

Persian Gulf

Sinjar
Mosul
Al Qayyarah
Bayji
Al Mishraq
Kirkuk
Sulaimaniya
Khanaqin
Samarra
Baghdad
Al Iskandariyah
Al Mussayyib
Karbala
As Samawa
Nasiriyah
Basra
Khawr az Zubayr
Akashat
Al-Qa'im
Haditha
closed
closed

Tigris R.
Euphrates R.

0   100   200 Miles
0   100   200 Kilometers

Iraq: Prior to the 1990–91 invasion of Kuwait by Iraq and the subsequent United Nations coalition's military expulsion of Iraq from its neighbor, Iraq was one of the most prosperous countries in the Middle East and the only one with full capacity to feed itself, even without the vast oil revenues generated by the country's immense reserves. Despite the inefficiencies of the Baathist dictatorship of Saddam Hussein, the country has a solid agricultural base and a burgeoning industry. The combination of military adventurism and conflict, in the form of a lengthy war with Iran and the ill-advised invasion of Kuwait, limited further economic development, however. Development was also problematic given the country's internal tensions between Arabic Sunni Muslims and Arabic Shiite Muslims, and between Arabs and Kurds and a few other minority populations in the northern parts of the country.

-98-

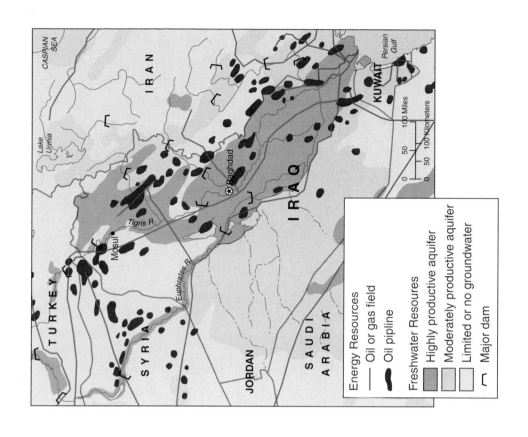

Energy Resources
— Oil or gas field
◖ Oil pipline
Freshwater Resoures
▨ Highly productive aquifer
▨ Moderately productive aquifer
▨ Limited or no groundwater
⌐ Major dam

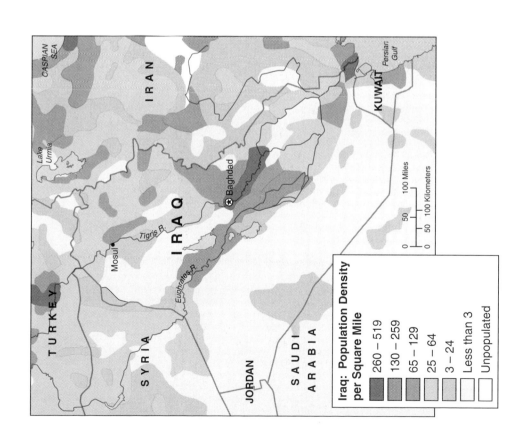

Iraq: Population Density per Square Mile
260 – 519
130 – 259
65 – 129
25 – 64
3 – 24
Less than 3
Unpopulated

Map 77 International Terrorism Incidents, 2000–2002

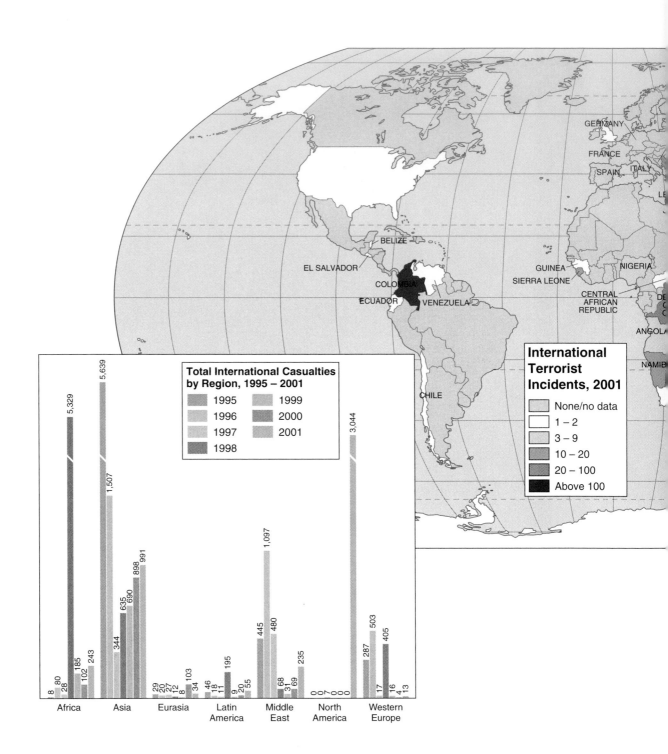

International Terrorist Incidents, 2001

- None/no data
- 1 – 2
- 3 – 9
- 10 – 20
- 20 – 100
- Above 100

Total International Casualties by Region, 1995 – 2001

- 1995
- 1996
- 1997
- 1998
- 1999
- 2000
- 2001

Africa: 8, 80, 28, 185, 102, 243, 5,329
Asia: 344, 635, 690, 898, 991, 1,507, 5,639
Eurasia: 29, 20, 27, 12, 8, 34, 103
Latin America: 46, 18, 11, 195, 9, 20, 55
Middle East: 445, 1,097, 480, 68, 31, 69, 235
North America: 0, 0, 7, 0, 0, 0, 3,044
Western Europe: 287, 503, 17, 405, 16, 4, 13

Americans have made a virtual mantra of the saying "The world has changed," as a consequence of the terrorist attacks on the World Trade Center and the Pentagon on September 11, 2001. As this map and the accompanying bar graphs point out, the world did not change, but the focus of a major terrorist attack shifted from Asia to North America. Many other areas of the world have lived with terrorism and terrorist activity for years. In 2000

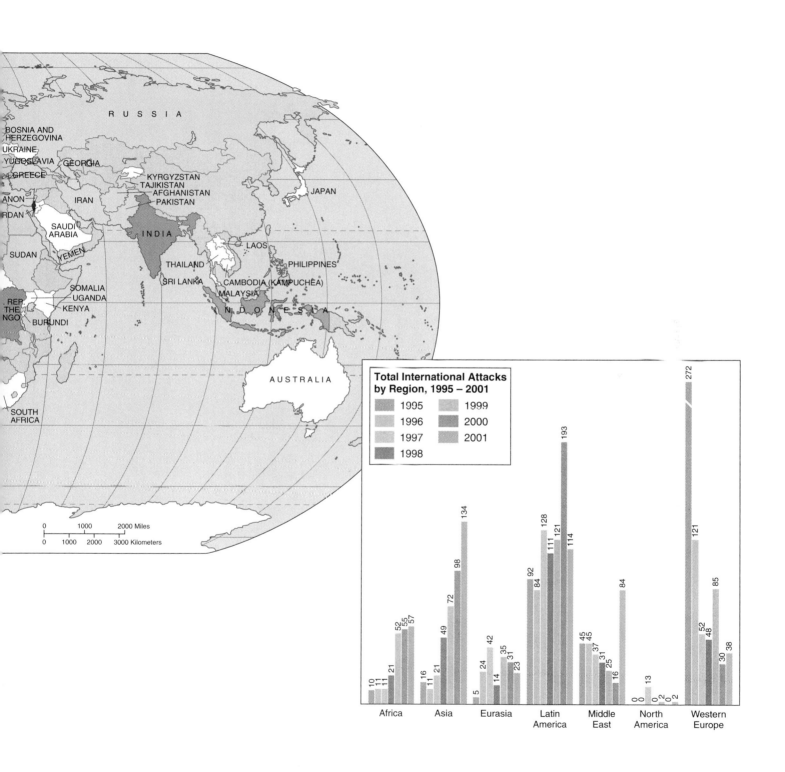

Total International Attacks
by Region, 1995 – 2001

1995	1999		
1996	2000		
1997	2001		
1998			

and 2001, despite the enormous losses in the United States in the 9/11 attacks, more lives were lost in Asia and Africa as a result of terrorism than were lost in North America. The world did not change, but Americans' perception of that world and their place in it has certainly changed.

Map 78 Nations With Nuclear Weapons

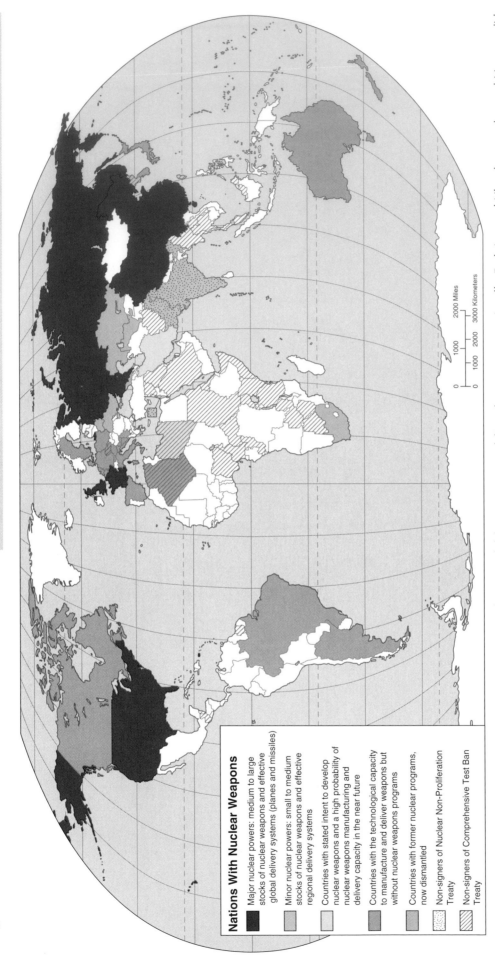

Nations With Nuclear Weapons

Major nuclear powers: medium to large stocks of nuclear weapons and effective global delivery systems (planes and missiles)

Minor nuclear powers: small to medium stocks of nuclear weapons and effective regional delivery systems

Countries with stated intent to develop nuclear weapons and a high probability of nuclear weapons manufacturing and delivery capacity in the near future

Countries with the technological capacity to manufacture and deliver weapons but without nuclear weapons programs

Countries with former nuclear programs, now dismantled

Non-signers of Nuclear Non-Proliferation Treaty

Non-signers of Comprehensive Test Ban Treaty

Since 1980, the number of countries possessing the capacity to manufacture and deliver nuclear weapons has grown dramatically, increasing the chances of accidental or intentional nuclear exchanges. In addition to the traditional nuclear powers of the United States, Russia, China, the United Kingdom, and France, must now be added Israel, India, and Pakistan as countries that, without possessing the large stocks of weapons of the major powers, nor the extensive delivery systems of the United States and Russia, still have effective regional (and possibly global) delivery systems and medium stocks of warheads. Countries such as Kazakhstan, Ukraine, Georgia, and Belarus that were created out of what had been the Soviet Union did have some nuclear capacity in the 1991–1995 period but have since had all nuclear weapons removed from their territories. However, North Korea has recently announced the re-establishment of its suspended nuclear weapons programs and may possess a small stock of nuclear warheads, along with the capacity to deliver those weapons regionally. Both Iran and Libya have nuclear ambitions, as did Iraq until the overthrow of the Baathist regime of Saddam Hussein by a U.S.-led military coalition in 2003. The proliferation of nuclear states threatens global security, and the objective of the Nuclear Non-Proliferation Treaty was to reduce the chances for expanding nuclear arsenals worldwide. This treaty has been partially successful in that a number of countries in the developed world certainly have the capacity to manufacture and deliver nuclear weapons but have chosen not to do so. These countries include Canada, European countries other than the United Kingdom and France, South Korea, Japan, Australia, and New Zealand, and Brazil and Argentina in South America. On the other side of the coin, the intent of North Korea to emerge as a nuclear power may force countries such as South Korea and Japan to re-think their positions as non-nuclear countries.

Map 79 Distribution of Minority Populations

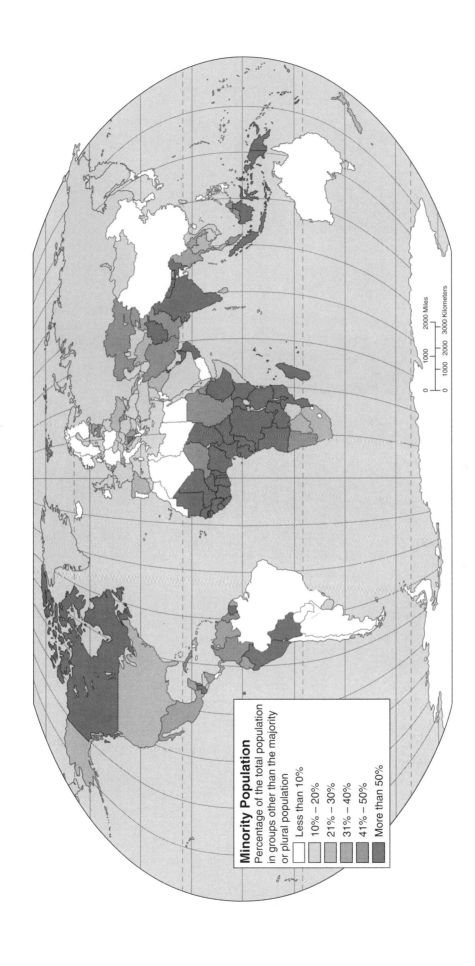

Minority Population
Percentage of the total population in groups other than the majority or plural population

- Less than 10%
- 10% – 20%
- 21% – 30%
- 31% – 40%
- 41% – 50%
- More than 50%

0   1000   2000 Miles
0 1000 2000 3000 Kilometers

The presence of minority ethnic, national, or racial groups within a country's population can add a vibrant and dynamic mix to the whole. Plural societies with a high degree of cultural and ethnic diversity should, according to some social theorists, be among the world's most healthy. Unfortunately, the reality of the situation is quite different from theory or expectation. The presence of significant minority populations played an important role in the disintegration of the Soviet Union; the continuing existence of minority populations within the new states formed from former Soviet republics threatens the viability and stability of those young political units. In Africa, national boundaries were drawn by colonial powers without regard for the geographical distribution of ethnic groups, and the continuing tribal conflicts that have resulted hamper both economic and political development. Even in the most highly developed regions of the world, the presence of minority ethnic populations poses significant problems: witness the separatist movement in Canada, driven by the desire of some French-Canadians to be independent of the English majority, and the continuing ethnic conflict between Flemish-speaking and Walloon-speaking Belgians. This map, by arraying states on a scale of homogeneity to heterogeneity, indicates areas of existing and potential social and political strife.

Map 80 Refugee Population, 2002

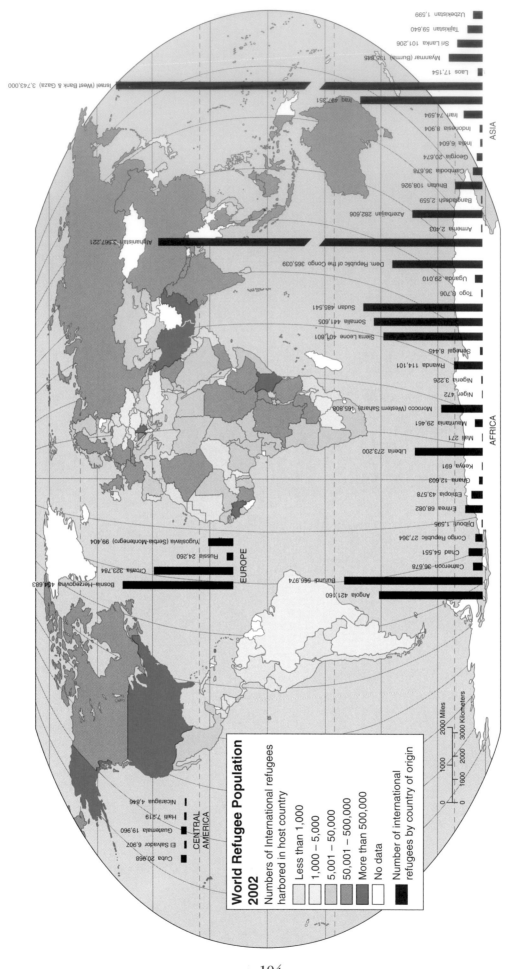

World Refugee Population 2002

Numbers of International refugees harbored in host country

- Less than 1,000
- 1,000 – 5,000
- 5,001 – 50,000
- 50,001 – 500,000
- More than 500,000
- No data

Number of international refugees by country of origin

CENTRAL AMERICA
- Cuba 20,968
- El Salvador 6,907
- Guatemala 19,960
- Haiti 7,219
- Nicaragua 4,846

EUROPE
- Bosnia-Herzegovina 454,683
- Croatia 323,784
- Russia 24,260
- Yugoslavia (Serbia-Montenegro) 99,404

AFRICA
- Angola 421,160
- Burundi 566,974
- Cameroon 36,678
- Chad 54,551
- Congo Republic 27,364
- Djibouti 1,595
- Eritrea 68,082
- Ethiopia 43,578
- Ghana 12,603
- Kenya 691
- Liberia 273,200
- Mali 271
- Mauritania 29,461
- Morocco (Western Sahara) 165,808
- Niger 472
- Nigeria 3,226
- Rwanda 114,101
- Senegal 8,445
- Sierra Leone 407,801
- Somalia 441,605
- Sudan 485,541
- Togo 8,706
- Uganda 29,010
- Dem. Republic of the Congo 365,039

ASIA
- Afghanistan 3,567,221
- Armenia 2,403
- Azerbaijan 282,606
- Bangladesh 2,559
- Bhutan 108,926
- Cambodia 36,678
- Georgia 20,674
- India 5,604
- Indonesia 8,904
- Iran 74,594
- Iraq 497,351
- Israel (West Bank & Gaza) 3,743,000
- Laos 17,154
- Myanmar (Burma) 135,845
- Sri Lanka 101,206
- Tajikistan 59,640
- Uzbekistan 1,599

Refugees are persons who have been driven from their homes, normally by armed conflict, and have sought refuge by relocating. The most numerous refugees have traditionally been international refugees, who have crossed the political boundaries of their homelands into other countries. This refugee population is recognized by international agencies, and the countries of refuge are often rewarded financially by those agencies for their willingness to take in externally displaced persons. In recent years, largely because of an increase in civil wars, there have been growing numbers of internally displaced persons—those who leave their homes but stay within their country of origin. There are no rewards for harboring such internal refugee populations.

-104-

Map 81 Political and Civil Liberties, 2004

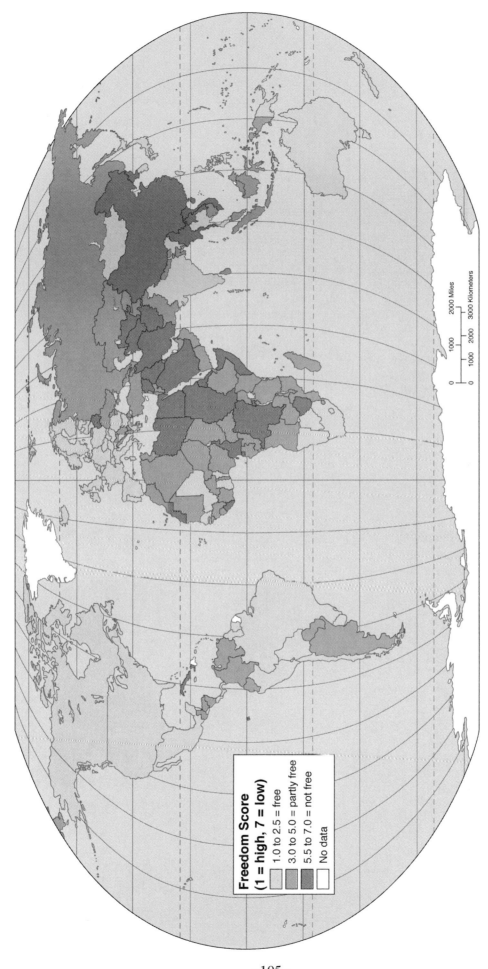

Freedom Score
(1 = high, 7 = low)

- 1.0 to 2.5 = free
- 3.0 to 5.0 = partly free
- 5.5 to 7.0 = not free
- No data

0   1000   2000 Miles
0   1000   2000   3000 Kilometers

Although measures of political and civil liberty are somewhat difficult to obtain and assess, there are some generally accepted standards that can be evaluated: open elections and competitive political parties, the rule of law, freedoms of speech and press, judicial systems separate from other branches of government, and limits on the power of elected or appointed governmental officials. Interestingly, there appear to be correlations between "degrees of freedom" and such other characteristics of a state as per cap-ita wealth, environmental quality, and healthy economic growth—characteristics that may be mutually contradictory. There is no empirical evidence of a causal link between democratic institutions and consumption; on the other hand, there is clear evidence of a positive relationship between wealth and consumption. Therefore, the three variables are closely correlated and should be used in assessing the nature of the state in any part of the world.

Part VII

World Regions

Map 82 North America: Physical Features

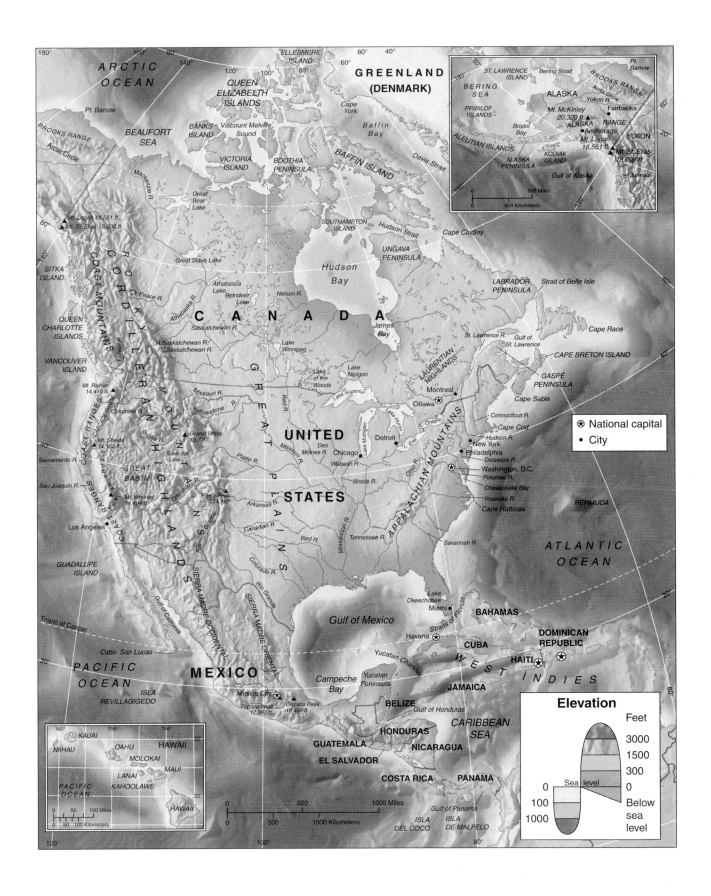

Elevation

	Feet
	3000
	1500
	300
Sea level	0
0	Below
100	sea
1000	level

⊛ National capital
• City

Map 83 North America: Political Divisions

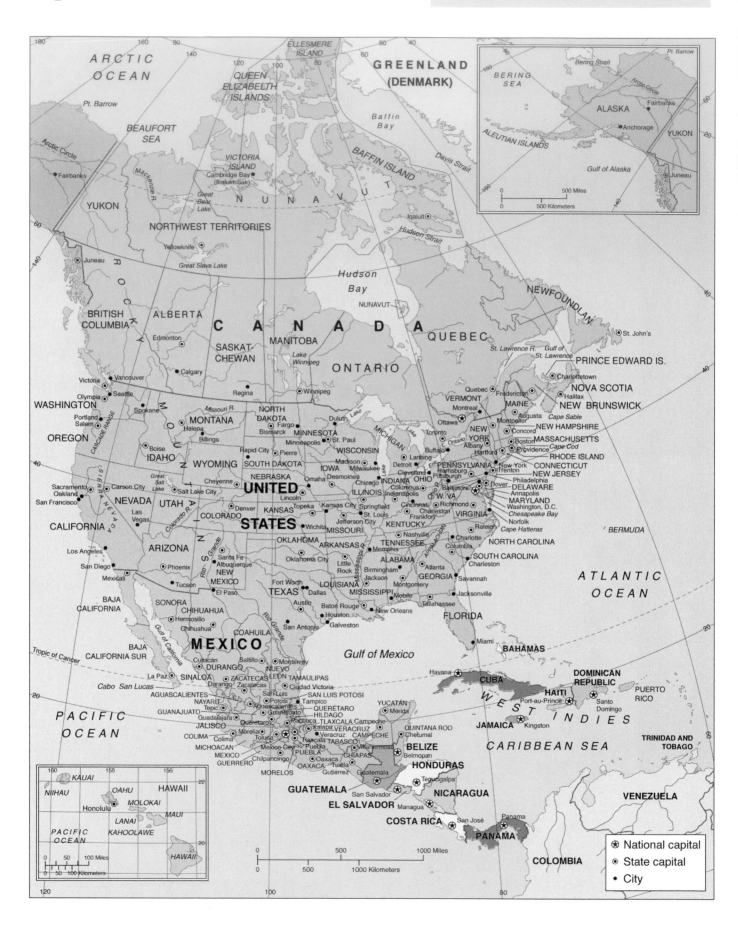

Map 84a Environment and Economy: The Use of Land

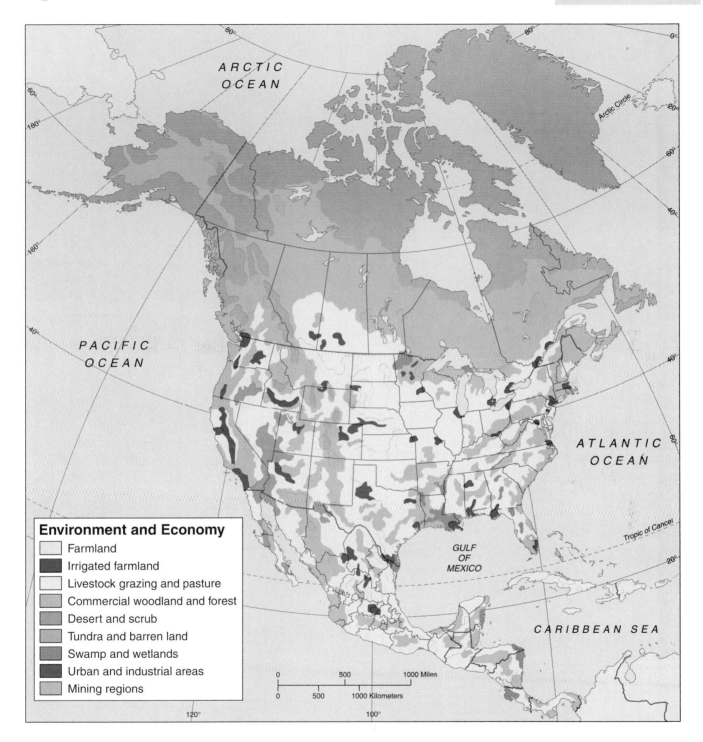

Environment and Economy
- Farmland
- Irrigated farmland
- Livestock grazing and pasture
- Commercial woodland and forest
- Desert and scrub
- Tundra and barren land
- Swamp and wetlands
- Urban and industrial areas
- Mining regions

The use of land in North America represents a balance between agriculture, resource extraction, and manufacturing that is unmatched. The United States, as the world's leading industrial power, is also the world's leader in commercial agricultural production. Canada, despite its small population, is a ranking producer of both agricultural and industrial products and Mexico has begun to emerge from its developing nation status to become an important industrial and agricultural nation as well. The countries of Middle America and the Caribbean are just beginning the transition from agriculture to modern industrial economies. Part of the basis for the high levels of economic productivity in North America is environmental: a superb blend of soil, climate, and raw materials. But just as important is the cultural and social mix of the plural societies of North America, a mix that historically aided the growth of the economic diversity necessary for developed economies.

Map **84c** Population Distribution: Clustering

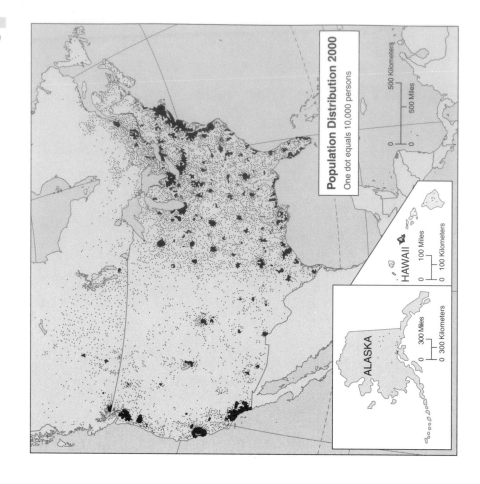

Population Distribution 2000
One dot equals 10,000 persons

ALASKA

HAWAII

Although population clustering is characteristic of highly economically developed regions, the Anglo-American population exhibits a remarkably clustered pattern. The primary reasons for this remarkable development are agricultural technology and affluence. A highly developed agricultural technology allows a small number of farmers, using sophisticated machinery, to grow and harvest enormous quantities of agricultural produce on large farms. In the United States, the world's leader in commercial agricultural production, only 2 percent of the population are farmers and that population is thinly distributed over wide areas. The vast majority of Americans live and work in city-suburb systems in which the widespread availability of private automobiles allows people to live considerable distances from where they work, keeping overall urban population densities relatively low but allowing for extensive urbanization—with cities large enough in area to be visible as population clusters on maps at this scale.

Map **84b** Population Density

Population Density
Persons per square mile (km)

Uninhabited area
Less than 2 (1)
2 – 25 (1–10)
25 – 50 (10 – 20)
50 – 150 (20 – 60)
150 – 300 (60 – 120)
More than 300 (120)

North America contains nearly 500 million people and the United States, with over 250 million inhabitants, is the third most populous country in the world, after China and India. Most of the present North American population has roots in the Old World. The native populations of the Americas had little or no resistance to Old World diseases in 1500, and within a couple of centuries of first contact with Europeans, most of the native peoples had either died out or preserved their genetic heritage by mixing with disease-resistant Old World populations. This left a North American population that is largely European, but with significant minorities resulting from the slave laborers imported from Africa, and from the mixture of native Americans with Europeans and/or Africans. The density of that population is largely the consequence of environmental factors (good soil, the availability of water, the presence of other resources) and cultural/economic ones (agricultural production, urbanization, industrialization).

Map **84d** Deforestation in the Americas

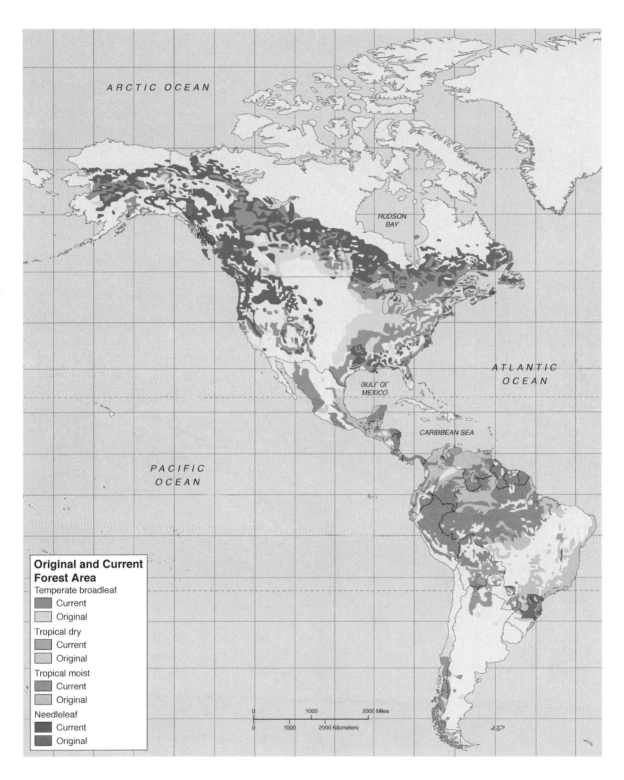

Original and Current Forest Area

Temperate broadleaf
- Current
- Original

Tropical dry
- Current
- Original

Tropical moist
- Current
- Original

Needleleaf
- Current
- Original

This map shows current forest cover adjacent to the estimated forest cover that would be present had there been no human intervention over the past several thousand years since the end of the last ice age, and assuming that climatic conditions would have been pretty much the same as they have been during that time. While in the popular imagination, it is the tropical forest regions of South America and Middle America that have been "deforested," the map shows otherwise. The greatest losses of forest cover have been in the now-agricultural regions of North America and the more temperate regions of South America (most in southern Brazil). Beside these forest losses in the temperate zones, the forest losses of the Amazon Basin seem relatively minor. This does not suggest that the current deforestation of Amazonia and other tropical forest regions is something about which we should not be concerned, but it does put that current deforestation in a more accurate historical context.

Map **85** Canada

-112-

Map 86 United States

-113-

Map 87 Middle America

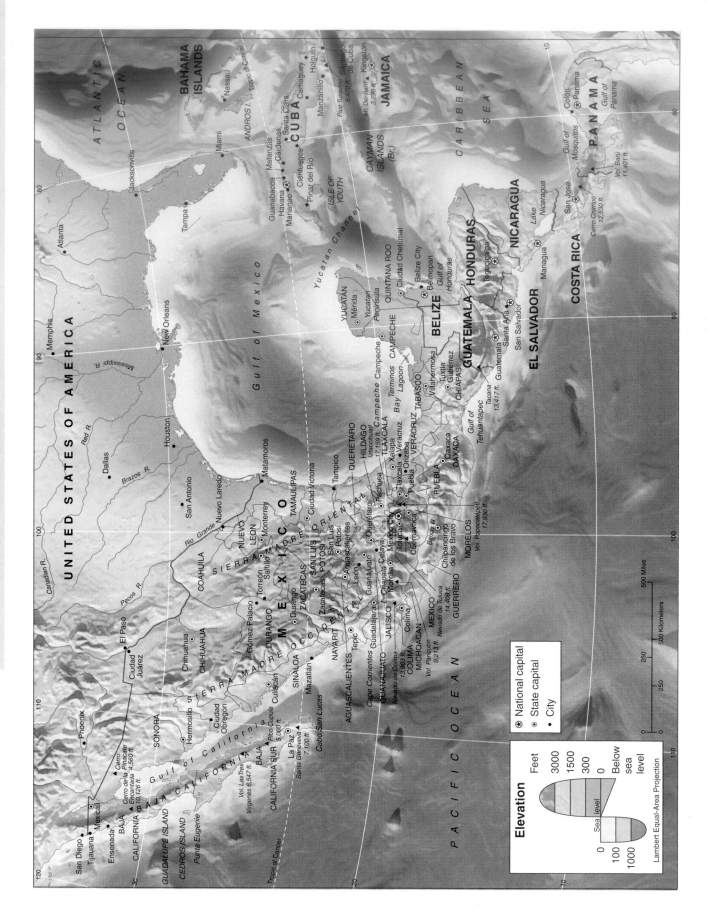

Map 88 The Caribbean

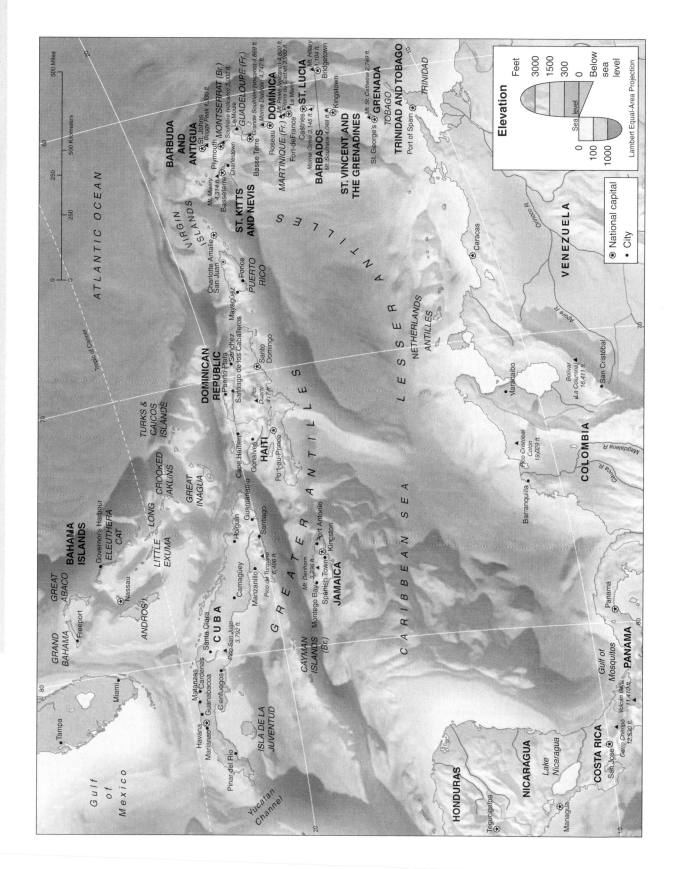

Elevation

Feet
3000
1500
300
0
Below sea level

Sea level

0 100 1000

Lambert Equal-Area Projection

⊛ National capital
• City

500 Miles
500 Kilometers

ATLANTIC OCEAN

Tropic of Cancer

BAHAMA ISLANDS
GRAND BAHAMA
GREAT ABACO
Freeport
Nassau
Governors Harbour
ELEUTHERA
CAT
ANDROS I.
LITTLE
LONG
EXUMA
CROOKED
AKLINS
GREAT INAGUA

TURKS & CAICOS ISLANDS

Miami
Tampa
Gulf of Mexico

Pinar del Rio
Havana ⊛
Marianao
Matanzas
Cardenas
Guanabacoa
Cienfuegos
Santa Clara
ISLA DE LA JUVENTUD
CUBA
Pico San Juan 3,792 ft.
Camagüey
Manzanillo
Pico de Turquino 6,496 ft.
Holguín
Guantánamo
Santiago
Yucatan Channel

CAYMAN ISLANDS (Br.)

Cape Haitien
Gonaives
HAITI
Port-au-Prince
Port Antonio
Mt. Denham 3,236 ft.
JAMAICA
Montego Bay
Spanish Town
Kingston

Santiago de los Caballeros
Sánchez
Puerto Plata
DOMINICAN REPUBLIC
Santo Domingo ⊛
Pico Duarte 4,174 ft.

Mayagüez
San Juan ⊛
Charlotte Amalie
Ponce
PUERTO RICO

VIRGIN ISLANDS

BARBUDA AND ANTIGUA
St. Johns ⊛
Boggy Peak 1,330 ft.
Plymouth
Mt. Misery 4,314 ft.
ST. KITTS AND NEVIS
Basseterre
Charlestown
MONTSERRAT (Br.)
Soufrière (volcano) 3,002 ft.

GUADELOUPE (Fr.)
Basse Terre
Grande Soufrière (volcano) 4,869 ft.
Pointe du Carbet 3,960 ft.
Morne Diablotin 4,747 ft.
Roseau ⊛
DOMINICA

MARTINIQUE (Fr.)
Mt. Pelee (volcano) 4,800 ft.
Fort-de-France
Le Marin

ST. LUCIA
Castries ⊛
Morne Gimie 3,145 ft.
Mt. Hillary 1,104 ft.

BARBADOS
Bridgetown ⊛
Mt. Soufrière 4,048 ft.

ST. VINCENT AND THE GRENADINES
Kingstown ⊛

GRENADA
St. George's ⊛
Mt. St. Catherine 2,749 ft.

TRINIDAD AND TOBAGO
TOBAGO
TRINIDAD
Port of Spain ⊛

G R E A T E R   A N T I L L E S

L E S S E R   A N T I L L E S

NETHERLANDS ANTILLES

C A R I B B E A N   S E A

HONDURAS
Tegucigalpa ⊛

NICARAGUA
Managua ⊛
Lake Nicaragua

COSTA RICA
San Jose ⊛
Cerro Chirripó 12,530 ft.
Volcán Baru 11,410 ft.

PANAMA
Panama ⊛
Gulf of Mosquitos

COLOMBIA
Barranquilla
Cartagena
Colón
Pico Cristóbal Colón 19,029 ft.
Magdalena R.
Cauca R.

VENEZUELA
Caracas ⊛
Maracaibo
Bolívar (La Columna) 16,411 ft.
San Cristóbal
Orinoco R.
Apure R.

-115-

Map 89 South America: Physical Features

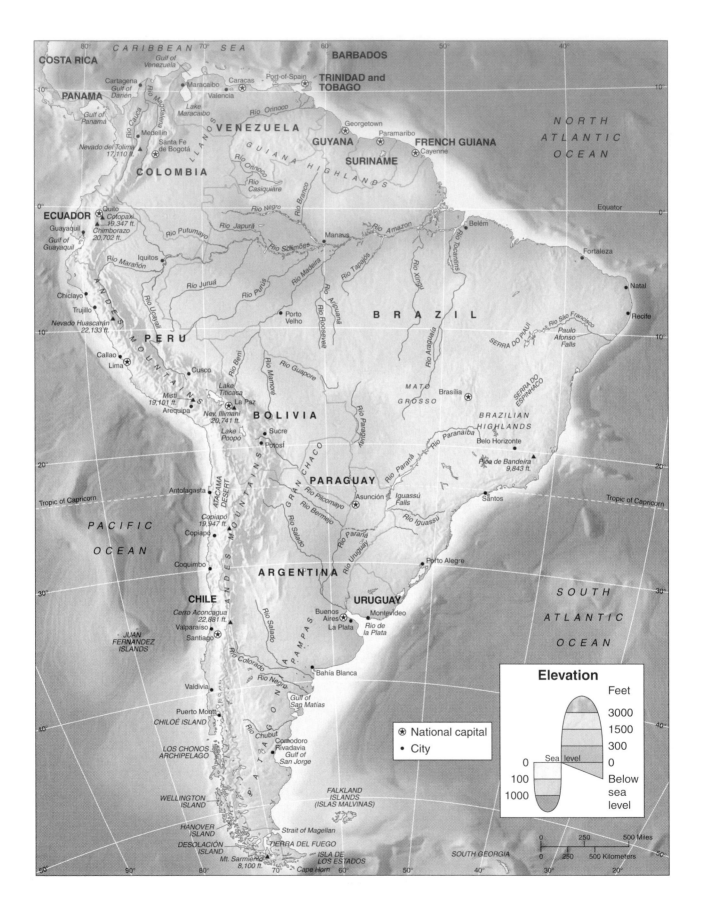

Map 90 South America: Political Divisions

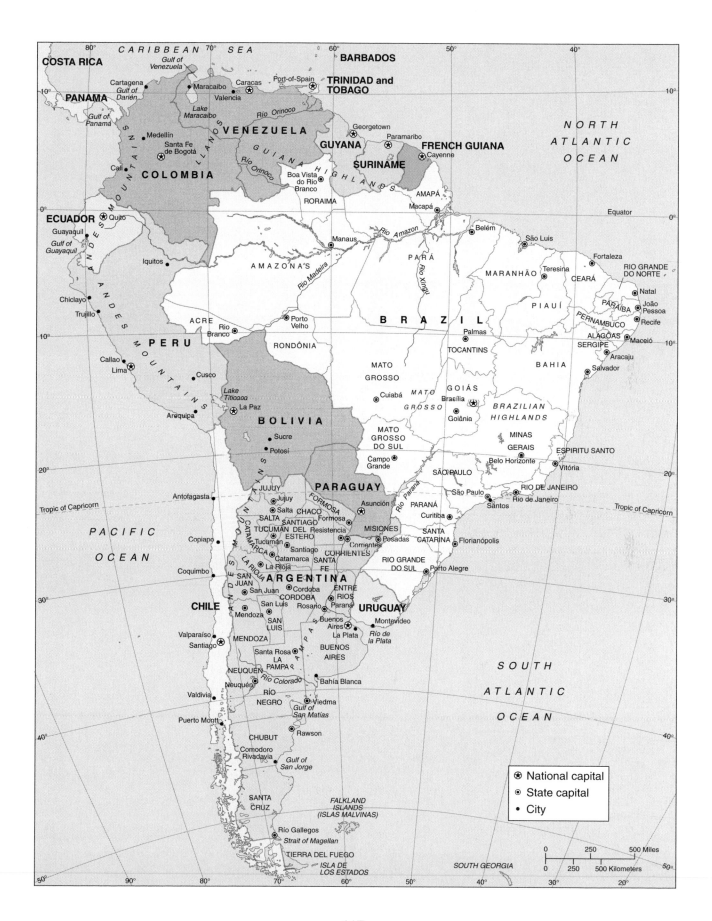

Map 91a Environment and Economy

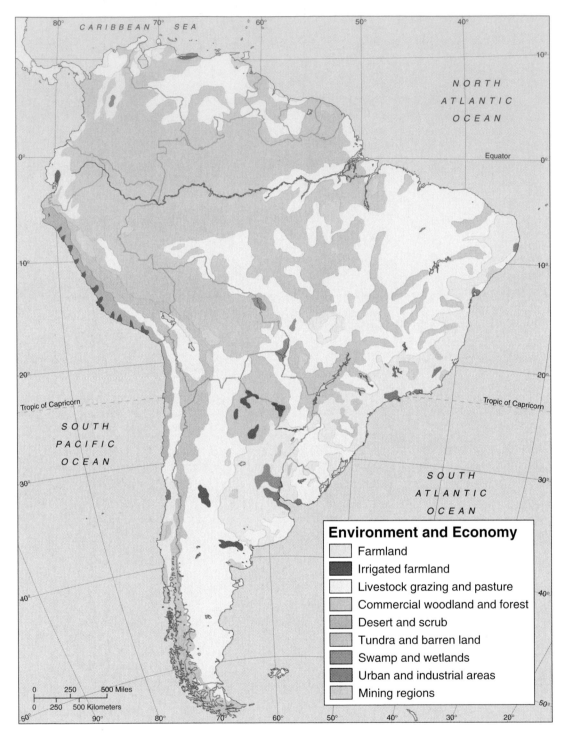

Environment and Economy
- Farmland
- Irrigated farmland
- Livestock grazing and pasture
- Commercial woodland and forest
- Desert and scrub
- Tundra and barren land
- Swamp and wetlands
- Urban and industrial areas
- Mining regions

South America is a region just beginning to emerge from a colonial-dependency economy in which raw materials flowed from the continent to more highly developed economic regions. With the exception of Brazil, Argentina, Chile, and Uruguay, most of the continent's countries still operate under the traditional mode of exporting raw materials in exchange for capital that tends to accumulate in the pockets of a small percentage of the population. The land use patterns of the continent are, therefore, still dominated by resource extraction and agriculture. A problem posed by these patterns is that little of the continent's land area is actually suitable for either commercial forestry or commercial crop agriculture without extremely high environmental costs. Much of the agriculture, then, is based on high value tropical crops that can be grown in small areas profitably, or on extensive livestock grazing. Even within the forested areas of the Amazon Basin where forest clearance is taking place at unprecedented rates, much of the land use that replaces forest is grazing.

Map 91b Population Density

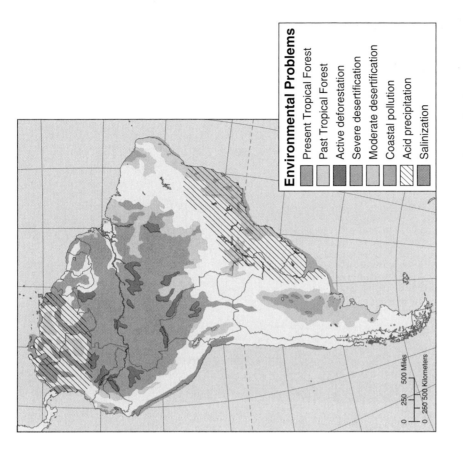

Population Density

Persons per square mile (kilometer)

- Uninhabited area
- Less than 2 (1)
- 2 – 50 (1 – 20)
- 50 – 150 (20 – 60)
- 150 – 300 (60 – 120)
- More than 300 (100)

0   250   500 Miles
0   250 500 Kilometers

Since so much of interior South America is uninhabitable (the high Andes) or only sparsely populated (the interior of the Amazon Basin), the continent's population tends to be peripheral—approximately 90 percent of the continent's nearly 375 million people live within 150 miles of the sea. This population also tends to be heavily urbanized. Over 80 percent of South America's population lives in cities and the continent has three of the world's 15 largest cities—Rio de Janeiro, São Paulo, and Buenos Aires. São Paulo is the world's third largest urban agglomeration after Tokyo and New York. As in North America, most of the population of South America can trace at least part of its ancestry to the Old World. Throughout the Spanish-speaking parts of the continent the population is predominantly *mestizo* or mixed European and native South American. In Portuguese-speaking Brazil, in addition to a *mestizo* population, there is a significant admixture of African blood, the result of a large slave labor force imported from Africa to work the sugar, cotton, and other plantations of the colonial period.

Map 91c Environmental Problems

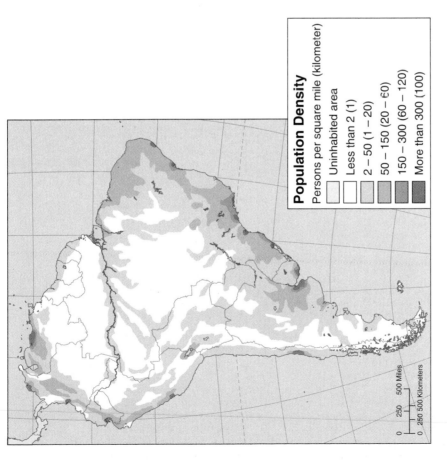

Environmental Problems

- Present Tropical Forest
- Past Tropical Forest
- Active deforestation
- Severe desertification
- Moderate desertification
- Coastal pollution
- Acid precipitation
- Salinization

0   250   500 Miles
0   250 500 Kilometers

The drainage basin of the Amazon River and its tributaries, along with adjacent regions, is the world's largest remaining area of tropical forest. Much of the periphery of this vast forested region has already been cleared for farming and grazing and the ax and chainsaw and the flames are working their way steadily toward the interior. Tropical deforestation produces a loss of the biological diversity represented by the world's most biologically productive ecosystem, along with changes in soil and soil-water systems. South America has other environmental problems: *desertification* in which grassland and/or scrub vegetation is converted to desert through overgrazing or other unwise agricultural practices; soil *salinization* in which soils become increasingly salty as the consequence of the over-application of irrigation water; *coastal* and *estuarine pollution* resulting from unregulated or unchecked use of coastal waters for industrial, commercial, and transportation purposes; and *acid precipitation* resulting from the combination of airborne industrial wastes and automobile-truck exhausts with water vapor to produce dry or wet acidic fallout.

Map 91d Protection of Natural Areas

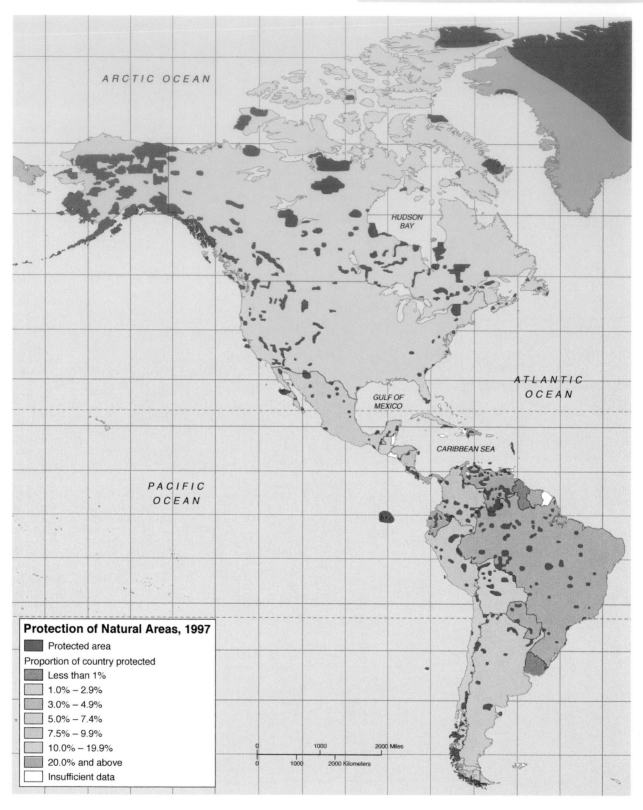

ARCTIC OCEAN

HUDSON BAY

ATLANTIC OCEAN

GULF OF MEXICO

CARIBBEAN SEA

PACIFIC OCEAN

Protection of Natural Areas, 1997

- Protected area

Proportion of country protected

- Less than 1%
- 1.0% – 2.9%
- 3.0% – 4.9%
- 5.0% – 7.4%
- 7.5% – 9.9%
- 10.0% – 19.9%
- 20.0% and above
- Insufficient data

0   1000   2000 Miles
0   1000   2000 Kilometers

There have been expressed concerns among international conservation organizations that the protection of natural areas is largely an empty measure and that many of the world's national, provincial, regional, or state parks are "paper parks"—more important in conception than in reality. Particularly in Africa, the park systems designated largely for the purposes of wildlife preservation have come under international scrutiny. In the Americas, however, there does seem to be a close correlation between the designation of protected natural areas and the actuality of ecological preservation. With an increasing demand for natural resources and with governmental policies in both the United States and Brazil (the two largest population and economic entities in the Americas) that seem directed against preservation measures, there is fear that American protected areas could become, like their counterparts in many areas of Africa and Asia, simply parks on paper.

Map 92 Northern South America

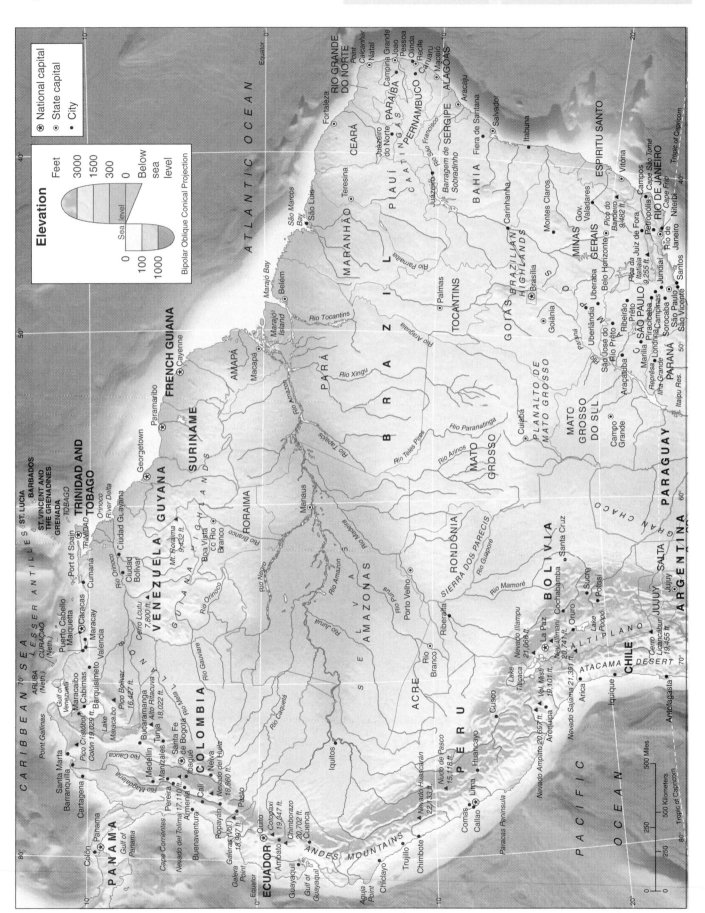

Map 93 Southern South America

Map **94** Europe: Physical Features

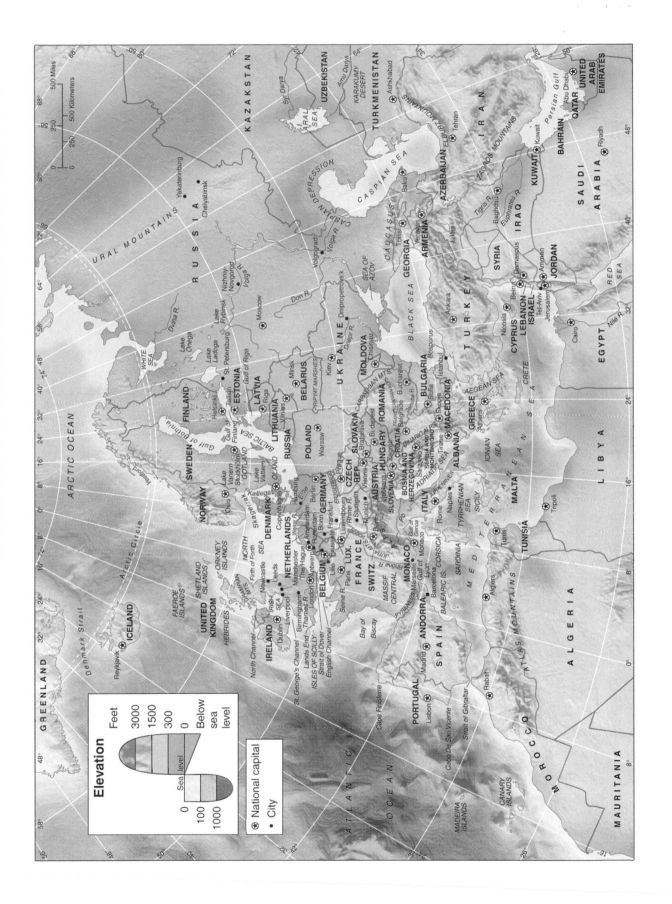

Europe: Physical Features

Elevation

Feet	
3000	
1500	
300	
0	Below sea level
Sea level	

0	
100	
1000	

⊛ National capital
• City

Map 95 Europe: Political Divisions

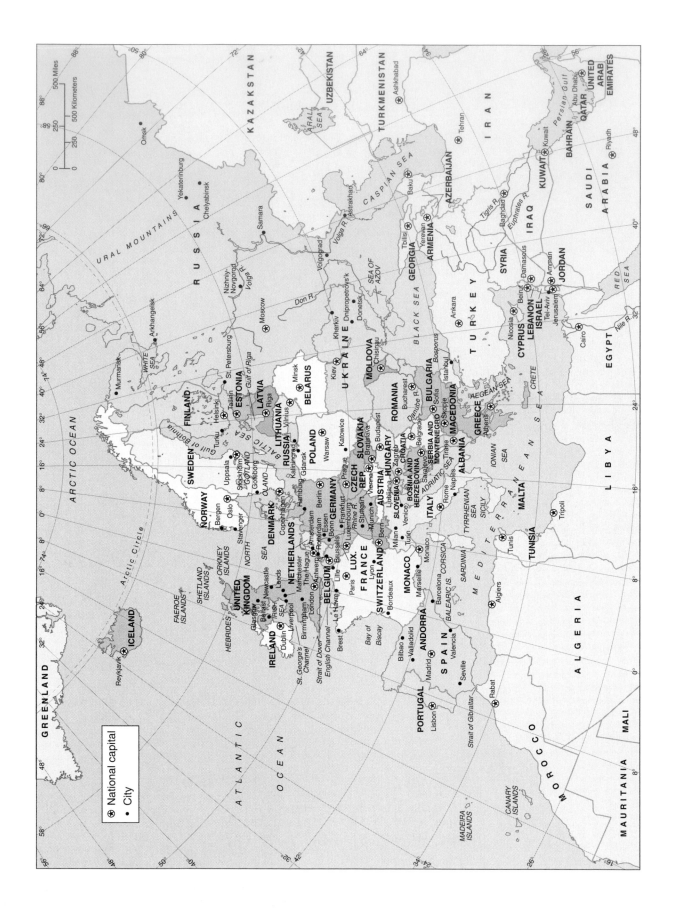

National capital
City

Europe: Thematic Features

Map 96a Environment and Economy: The Use of Land

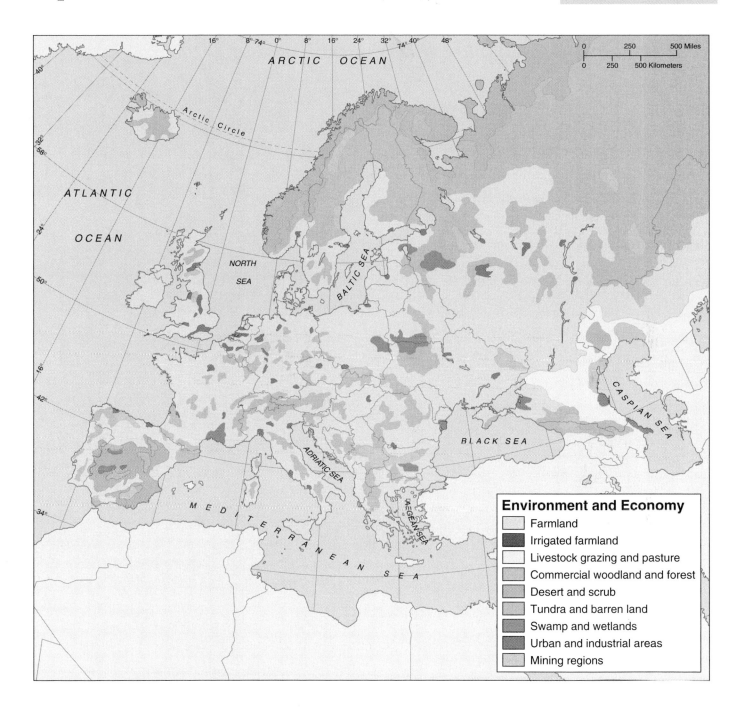

Environment and Economy
- Farmland
- Irrigated farmland
- Livestock grazing and pasture
- Commercial woodland and forest
- Desert and scrub
- Tundra and barren land
- Swamp and wetlands
- Urban and industrial areas
- Mining regions

More than any other continent, Europe bears the imprint of human activity—mining, forestry, agriculture, industry, and urbanization. Virtually all of western and central Europe's natural forest vegetation is gone, lost to clearing for agriculture beginning in prehistory, to lumbering that began in earnest during the Middle Ages, or more recently, to disease and destruction brought about by acid precipitation. Only in the far north and the east do some natural stands remain. The region is the world's most heavily industrialized and the industrial areas on the map represent only the largest and most significant. Not shown are the industries that are found in virtually every small town and village and smaller city throughout the industrial countries for Europe. Europe also possesses abundant raw materials and a very productive agricultural base. The mineral resources have long been in a state of active exploitation and the mining regions shown on the map are, for the most part, old regions in upland areas that are somewhat less significant now than they may have been in the past. Agriculturally, the northern European plain is one of the world's great agricultural regions but most of Europe contains decent land for agriculture.

Map 96b Population Density

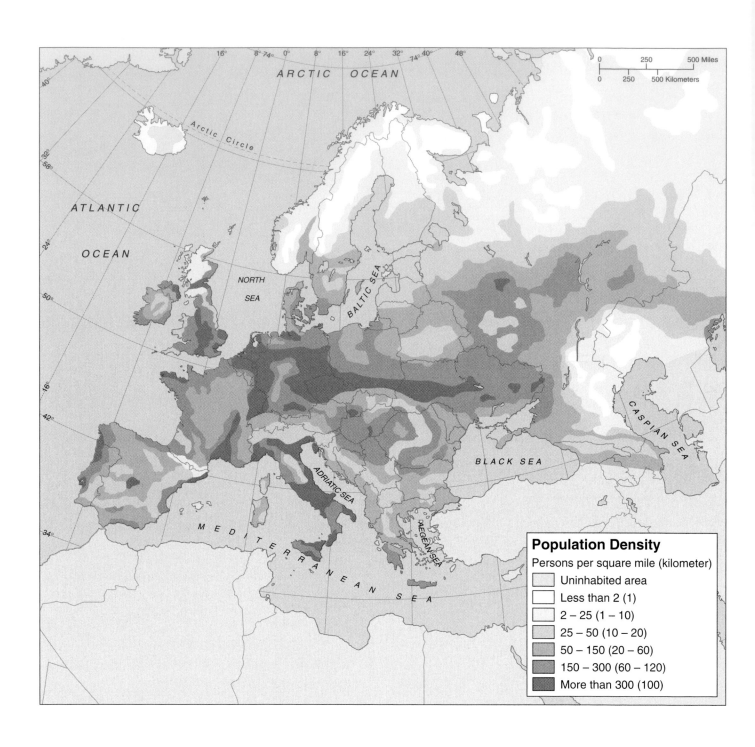

Population Density

Persons per square mile (kilometer)

- ☐ Uninhabited area
- ☐ Less than 2 (1)
- ☐ 2 – 25 (1 – 10)
- ☐ 25 – 50 (10 – 20)
- ☐ 50 – 150 (20 – 60)
- ☐ 150 – 300 (60 – 120)
- ☐ More than 300 (100)

Europe is one of the most densely settled regions of the world with an overall population density nearing 200 persons per square mile (80 per square kilometer), the consequence of a high level of urbanization and an economic system that is heavily industrialized. Even in agricultural regions, the population density is high. Beyond high density, the two chief identifying marks of the European population are remarkable diversity and unusual dynamics. For a small part of the world, Europe has cultural and ethnic diversity that is rarely matched elsewhere; more than 60 languages are spoken in an area not much larger than the United States. The population dynamics of Europe show a mature population that has passed through the "Demographic Transition"—a remarkable increase and then decline in growth rates resulting from the rise of an urban-industrial society. Only in other heavily industrialized regions of the world are found the very small overall growth rates characteristic of Europe.

Map 96c European Political Boundaries, 1914–1948

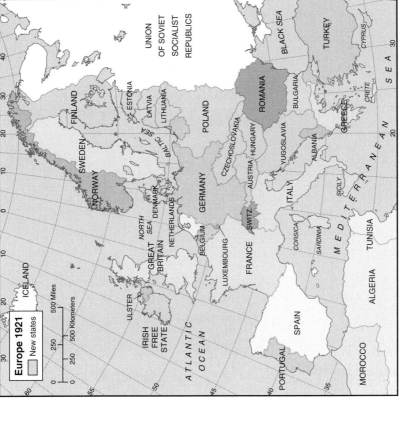

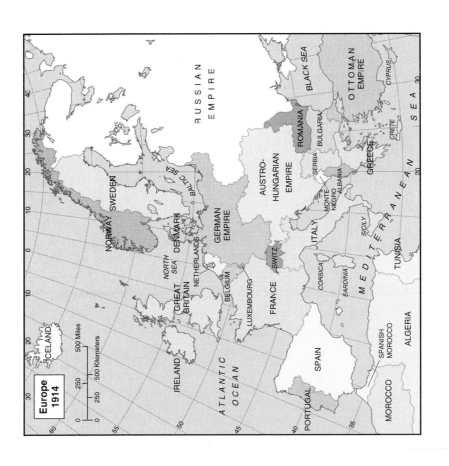

In 1914, on the eve of the First World War, Europe was dominated by the United Kingdom and France in the west, the German Empire and the Austro-Hungarian Empire in central Europe, and the Russian Empire in the east. Battle lines for the conflict that began in 1914 were drawn when the United Kingdom, France, and the Russian Empire joined together as the Triple Entente. In the view of the Germans, this coalition was designed to encircle Germany and its Austrian ally, which, along with Italy, made up the Triple Alliance. The German and Austrian fears were heightened in 1912–14 when a Russian-sponsored "Balkan League" pushed the Ottoman Turkish Empire from Europe, leaving behind the weak and mutually antagonistic Balkan states Serbia and

Montenegro. In August 1914, Germany and Austria-Hungary attacked in several directions and World War I began. Four years later, after massive loss of life and destruction, the central European empires were defeated. The victorious French, English, and Americans (who had entered the war in 1917) restructured the map of Europe in 1919, carving nine new states out of the remains of the German and Austro-Hungarian empires and the westernmost portions of the Russian Empire which, by the end of the war, was deep in the Revolution that deposed the czar and brought the Communists to power in a new Union of Soviet Socialist Republics.

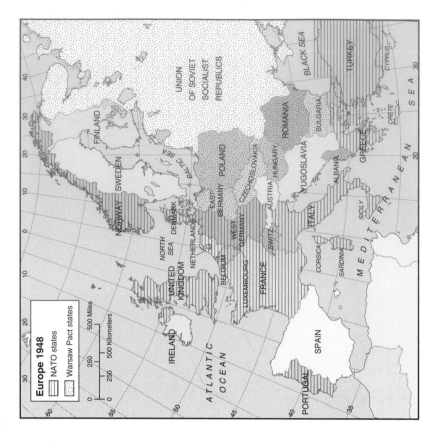

Europe 1943

- Axis powers
- Under German rule
- Axis military occupation
- Greater German Reich

- Axis satellites
- Allied territory
- Allied occupied
- Neutral powers

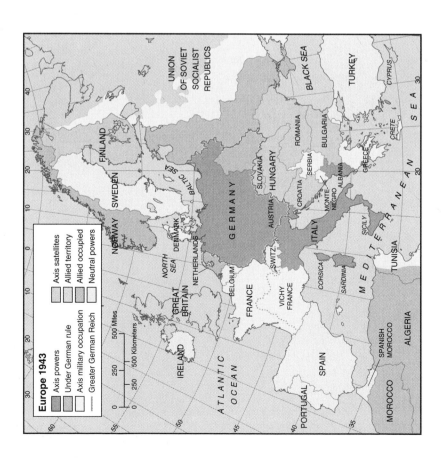

Europe 1948

- NATO states
- Warsaw Pact states

When the victorious Allies redrew the map of central and eastern Europe in 1919, they caused as many problems as they were trying to solve. The interval between the First and Second World Wars was really just a lull in a long war that halted temporarily in 1918 and erupted once again in 1939. Defeated Germany, resentful of the terms of the 1918 armistice and 1919 Treaty of Versailles and beset by massive inflation and unemployment at home, overthrew the Weimar republican government in 1933 and installed the National Socialist (Nazi) party led by Adolf Hitler in Berlin. Hitler quickly began making good on his promises to create a "thousand year realm" of German influence by annexing Austria and the Czech region of Czechoslovakia and allying Germany with a fellow fascist state in Mussolini's Italy. In September 1939 Germany launched the lightning-quick combined infantry, artillery, and armor attack known as *der Blitzkrieg* and took Poland to the east and, in quick succession, the Netherlands, Belgium, and France to the west. By 1943 the greater German Reich extended from the Russian Plain

to the Atlantic and from the Black Sea to the Baltic. But the Axis powers of Germany and Italy could not withstand the greater resources and manpower of the combined United Kingdom–United States–USSR-led Allies and, in 1945, Allied armies occupied Germany. Once again, the lines of the central and eastern European map were redrawn. This time, a strengthened Soviet Union took back most of the territory the Russian Empire had lost at the end of the First World War. Germany was partitioned into four occupied sectors (English, French, American, and Russian) and later into two independent countries, the Federal Republic of German (West Germany) and the German Democratic Republic (East Germany). Although the Soviet Union's territory stopped at the Polish, Hungarian, Czechoslovakian, and Romanian borders, the eastern European countries (Poland, East Germany, Czechoslovakia, Hungary, Romania, Yugoslavia, Albania, and Bulgaria) became Communist between 1945 and 1948 and were separated from the West by the Iron Curtain.

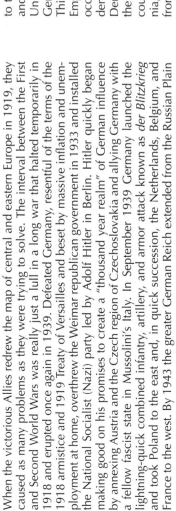

Map 96d Political Changes 1989–2004

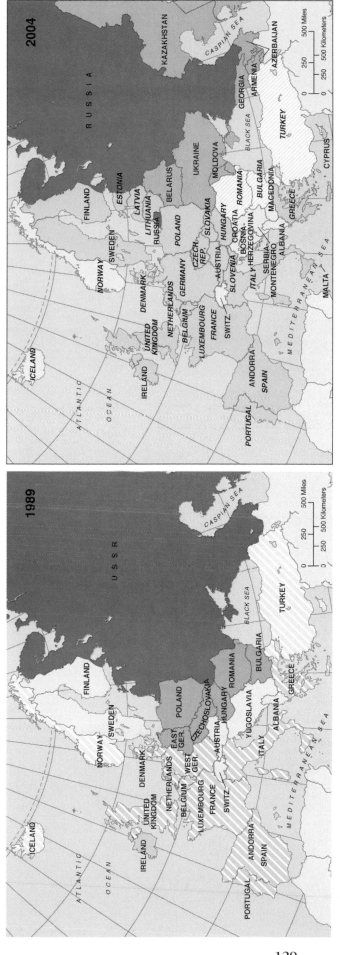

Europe: Political Changes 1989–2004

1989

- Union of Soviet Socialist Republics
- Warsaw Pact Countries (excluding the USSR)
- North Atlantic Treaty Organization Countries
- European Community (formerly the EEC)

2004

- Former Republics of the USSR, now independent countries
- Russian Federation
- *FRANCE* NATO Countries shown in italics
- European Union (formerly the EC)
- European Union applicant country

During the last decade of the twentieth century, one of the most remarkable series of political geographic changes of the last 500 years took place. The bi-polar "East-West" structure that had characterized Europe's political geography since the end of World War II altered in the space of a very few years. In the mid-1980s, as Soviet influence over eastern and central Europe weakened, those countries began to turn to the capitalist West. Between 1989, when the country of Hungary was the first Soviet satellite to open its borders to travel, and 1991, when the Soviet Union dissolved into 15 independent countries, abrupt change in political systems occurred. The result is a new map of Europe that includes a number of countries not present on the map of 1989. These countries have emerged as the result of reunification, separation, or independence from

the old Soviet Union. The new political structure has been accompanied by growing economic cooperation. In May, 2004, ten new applicant countries (the Greek part of Cyprus, Czech Republic, Estonia, Hungary, Latvia, Lithuania, Malta, Poland, Slovakia, and Slovenia) were added to the European Union, increasing the economic strength of the EU, which now has its own currency (the euro) and is presently attempting to develop a constitution acceptable to the 25 member nations. The only remaining formal applicant country to EU membership is Turkey. Concerned about Turkish migration and the potential political problems of adding a country with a population of 60 million Muslims to the European Union, the majority of EU countries have thus far failed to act positively on Turkey's application for membership.

Map 97 Western Europe

Elevation

	Feet
	3000
	1500
	300
0 Sea level	0
100	Below
1000	sea level

Azimuthal Equidistant Projection

⊛ National capital
• City

FAEROES

NORWAY
Trondheim
Östersund
Helagsfjället 5,892 ft.
Sånfjället 4,190 ft.
Sundsvall
Glittertinden 8,110 ft.
SWEDEN
Bergen
Gävle
Falun
Hönefoss
Oslo
Uppsala
Haugesund
Drammen
Moss Karlstad
Västerås
Stockholm
Stavanger
Skien
Fredrikstad
VÄNERN
Eskilstuna
Södertälje
Örebro

SHETLAND IS.
Kristiansand
Skagerrak
Uddevalla
Trollhättan
Linköping
HEBRIDES
ORKNEY IS.
58°
Frederikshavn
Göteborg
Jönköping
GOTLAND
UNITED KINGDOM
Inverness
Ålborg
Borås
Aberdeen
Ben Nevis 4,406 ft.
SCOTLAND
Holstebro
Randers
Århus
Kalmar
ÖLAND
Karlskrona
NORTHERN IRELAND
Perth
Dundee
Kristiansand
Helsingør Helsingborg
Kristianstad
Greenock
Paisley
Edinburgh
NORTH SEA
DENMARK
Vejle
Copenhagen
Malmö
BORNHOLM
Londonderry
Glasgow
Motherwell
Esbjerg
Odense
Næstved
Scafell Pikes 3,210 ft.
Newcastle upon Tyne
Flensburg
Sønderborg
BALTIC SEA
Belfast
Carlisle
Sunderland
Schleswig
Newmünster
Kiel
Rostock
Stralsund
Gdynia
ISLE OF MAN
York Leeds
Cuxhaven
Lübeck
Koszalin
Gdansk
IRELAND
Limerick
IRISH SEA
Liverpool
Manchester
Sheffield
Hamburg
Szczecin
Bydgoszcz
Carrauntoohil 3,414 ft.
Dublin
Dub Laoghaire
Stoke-on-Trent
Lincoln
NETHERLANDS
Groningen
Bremerhaven
Bremen
Hannover
Berlin
Frankfurt an der Oder
Poznan
Waterford
Snowdon 3,560 ft.
Nottingham
Leeuwarden
Oldenburg
POLAND
Cork
WALES
Birmingham
Norwich
Amsterdam
Zwolle
Apeldoorn
Münster
Braunschweig
Magdeburg
Potsdam
Dessau
Kalisz
ENGLAND
Worcester
Cambridge
The Hague
Utrecht
Duisburg
Bielefeld
Hildesheim
GERMANY
Swansea
Cardiff
Bristol
Bath
Oxford
London
Ipswich
Rotterdam
Brugge
Düsseldorf
Essen
Dortmund
Wuppertal
Kassel
Halle
Leipzig
Dresden
Legnica
Wrocław
Görlitz
Liberec
Opole
Reading
Thames R.
Dover
BELGIUM
Antwerp
Gent
Brussels
Cologne
Bonn
Erfurt
Weimar
Zwickau
Karlovy Vary
Prague
Hradec Králové
Plymouth
Southampton
Brighton
Portsmouth
English Channel
Boulogne-sur-Mer
Lille
Charleroi
Aachen
Koblenz
Wiesbaden
Gera
Pilsen
Havlíčkův
Brod
Olomouc
Cherbourg
Le Havre
Arras
Mons
St. Quentin
LUXEMBOURG
Mainz
Frankfurt
Würzburg
CZECH REPUBLIC
Brno
Žilina
Brest
St. Brieuc
St. Malo
Caen
Rouen
Amiens
Rheims
Châlons-sur-Marne
Kaiserslautern
Saarbrücken
Darmstadt
Nürnberg
Regensburg
České Budějovice
SLOVAKIA
Trenčín
Nové Zámky
Quimper
Fougères
St. Denis
Paris
Metz
Nancy
Mannheim
Sarreaguemines
Karlsruhe
Stuttgart
Ulm
Augsburg
Passau
Linz
Wels
Wiener Neustadt
Bratislava
Győr
Pápa
Lorient
Vannes
Rennes
Laval
Versailles
Troyes
Épinal
Colmar
Mulhouse
Baden-Baden
Strasbourg
Freiburg
Danube R.
Munich
Salzburg
Vienna
AUSTRIA
Szombathely
St. Nazaire
Angers
Le Mans
Orléans
Auxerre
Dijon
Belfort
Basel
Schaffhausen
Zugspitze 9,721 ft.
Innsbruck
ALPS
Graz
Székesfehérvár
HUNGARY
Nantes
Blois
Tours
Loire R.
Nevers
Besançon
Zürich
SWITZ.
Sankt Gallen
LIECHTENSTEIN
Grossglockner 12,461 ft.
Villach
Klagenfurt
Nagykanizsa
Pécs
Chalet
Bourges
Mâcon
Lausanne
Luzern
Jungfrau 13,642 ft.
Triglav 9,393 ft.
SLOVENIA
Maribor
Varaždin
Châtellerault
Poitiers
Moulins
Geneva
Bourg-en-Bresse
Matterhorn 14,692 ft.
Finsteraarhorn 14,022 ft.
Bolzano
Udine
Ljubljana
Celje
Zagreb
Osijek
Slavonsky Brod
La Rochelle
Rochefort
Niort
Limoges
Vichy
Roanne
Lyon
Monte Rosa 15,203 ft.
Bergamo
Brescia
Novara
Milan
Pavia
Vicenza
Padua
Trieste
Novo Mesto
Rijeka
Sisak
CROATIA
Bordeaux
Angoulême
Périgueux
Clermont-Ferrand
Puy de Sancy 6,185 ft.
St. Étienne
Villeurbanne
Chambéry
Grenoble
Mont Blanc 15,771 ft.
Turin
Alessandria
Cremona
Verona
Parma
Modena
Genoa
Po R.
Venice
Banja Luka
BOSNIA-HERZEGOVINA
Tuzla
Bay of Biscay
El Ferrol
La Coruña
Santiago de la Compostela
Pontevedra
Vigo
Lugo
Oviedo
Avilés
Gijón
Ponferrada
León
Santander
Bilbao
San Sebastián
Vitoria
Pamplona
Bayonne
Pau
Tarbes
Montauban
Toulouse
Plomb du Cantal 6,096 ft.
MASSIF
CENTRAL
Mt. Pelat 10,010 ft.
Montpellier
Nîmes
Avignon
Aix-en-Provence
Marseilles
Toulon
Mt. Viso 12,602 ft.
Savona
La Spezia
Prato
Bologna
Florence
Livorno
Ravenna
Zadar
Šibenik
Split
Sarajevo
Mostar
Nikšić
Braga
Porto
Palencia
Burgos
Logroño
Zaragoza
Soria
PYRENEES
Pic de Montcalm 10,305 ft.
Mt. Perdido 11,000 ft.
Pica de Aneto 11,168 ft.
Andorra
La Vella
ANDORRA
Perpignan
Carcassonne
Nice
Cannes
MONACO
SAN-MARINO
ITALY
Siena
Arezzo
Ancona
ADRIATIC SEA
Dubrovnik
PORTUGAL
Zamora
Valladolid
Duero R.
Salamanca
Segovia
Guadalajara
Madrid
Lérida
Sabadell
Barcelona
Tarragona
Gerona
Mataró
Bastia
CORSICA (France)
ELBA
Mt. Cinto 8,878 ft.
Piombino
Perugia
Terni
Mt. Corno 9,554 ft.
Chieti
Mt. Amaro 9,163 ft.
42°
Coimbra
Covilhã
Ávila
Aranjuez
Cuenca
Castellón de la Plana
Ajaccio
Vatican City
Rome
L'Aquila
Lisbon
Portalegre
Cáceres
Mérida
Toledo
Ciudad Real
SPAIN
Sagunto
Valencia
Tiber R.
Terracina
Vesuvio 4,190 ft.
Foggia
Cerignola
Barletta
Barreiro
Setúbal
Elvas
Badajoz
Guadiana R.
Tomelloso
Valdepeñas
Albacete
BALEARIC IS.
MINORCA
Sassari
Alghero
Punta la Marmora 6,016 ft.
Naples
Molfetta
Bari
Altamura
Brindisi
Évora
Puertollano
Córdoba
Linares
La Sagra 7,999 ft.
Alcoy
Valencia
Alicante
Palma
MAJORCA
IBIZA
Salerno
Potenza
Taranto
Lecce
Lagos
Seville
Éxija
Guadalquivir R.
Jaén
Murcia
Oristano
SARDINIA (Italy)
TYRRHENIAN SEA
Gulf of Taranto
Huelva
Utrera
Antequera
Granada
Lucena
Lorca
Cartagena
Igelsias
Cagliari
Jerez de la Frontera
Cádiz
Strait of Gibraltar
La Línea
Gibraltar
Málaga
Motril
Mulhacén 11,424 ft.
Almería
MEDITERRANEAN SEA
Cosenza
Nicastro
Catanzaro
IONIAN SEA
Tangier
Ceuta
Algiers
Mt. Etna (volcano) 10,902 ft.
Messina
Reggio di Calabria
Rabat
Melilla
Oran
'Annaba
Trapani
Palermo
Marsala
Caltanissetta
Bizerte
SICILY
Catania
Caltagirone
Siracusa
Casablanca
MIDDLE ATLAS
MOROCCO
Vittoria
Modica
Ragusa
ALGERIA
Tunis
34°
SAHARAN ATLAS
TUNISIA
MALTA

ATLANTIC OCEAN

0 250 500 Miles
0 250 500 Kilometers

Map 98 Eastern Europe

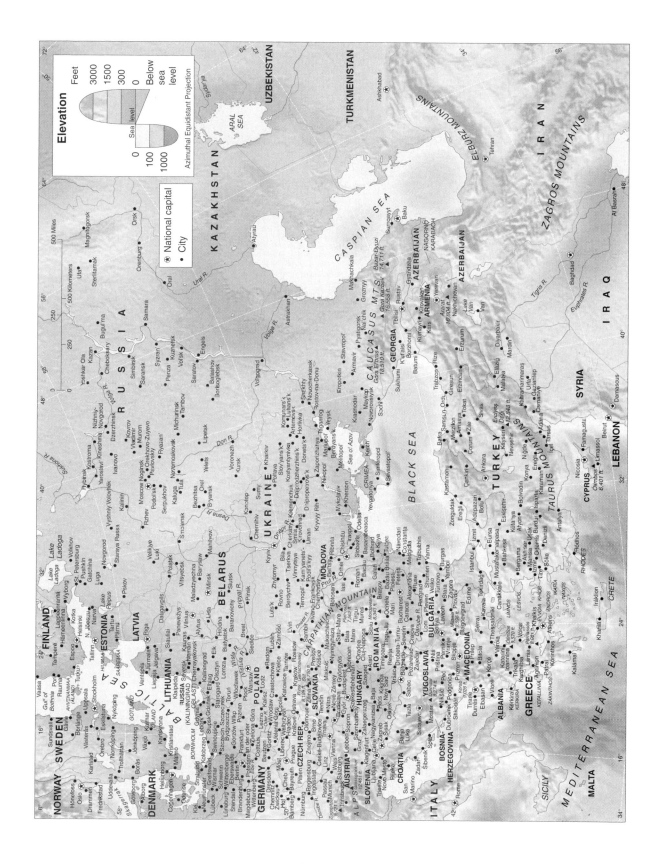

Map 99 Northern Europe

Map 100 Africa: Physical Features

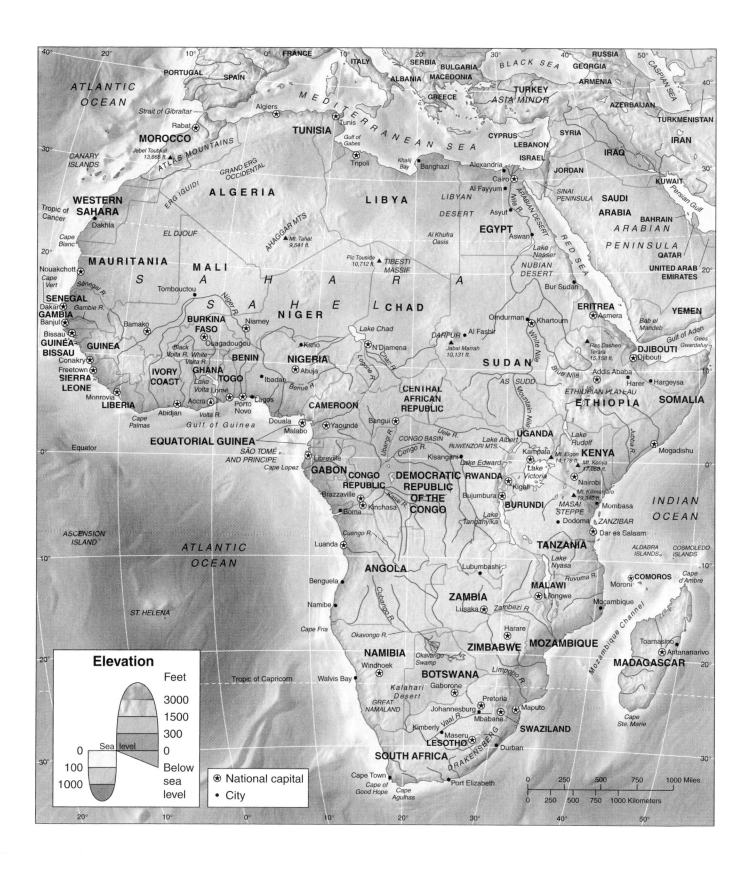

Map **101** Africa: Political Divisions

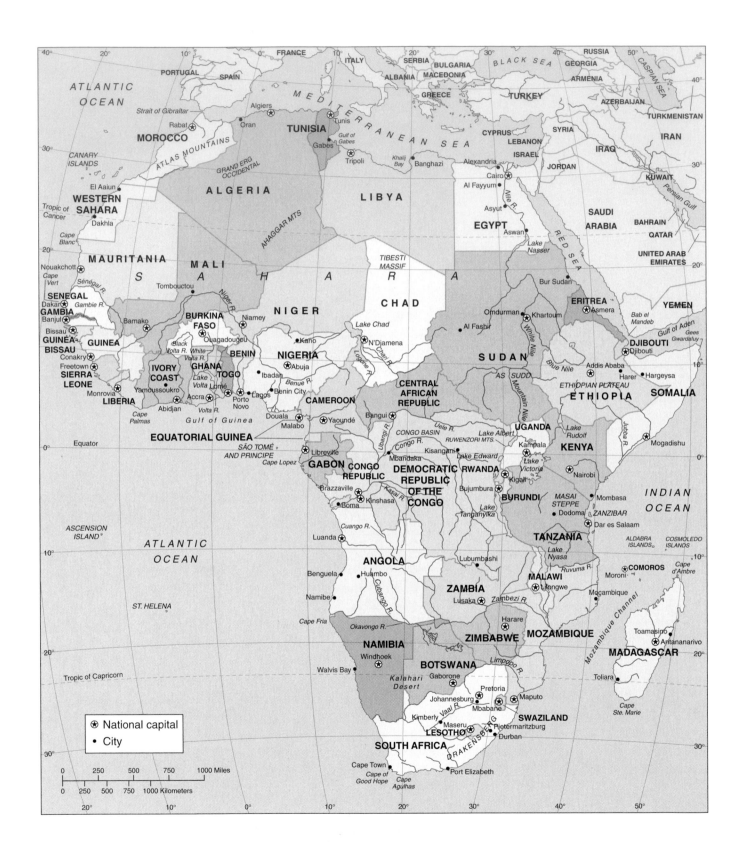

Africa: Thematic Features

Map 102a Environment and Economy

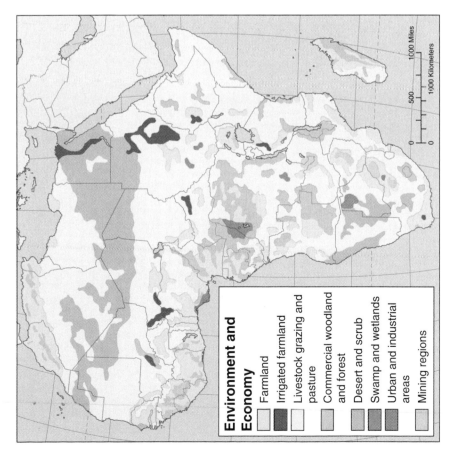

Environment and Economy

- Farmland
- Irrigated farmland
- Livestock grazing and pasture
- Commercial woodland and forest
- Desert and scrub
- Swamp and wetlands
- Urban and industrial areas
- Mining regions

1000 Miles
0 500 1000 Kilometers

Africa's economic landscape is dominated by subsistence, or marginally-commercial agricultural activities and raw material extraction, engaging three-fourths of Africa's workers. Much of this grazing land is very poor desert scrub and bunch grass that is easily impacted by cattle, sheep, and goats. Growing human and livestock populations place enormous stress on this fragile support capacity and the result is desertification: the conversion of even the most minimal of grazing environments to a small quantity of land suitable for crop farming. Although the continent has approximately 20 percent of the world's total land area, the proportion of Africa's arable land is small. The agricultural environment is also uncertain; unpredictable precipitation and poor soils hamper crop agriculture.

Map 102b Population Density

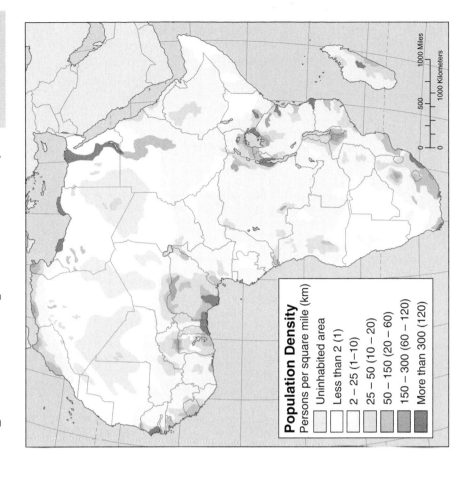

Population Density

Persons per square mile (km)

- Uninhabited area
- Less than 2 (1)
- 2 – 25 (1–10)
- 25 – 50 (10 – 20)
- 50 – 150 (20 – 60)
- 150 – 300 (60 – 120)
- More than 300 (120)

1000 Miles
0 500 1000 Kilometers

Nearly 800 million people occupy the African continent, approximately one-eighth of the world's population. In general, this population has two chief characteristics: a low level of quality of life and growth rates that are among the world's highest. On a continent beset by poverty, recurrent internal civil and tribal war, and a host of environmental problems, the populations of many African countries are nevertheless increasing at a rate above 3 percent per year. The bulk of the population is concentrated in relatively small areas of the Mediterranean coastal regions of the north, the bulge of West Africa, and the eastern coastal and highland zone stretching from South Africa to Kenya. Where populations in other parts of the world tend to avoid highland locations, in Africa highlands often tend to be the most densely settled regions: moister with better soils, freer of insect pests, and somewhat cooler.

Map 102c Colonial Patterns

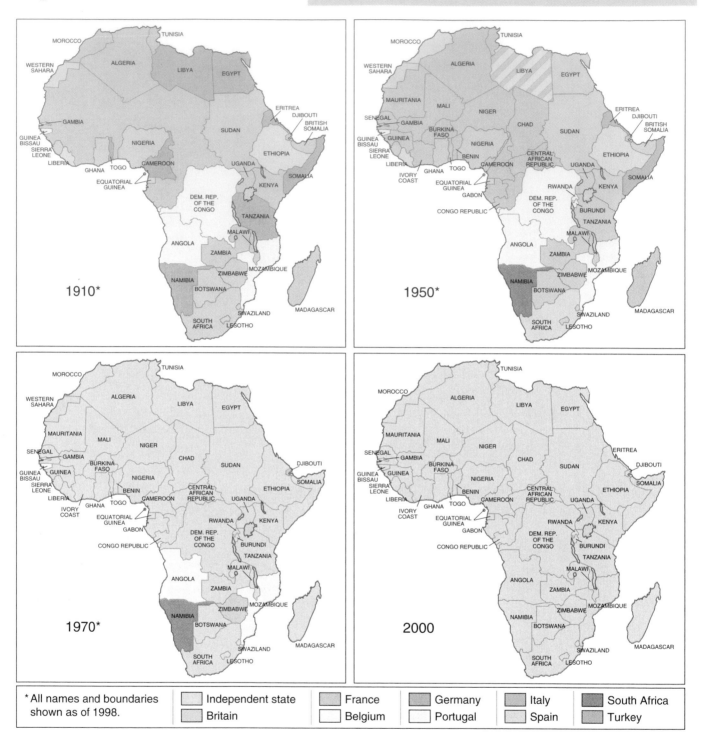

1910*

1950*

1970*

2000

* All names and boundaries shown as of 1998.

Independent state	France
Britain	Belgium

Germany	Italy
Portugal	Spain

South Africa
Turkey

In few parts of the world has the transition from colonialism to independence been as abrupt as on the African continent. Most African states did not become colonies until the nineteenth century and did not become independent until the twentieth, nearly all of them after World War II. Much of the colonial power in Africa is social and economic. The African colony provided the mother country with raw materials in exchange for marginal economic returns, and many African countries still exist in this colonial dependency relationship. An even more important component of the colonial legacy of Europe in Africa is geopolitical. When the world's colonial powers joined at the Conference of Berlin in 1884, they divided up Africa to fit their own needs, drawing boundary lines on maps without regard for terrain or drainage features, or for tribal/ethnic linguistic, cultural, economic, or political borders. Traditional Africa was enormously disrupted by this process. After independence, African countries retained boundaries that are legacies of the colonial past and African countries today are beset by internal problems related to tribal and ethnic conflicts, the disruption of traditional migration patterns, and inefficient spatial structures of market and supply.

Map 102d   African Cropland and Dryland Degradation

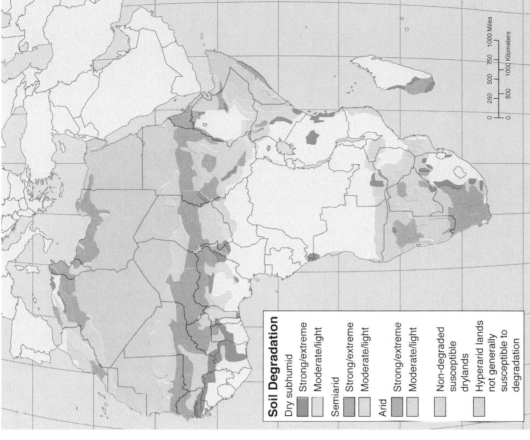

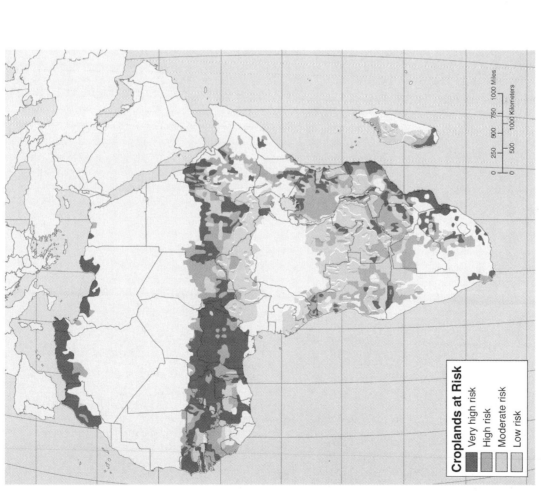

Croplands at Risk
- Very high risk
- High risk
- Moderate risk
- Low risk

Soil Degradation

Dry subhumid
- Strong/extreme
- Moderate/light

Semiarid
- Strong/extreme
- Moderate/light

Arid
- Strong/extreme
- Moderate/light

- Non-degraded susceptible drylands
- Hyperarid lands not generally susceptible to degradation

The economy of the African continent is largely agricultural and given an African population in excess of 500 million people, much of the agricultural environment is degraded. Two forms of degradation exist. The first of these is in cropland areas where susceptible tropical soils and high population densities have produced major cropland degradation, largely in the form of loss of fertility. Irrigation and the advent of artificial fertilizer have not helped soils to restore their natural chemical balances after genera-

tions of misuse. The second form of degradation is in the dryland areas where livestock grazing rather than cropping is the dominant agricultural form. Populations of domesticated stock that are too large for the carrying capacity of the environment, exacerbated by the view of livestock as wealth and worsened by increasing human populations, have all contributed to the conversion of semi-arid grasslands into desert.

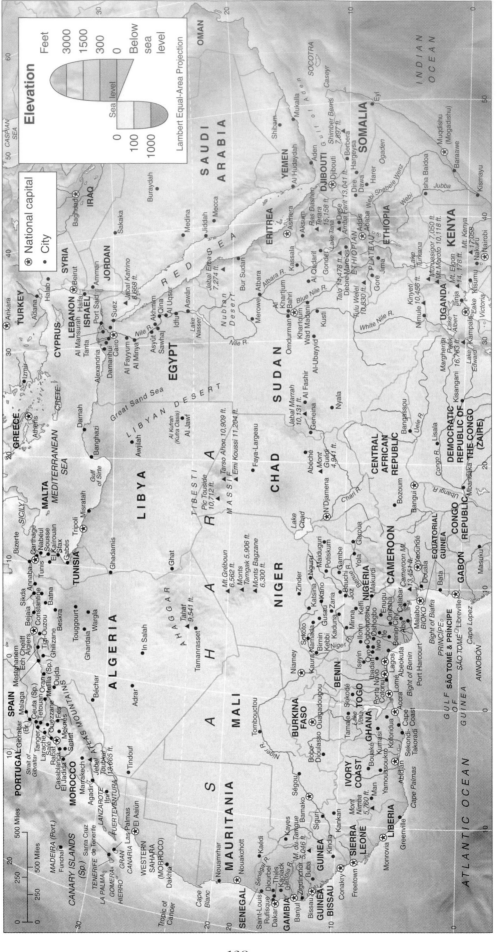

Map 103 Northern Africa

Map 104 Southern Africa

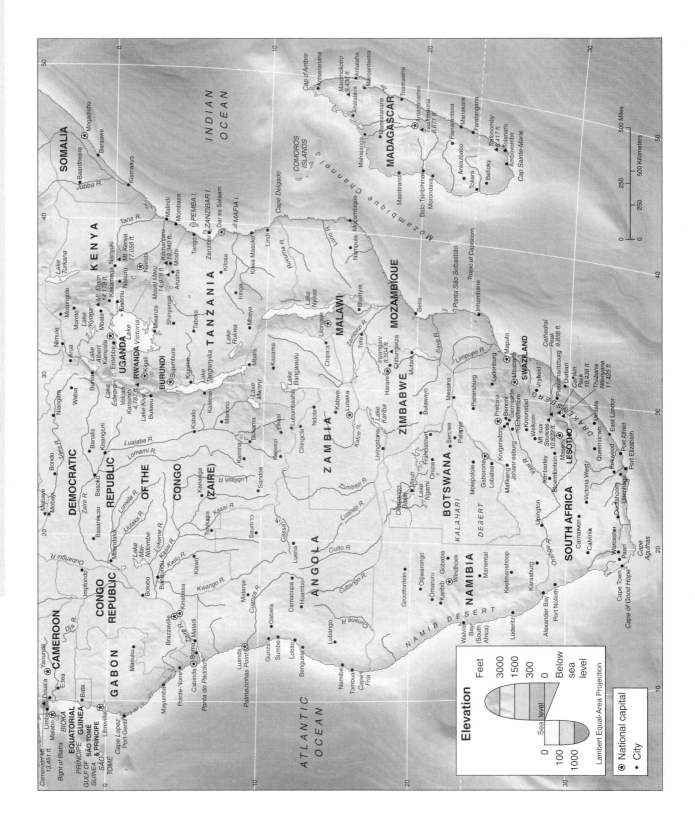

Map 105 Asia: Physical Features

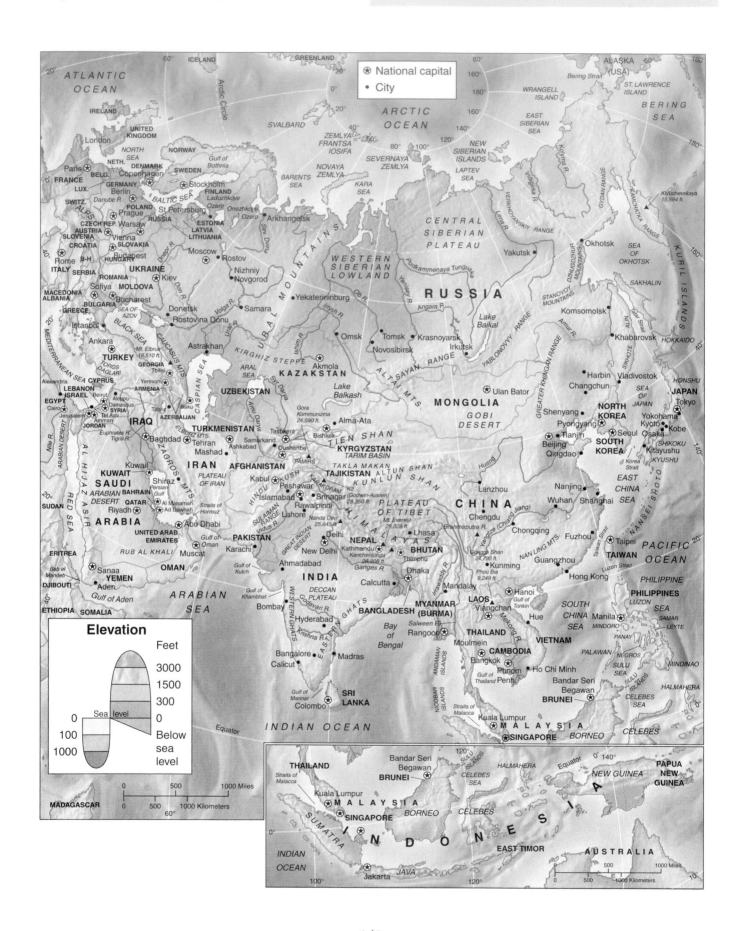

Elevation

Feet
3000
1500
300
0
Below sea level

⊛ National capital
• City

Map 106 Asia: Political Divisions

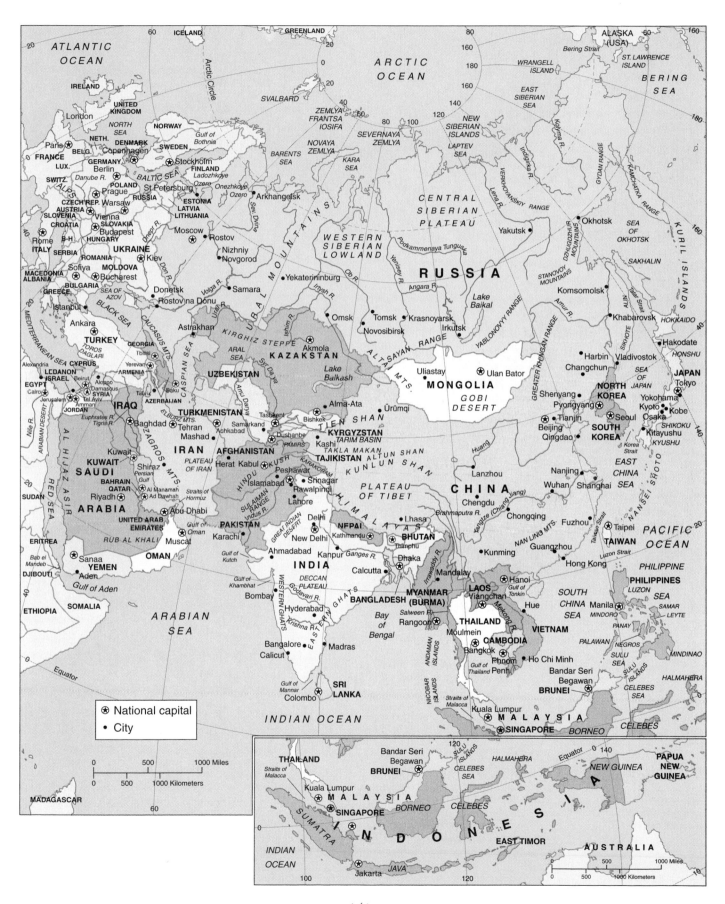

ATLANTIC OCEAN

ICELAND

GREENLAND

ALASKA (USA)

ARCTIC OCEAN

Bering Strait

ST. LAWRENCE ISLAND

BERING SEA

IRELAND

UNITED KINGDOM
London

NORTH SEA

NORWAY

SVALBARD

ZEMLYA FRANTSA IOSIFA

NOVAYA ZEMLYA

SEVERNAYA ZEMLYA

NEW SIBERIAN ISLANDS

EAST SIBERIAN SEA

WRANGELL ISLAND

Paris
FRANCE
NETH.
BELG.
LUX.
DENMARK
Copenhagen
SWEDEN
Stockholm
FINLAND
Gulf of Bothnia

BALTIC SEA

BARENTS SEA

KARA SEA

LAPTEV SEA

Kolyma R.

GYDAN RANGE

KAMCHATKA RANGE

SWITZ.
GERMANY
Berlin
ALPS
Danube R.
POLAND
Prague
CZECH REP.
Warsaw
RUSSIA
St Petersburg
Ladozhkoye Ozero
Onezhskoye Ozero
Arkhangelsk

Sev. Dvina

VERKHOYANSKIY RANGE

Indigirka R.

Okhotsk

SEA OF OKHOTSK

KURIL ISLANDS

AUSTRIA
SLOVENIA
CROATIA
Vienna
Budapest
SLOVAKIA
HUNGARY
B-H
Rome
ITALY
SERBIA
ROMANIA
Sofiya
MACEDONIA
ALBANIA
BULGARIA
GREECE
Istanbul

Moscow
Rostov
Nizhniy Novgorod
Kiev
UKRAINE
MOLDOVA
Bucharest

Dnepr R.
Don R.
SEA OF AZOV

Yekaterininburg
Ob. R.
Donetsk
Volga R.
Samara
Rostov na Donu
Astrakhan
Irtysh R.
Ishim R.

URAL MOUNTAINS

WESTERN SIBERIAN LOWLAND

Yenisey R.
Omsk
Tomsk
Novosibirsk

CENTRAL SIBERIAN PLATEAU

Podkammenaya Tunguska

Yakutsk

RUSSIA

Komsomolsk

SAKHALIN

Tatar Strait

ALIN

SIKHOTE

Khabarovsk

HOKKAIDO

Hakodate

MEDITERRANEAN SEA

BLACK SEA

TURKEY
Ankara
TOROS DAGLARI
CYPRUS

CAUCASUS MTS.

GEORGIA
Tbilisi
Yerevan
ARMENIA

KIRGHIZ STEPPE
Akmola
KAZAKSTAN

ALTAI

SAYAN RANGE

Krasnoyarsk
Irkutsk

Lake Baikal

Angara R.

YABLONOVY RANGE

GREATER KHINGAN RANGE

STANOVOY MOUNTAINS

Amur R.

Harbin
Changchun
Vladivostok

SEA OF JAPAN

HONSHU

JAPAN
Tokyo

Alexandria
LEBANON
ISRAEL
Beirut
EGYPT
Cairo
Jerusalem
Damascus
Aleppo
SYRIA
Tel Aviv
Amman
JORDAN

Nile R.

ARAL SEA

UZBEKISTAN

CASPIAN SEA

AZERBAIJAN
Baku
ELBURZ MTS.
Tabriz

Syr Darya
Lake Balkash

Tashkent
Samarkand
Ashkabad
TURKMENISTAN

Amu Darya

Bishkek
Alma-Ata
Ürümqi
TIEN SHAN
KYRGYZSTAN
TARIM BASIN

Uliastay

MONGOLIA
Ulan Bator
GOBI DESERT

Shenyang
Beijing
Tianjin
Qingdao
Pyongyang
NORTH KOREA
SOUTH KOREA
Seoul
Korea Strait

SEA OF JAPAN

Yokohama
Kyoto
Kobe
Osaka
SHIKOKU
Kitayushu
KYUSHU

IRAQ
Baghdad
Euphrates R.
Tigris R.
ZAGROS MTS.

IRAN
Tehran
Mashad
PLATEAU OF IRAN

Dushanbe
PAMIRS
TAJIKISTAN
Kashi
TAKLA MAKAN
ALTUN SHAN
KUNLUN SHAN

Lanzhou

CHINA

Nanjing
Wuhan
Shanghai

EAST CHINA SEA

NANSEI SHOTO

AFGHANISTAN
Herat
Kabul
HINDU KUSH
Peshawar
Islamabad
Rawalpindi
Srinagar
KARAKORAM
SULAIMAN RANGE
Indus R.
Lahore

PLATEAU OF TIBET

Chengdu
Chongqing
Lhasa
HIMALAYAS

Huang
Yangtze (Chiang Jiang)
Brahmaputra R.

Fuzhou

Kunming
Guangzhou

Taiwan Strait

Taipei
TAIWAN

PACIFIC OCEAN

KUWAIT
Kuwait
SAUDI
Shiraz
Persian Gulf
BAHRAIN
QATAR
Al Manamah
Ad Dawhah
Riyadh
ARABIA
AL HIJAZ

RED SEA

UNITED ARAB EMIRATES
Abu Dhabi
Gulf of Oman
Muscat
OMAN

Straits of Hormuz

RUB AL KHALI

PAKISTAN
Karachi
GREAT INDIAN DESERT
Delhi
New Delhi
Ahmadabad
Gulf of Kutch

NEPAL
Kathmandu
Kanpur
Ganges R.
BHUTAN
Thimphu

Dhaka
BANGLADESH

NAN LING MTS.

Hong Kong

PHILIPPINE SEA

PHILIPPINES
LUZON
Manila
MINDORO
SAMAR
LEYTE

SUDAN

ERITREA

Bab el Mandeb
DJIBOUTI

ETHIOPIA

SOMALIA

Sanaa
YEMEN
Aden
Gulf of Aden

ARABIAN SEA

INDIA
Bombay
DECCAN PLATEAU
WESTERN GHATS
Godavari R.
Hyderabad
Krishna R.
Bangalore
Calicut
Madras
EASTERN GHATS

Calcutta
Gulf of Khambhat

Bay of Bengal

ANDAMAN ISLANDS

Mandalay
MYANMAR (BURMA)
Irrawaddy R.
Salween R.
Rangoon
Moulmein
THAILAND
Bangkok

Viangchan
LAOS
Gulf of Tonkin
Hue
VIETNAM
Hanoi

SOUTH CHINA SEA

PANAY
NEGROS
PALAWAN

SULU SEA

MINDINAO

CELEBES SEA

HALMAHERA

SRI LANKA
Colombo
Gulf of Mannar

NICOBAR ISLANDS

Straits of Malacca

CAMBODIA
Phnom Penh
Gulf of Thailand
Ho Chi Minh
Bandar Seri Begawan
BRUNEI

Kuala Lumpur
MALAYSIA
SINGAPORE
BORNEO
CELEBES

MADAGASCAR

INDIAN OCEAN

⊛ National capital
• City

THAILAND
Straits of Malacca
Kuala Lumpur
MALAYSIA
SINGAPORE

Bandar Seri Begawan
BRUNEI
SULU ISLANDS
CELEBES SEA
HALMAHERA

Equator

BORNEO
CELEBES

SUMATRA
INDONESIA
Jakarta
JAVA

EAST TIMOR

PAPUA NEW GUINEA
NEW GUINEA

AUSTRALIA

0   500   1000 Miles
0   500   1000 Kilometers

-141-

Map 107a Environment and Economy

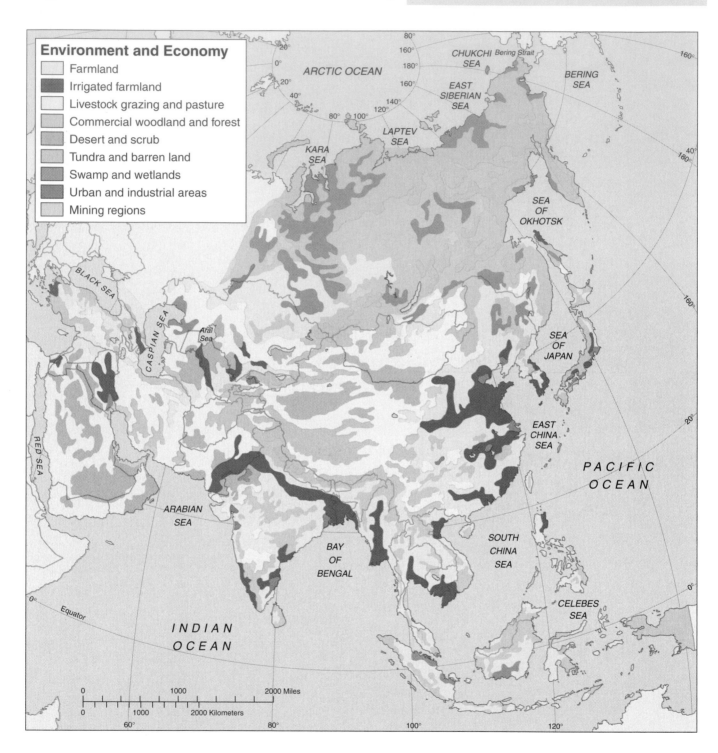

Environment and Economy
- Farmland
- Irrigated farmland
- Livestock grazing and pasture
- Commercial woodland and forest
- Desert and scrub
- Tundra and barren land
- Swamp and wetlands
- Urban and industrial areas
- Mining regions

Asia is a land of extremes of land use with some of the world's most heavily industrialized regions, barren and empty areas, and productive and densely populated farm regions. Asia is a region of rapid industrial growth. Yet Asia remains an agricultural region with three out of every four workers engaged in agriculture. Asian commercial agriculture and intensive subsistence agriculture is characterized by irrigation. Some of Asia's irrigated lands are desert requiring additional water. But most of the Asian irrigated regions have

sufficient precipitation for crop agriculture and irrigation is a way of coping with seasonal drought—the wet-and-dry cycle of the monsoon—often gaining more than one crop per year on irrigated farms. Agricultural yields per unit area in many areas of Asia are among the world's highest. Because the Asian population is so large and the demands for agricultural land so great, Asia is undergoing rapid deforestation and some areas of the continent have only small remnants of a once-abundant forest reserve.

Map 107b Population Density

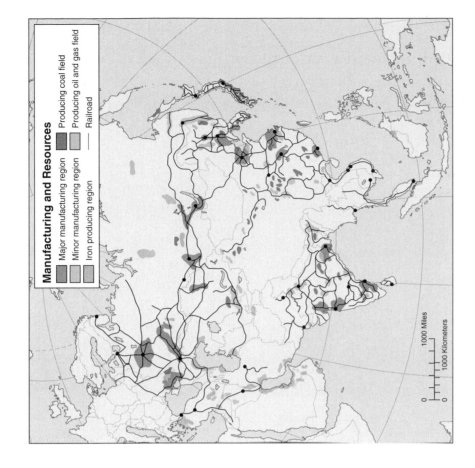

Population Density

Persons per square mile (km)

- Uninhabited area
- Less than 2 (1)
- 2 – 25 (1 – 10)
- 25 – 50 (10 – 20)
- 50 – 150 (20 – 60)
- 150 – 300 (60 – 120)
- More than 300 (120)

1000 Miles

0

1000 Kilometers

0

With one-third of the world's land area and nearly two-thirds of the world's population, Asia is more densely settled than any other region of the world. In some of the continent's farming regions, agricultural population density exceeds 2,000 persons per square mile. In some portions of the continent, particularly the Islamic areas of Central and Southwest Asia, this already large population is growing very rapidly with some countries having population growth rates above 3 percent per year and doubling times between 20 and 25 years. The populations of neither China nor India are growing particularly rapidly but since both countries have population bases that are enormous, the absolute number of Indians and Chinese added to the world's population each year is staggering. In spite of these massive populations, Asia also contains areas that are either completely uninhabited or have population densities that are as low as any on earth.

Map 107c Industrialization: Manufacturing and Resources

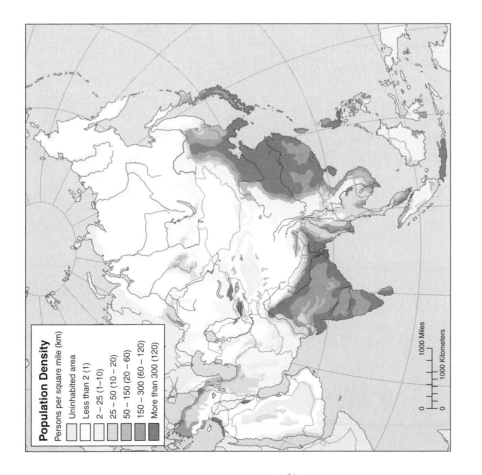

Manufacturing and Resources

- Major manufacturing region
- Minor manufacturing region
- Iron producing region
- Producing coal field
- Producing oil and gas field
- Railroad

1000 Miles

0

1000 Kilometers

0

International economists have predicted that the next century will be one of Asian economic dominance, as other Asian nations approach the economic levels of industrial giant Japan. Asian countries have begun to industrialize rapidly and have increased industrial production more than 100 percent in the last decade. But nothing guarantees industrial output, it must master the great distances separating critical raw material and production sites from the locations of the markets for them. Such a mastery is achieved only by the development of efficient transportation systems. Usually that means water travel and here Asia is remarkably deficient when compared with Europe and North America. Nevertheless, the mixed prospects for Asian economic growth into the next century are a great deal better than would have been predicted a decade ago.

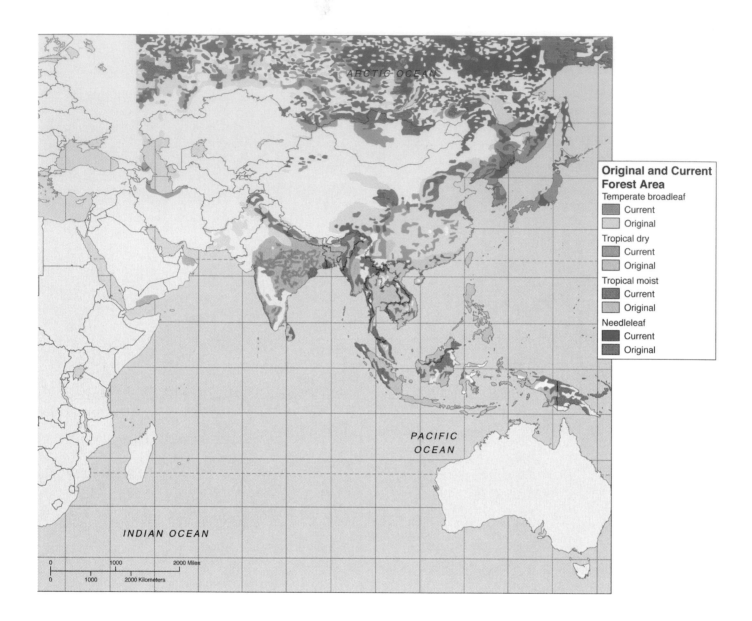

In no part of the world has deforestation produced such dramatic changes in the environment as in Asia. Here, the world's oldest continuous agricultural civilizations have placed enormous demands upon the vegetative and soil environments to the extent that precious little natural forest cover remains in two world's two most populous countries, China and India. While Southeast Asia has been historically less denuded than the major culture centers to its north and west, recent decades have seen significantly forest loss to commercial timber operations, stimulated by the high-market value of tropical hardwoods. Also, a fairly recent trend has been the massive deforestation in the northeastern parts of Russia where economic expansion over the last century has stimulated the rapid loss of softwood forests for construction lumber and for paper production.

Map **108** Northern Asia

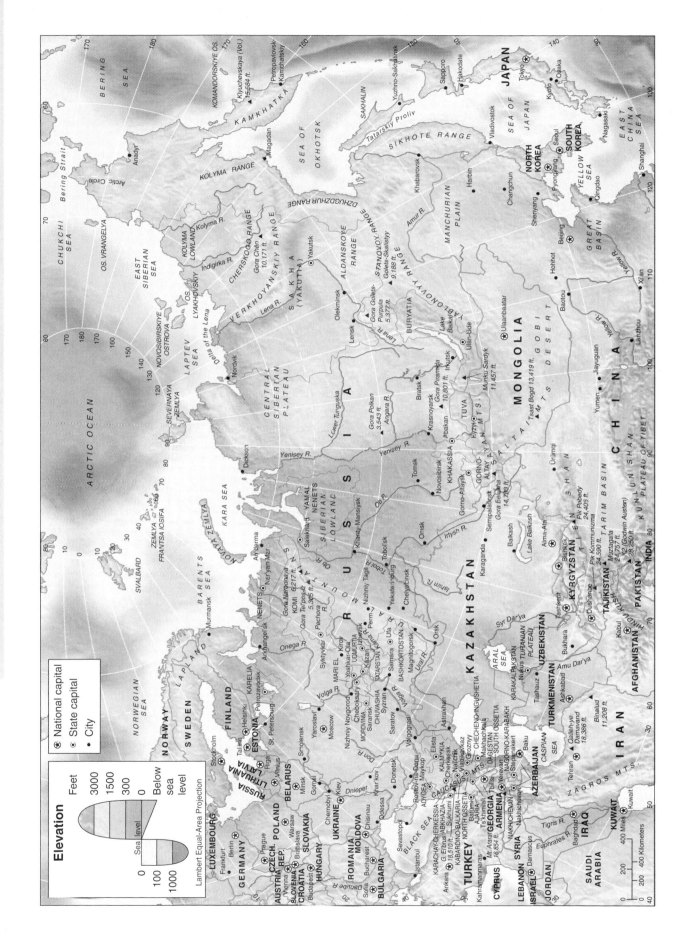

Map **109** Southwestern Asia

Elevation

	Feet
	3000
	1500
	300
0 Sea level	0
100	Below
1000	sea level

Simple Conic Projection

⊛ National capital
• City

Map 109a The Middle East: Territorial Changes, 1918-2004

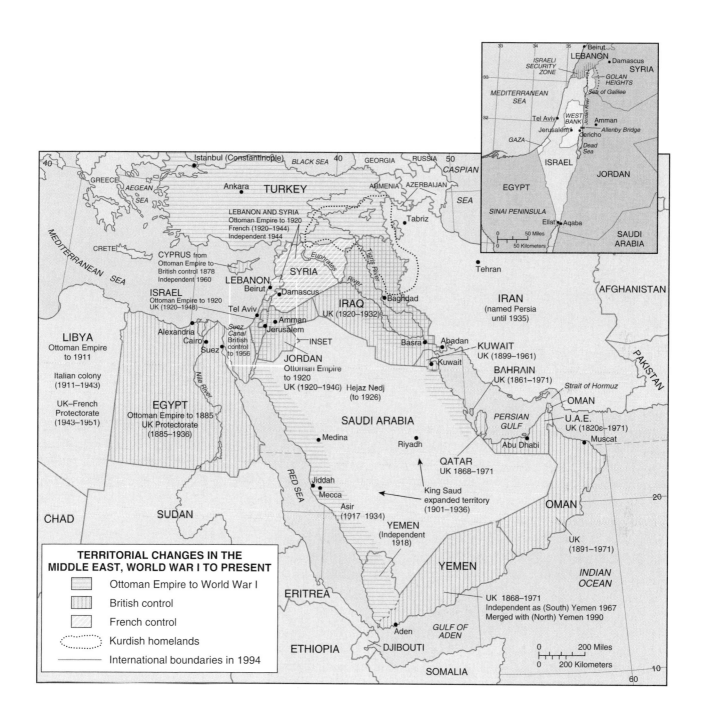

TERRITORIAL CHANGES IN THE MIDDLE EAST, WORLD WAR I TO PRESENT

- Ottoman Empire to World War I
- British control
- French control
- Kurdish homelands
- International boundaries in 1994

The Middle East, encompassing the northeastern part of Africa and southwestern Asia, has experienced a turbulent history. In the last century alone, many of the region's countries have gone from being ruled by the Turkish Ottoman Empire, to being dependencies of Great Britain or France, to being independent. Having experienced the Crusades and colonial domination by European powers, the region's predominantly Islamic countries are now resentful of interference in the region's affairs by countries with a European and/or Christian heritage. The tension between Israel (settled largely in the late nineteenth and twentieth centuries by Jews of predomi-

nantly European background) and its neighbors is a matter of European–Middle Eastern cultural stress as well as a religious conflict between Islamic Arab culture and Judaism. The political boundaries on the map, like those throughout most of Africa, are the invention of European colonial powers and often do not take into account the pre-existing lines of tribal control or authority. In Iraq, for example, three distinct cultural areas exist: Shiite Muslim (Arabic), Sunni Muslim (Arabic), and Kurd (also Islamic but distinctly not Arabic). These three cultures occupying the territory of a state makes stability in that tortured region a distinct problem for the future.

Map 109b South-Central Eurasia: An Ethnolinguistic Crazy Quilt

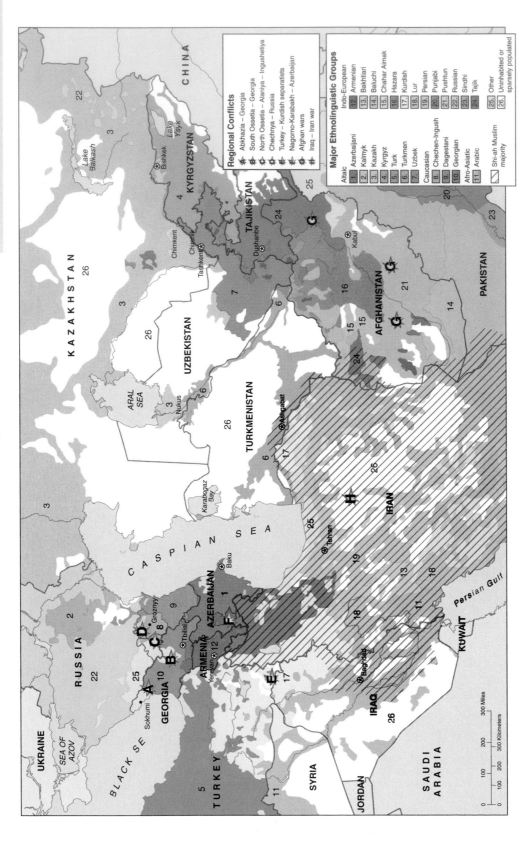

Regional Conflicts

✳ Abkhazia – Georgia
✳ South Ossetia – Georgia
✳ North Ossetia – Alaniya – Ingushetiya
✳ Chechnya – Russia
✳ Turkey – Kurdish separatists
✳ Nagorno-Karabakh – Azerbaijan
✳ Afghan wars
✳ Iraq – Iran war

Major Ethnolinguistic Groups

Altaic
1. Azerbaijani
2. Kalmyk
3. Kazakh
4. Kyrgyz
5. Turk
6. Turkmen
7. Uzbek

Caucasian
8. Chechen-Ingush
9. Dagestani
10. Georgian

Afro-Asiatic
11. Arabic

Indo-European
12. Armenian
13. Bakhtiari
14. Baluchi
15. Chahar Aimak
16. Hazara
17. Kurdish
18. Lur
19. Persian
20. Punjabi
21. Pushtun
22. Russian
23. Sindhi
24. Tajik

25. Other
26. Uninhabited or sparsely populated

▨ Shi-ah Muslim majority

South-Central Eurasia: The area of south-central Eurasia represents what is perhaps the world's most volatile area in terms of potential military conflict. A crazy quilt of ethnic and linguistic groups, this region contains politically-defined "states," but nothing approaching "nation-states" as they are understood elsewhere. Even reasonably well-consolidated states, such as Iran, are hampered by the mixture of languages and ethnic populations within their borders. For some countries, such as Afghanistan, the ethnolinguistic mix is so historically fixed as to render any attempts at modern state building nearly hopeless. This vast area—although nearly universally Muslim—is also split between the two primary Islamic sects, Sunni and Shia, with several smaller sectarian divisions as well. The division of one of the world's global religions into two principal factions occurred in the seventh century, originally over the question of the source of authority in the religious hierarchy. But it has since come to be a theological and metaphysical separation, and Sunni and Shia Muslims now bear somewhat the same relationship to one another as did Catholics and Protestants in seventeenth- and eighteenth-century Europe—a mutual antipathy that periodically flares into civil unrest and conflict that goes far beyond the bounds of religious debate. The area is rich in natural resources, but the mixture of ethnicity, language, and religion has and will continue to produce the human conflicts that inhibit human development.

Map **110** Southern Asia

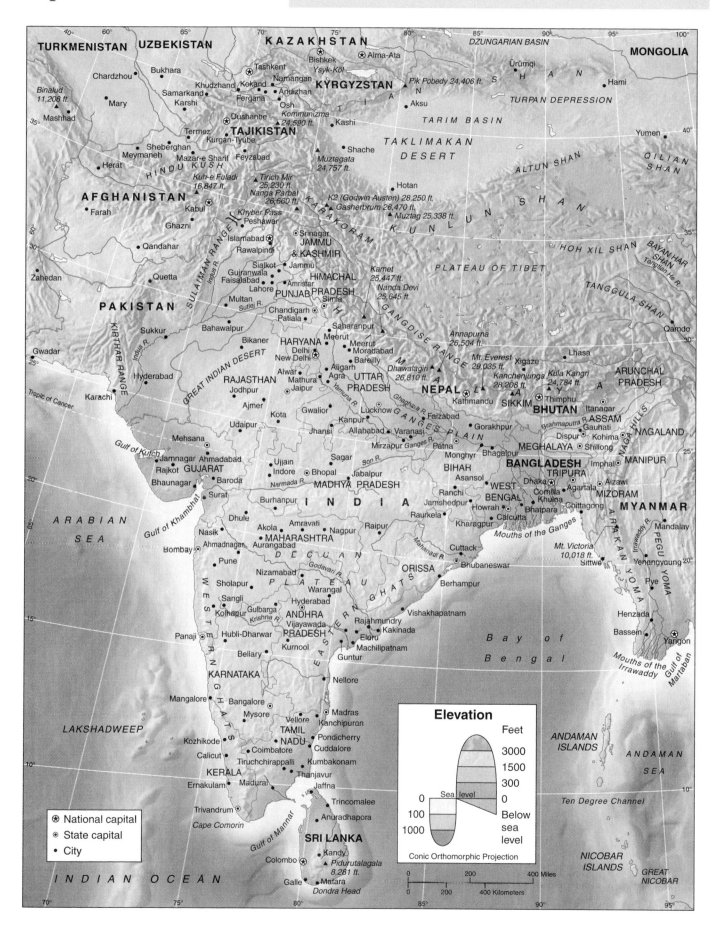

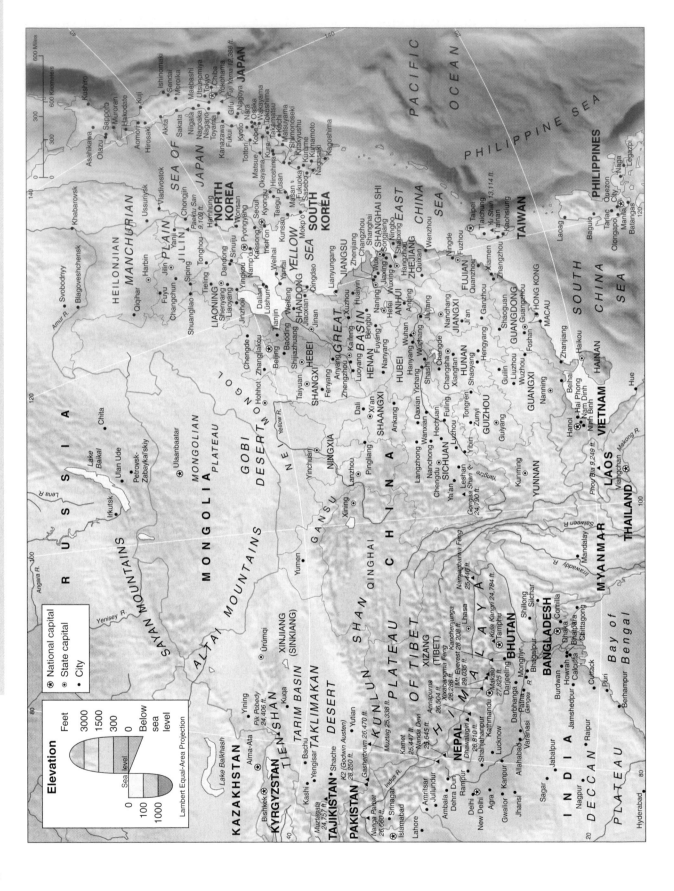

Map 111 East Asia

Elevation

Feet	
3000	
1500	
300	
0	Sea level
Below sea level	

Sea level

0
100
1000

Lambert Equal-Area Projection

⊛ National capital
⊙ State capital
• City

600 Miles
600 Kilometers
300
300
0

RUSSIA

KAZAKHSTAN

KYRGYZSTAN

TAJIKISTAN

PAKISTAN

INDIA

NEPAL

BHUTAN

BANGLADESH

MYANMAR

THAILAND

LAOS

VIETNAM

CHINA

MONGOLIA

NORTH KOREA

SOUTH KOREA

JAPAN

TAIWAN

PHILIPPINES

PACIFIC OCEAN

PHILIPPINE SEA

SOUTH CHINA SEA

EAST CHINA SEA

YELLOW SEA

SEA OF JAPAN

Bay of Bengal

SAYAN MOUNTAINS

ALTAI MOUNTAINS

TIEN SHAN

KUNLUN SHAN

HIMALAYA

GOBI DESERT

MONGOLIAN PLATEAU

PLATEAU OF TIBET

TARIM BASIN

TAKLIMAKAN DESERT

MANCHURIAN PLAIN

GREAT BASIN

PLATEAU OF TIBET (XIZANG) (TIBET)

DECCAN PLATEAU

Lake Baikal
Lake Balkhash

Amur R.
Lena R.
Yenisey R.
Angara R.
Yellow R.
Yangtze
Mekong R.
Salween R.
Irrawaddy R.
Ganges R.
Indus R.
Brahmaputra R.

Fuji Yama 12,388 ft.
Paektu San 9,100 ft.
Yu Shan 13,114 ft.
Phou Bia 9,249 ft.
Namjagbarwa Feng 25,446 ft.
Gongga Shan 24,790 ft.
Mt. Everest 29,035 ft.
Kanchenjunga 28,208 ft.
Lhotse 27,923 ft.
Makalu 27,825 ft.
Xixabangma Feng 26,286 ft.
Annapurna 26,504 ft.
Dhaulagiri 26,810 ft.
Kula Kangri 24,784 ft.
Nanda Devi 25,645 ft.
Kamet 25,447 ft.
Muztagata 24,757 ft.
Muztag 25,338 ft.
Gasherbrum 26,470 ft.
K2 (Godwin Austen) 28,250 ft.
Nanga Parbat 26,660 ft.
Pik Pobedy 24,406 ft.
Yutian

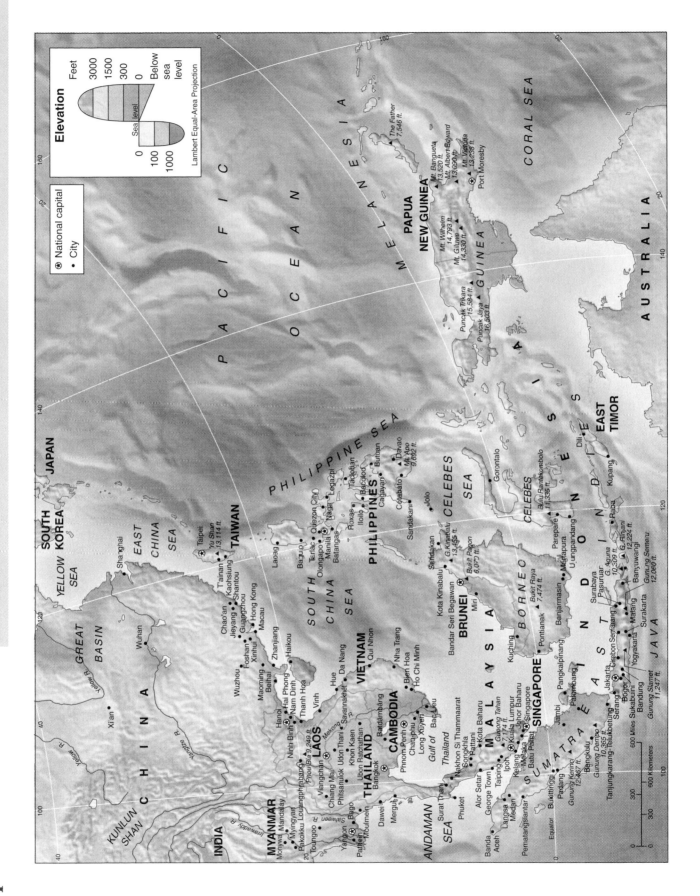

Map 112 Southeast Asia

Elevation

Feet
3000
1500
300
0
Below sea level

Sea level

| 0 |
| 100 |
| 1000 |

Lambert Equal-Area Projection

⊛ National capital
• City

INDIA

CHINA

KUNLUN SHAN

GREAT BASIN

Xian

Yellow R.

Yangtze R.

Wuhan

SOUTH KOREA

JAPAN

YELLOW SEA

Shanghai

EAST CHINA SEA

TAIWAN

Taipei ⊛
Yu Shan 13,114 ft.
Tainan
Kaohsiung
Shantou
Chao'an
Jieyang
Guangzhou
Xinhui
Macau
Hong Kong
Haikou
Zhanjiang
Belhai
Maoming
Foshan
Wuzhou

MYANMAR

Mandalay
Monywa
Myingyan
Pakokku
Loungphyalaung
Irrawaddy R.

Salween R.
Salween R. (Nu R.)

Toungoo
Bago
Yangon ⊛
Pathein
Moulmein
Dawei
Merguh

Chiang Mai

Phrai
Phitsanulok
Vangchan
Phouï Bia 9,249 ft.
LAOS
Vientiane ⊛
Udon Thani
Savannakhet
Khon Kaen
Ubon Ratchathani

THAILAND
Bangkok ⊛
Nakhon Ratchasima

CAMBODIA
Battambang
Phnom Penh ⊛
Chanthaburi
Long Xuyen
Bac Lieu

VIETNAM

Hanoi ⊛
Ninh Binh
Hai Phong
Nam Dinh
Thanh Hoa
Vinh
Hue
Da Nang
Qui Nhon
Nha Trang
Bien Hoa
Ho Chi Minh

Gulf of Thailand

ANDAMAN SEA

Surat Thani
Phuket
Nakhon Si Thammaarat
Songkhla
Pattani
Kota Baharu
Alor Setar
George Town
Taiping
Ipoh
Kelang
Kuala Lumpur
Jehor Baharu

MALAYSIA
Gunung Tahan 7,174 ft.
MELAKA
Batu Pahat

SINGAPORE ⊛

SOUTH CHINA SEA

Kota Kinabalu
BRUNEI
Bandar Seri Begawan ⊛
Miri
G. Kinabalu 13,455 ft.
Bukit Pagon 6,070 ft.

Kuching
BORNEO
Pontianak
Bukit Raya 7,474 ft.
Banjarmasin

PHILIPPINES

Laoag
Baguio
Tarlac
Quezon City
Manila ⊛
Olongapo
Batangas
Naga
Legazpi
Roxas
Iloilo
Bacolod
Tacloban
Cagayan
Butuan
Davao
Mt. Apo 9,692 ft.
Cotabato
Zamboanga
Jolo
Sandakan
Sandakan

PHILIPPINE SEA

PACIFIC OCEAN

MELANESIA

The Father 7,546 ft.
Mt. Bangueta 13,520 ft.
Mt. Albert Edward
Mt. Victoria 13,238 ft.
13,090 ft.

PAPUA NEW GUINEA

Mt. Wilhelm 14,793 ft.
Mt. Giluwe 14,330 ft.

NEW GUINEA

Port Moresby ⊛

Puncak Trikara 15,584 ft.
Puncak Jaya 16,503 ft.

CORAL SEA

CELEBES SEA

Gorontalo

Parepare
Ujungpandang

CELEBES
Bulu Rantekombolo 11,335 ft.

Manado
Ujungdang

INDONESIA

EAST TIMOR

Dili ⊛
Raba

Kupang

AUSTRALIA

Palembang
Pangkalpinang
Jambi

SUMATRA

Padang
Bengkulu
Gunung Kerinci 12,467 ft.
Tanjungkarang-Telukbetung
Gunung Dempo 10,365 ft.

Bukittinggi
Pematangsiantar
Medan
Langsa
Banda Aceh

Equator

JAVA

Jakarta ⊛
Bogor
Serang
Bandung
Cirebon
Sukabumi
Semarang
Yogyakarta
Surakarta
Surabaya
Pasuruan
Matang
Banyuwangi
Gunung Slamet 11,247 ft.
Gunung Semeru 12,060 ft.

G. Rinjani 12,224 ft.
G. Agung 10,309 ft.

EAST INDIES

600 Miles
600 Kilometers

0 300
0 300

100
300
600

-151-

Map 113 Australasia and Oceania: Physical Features

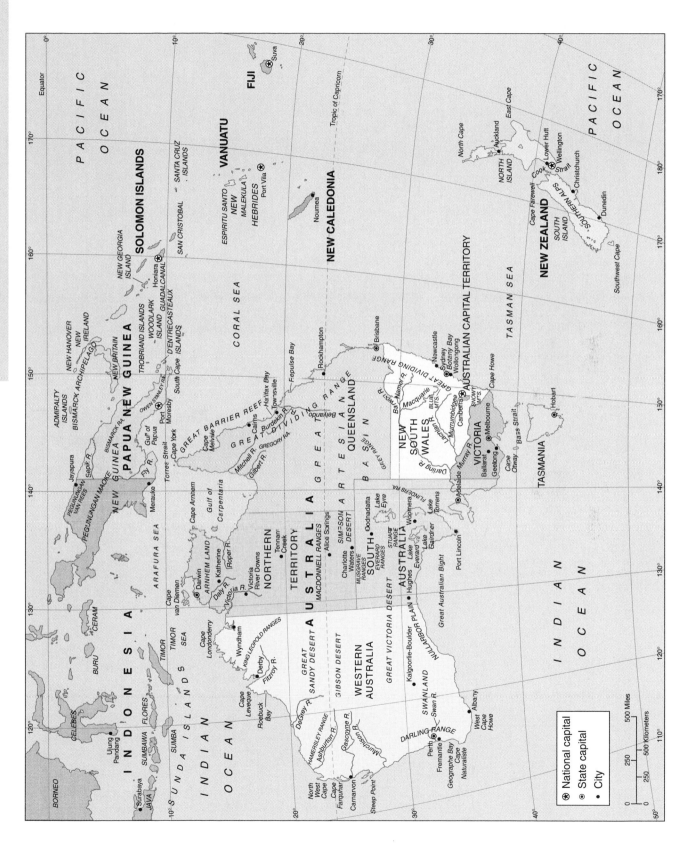

Map 114 Australasia and Oceania: Political Divisions

Australasia: Thematic Features

Map 115a Environment and Economy

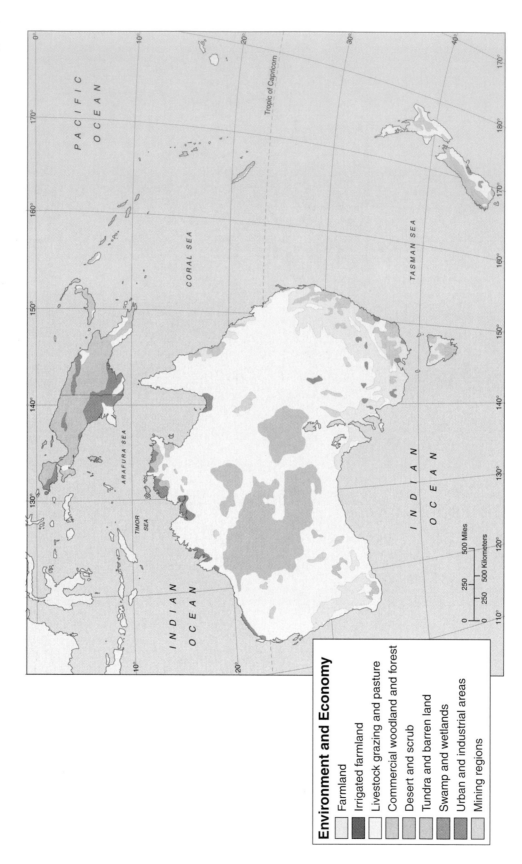

Environment and Economy

- Farmland
- Irrigated farmland
- Livestock grazing and pasture
- Commercial woodland and forest
- Desert and scrub
- Tundra and barren land
- Swamp and wetlands
- Urban and industrial areas
- Mining regions

Australasia is dominated by the world's smallest and most uniform continent. Flat, dry, and mostly hot, Australia has the simplest of land use patterns: where rainfall exists so does agricultural activity. Two agricultural patterns dominate the map: livestock grazing, primarily sheep, and wheat farming, although some sugar cane production exists in the north and some cotton is grown elsewhere. Only about 6 percent of the continent consists of arable land so the areas of wheat farming, dominant as they may be in the context of Australian agriculture, are small. Australia also supports a healthy mineral resource economy, with iron and copper and precious metals making up the bulk of the extraction. Elsewhere in the region, tropical forests dominate Papua New Guinea, with some subsistence agriculture and livestock. New Zealand's temperate climate with abundant precipitation supports a productive livestock industry and little else besides tourism—which is an important economic element throughout the remainder of the region as well.

-154-

Map 115c Climate Patterns

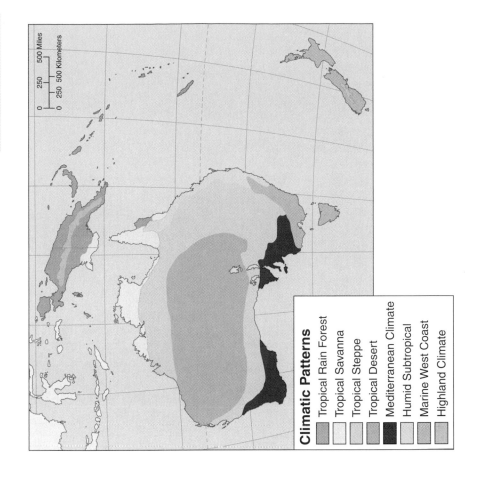

Climatic Patterns

- Tropical Rain Forest
- Tropical Savanna
- Tropical Steppe
- Tropical Desert
- Mediterranean Climate
- Humid Subtropical
- Marine West Coast
- Highland Climate

Because of its nearly uniform surface, with only a few low and scattered uplands, Australian climate is a consequence only of the two great climate controls—latitudinal position and location relative to continental margin and interior. The continent bestrides the 30th parallel of latitude and its climatic pattern is dominated by the subtropical high pressure system with dry air masses that are responsible for the existence of great deserts. Toward the equator, the desert grades into steppe, savanna, and tropical forest as the subtropical high gives way to equatorial low pressure and abundant precipitation. Toward the pole, arid land fades into more well-watered steppe grasslands, the Mediterranean type climate of the southern margins of the continent, and the marine west coast climate of the Australian southeast and New Zealand. This latter climate is where most of the region's people live.

Map 115b Population Density

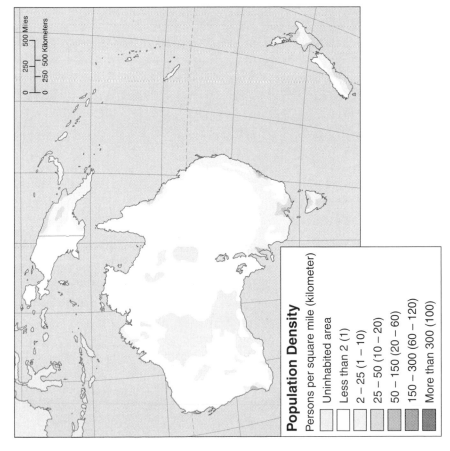

Population Density

Persons per square mile (kilometer)

- Uninhabited area
- Less than 2 (1)
- 2 – 25 (1 – 10)
- 25 – 50 (10 – 20)
- 50 – 150 (20 – 60)
- 150 – 300 (60 – 120)
- More than 300 (100)

The region's small population is remarkably diverse. To the north are the Melanesian New Guinea peoples, while Europeans dominate the populations of Australia and New Zealand, both of which have significant indigenous populations. Throughout most of the smaller island groups, the bulk of the population is Melanesian but with a scattering of Europeans. The distribution of population is extremely uneven with New Guinea, southeastern Australia, and New Zealand supporting the bulk of the region's people while the remainder of the region—meaning nearly all of Australia—is either sparsely populated or devoid of population altogether. Nowhere do population densities reach the levels they do in other major regions of the world and densities of 50 persons per square mile are the highest to be found. Population location is dependent on precipitation and population growth patterns are culturally variable.

Map **116** The Pacific Rim

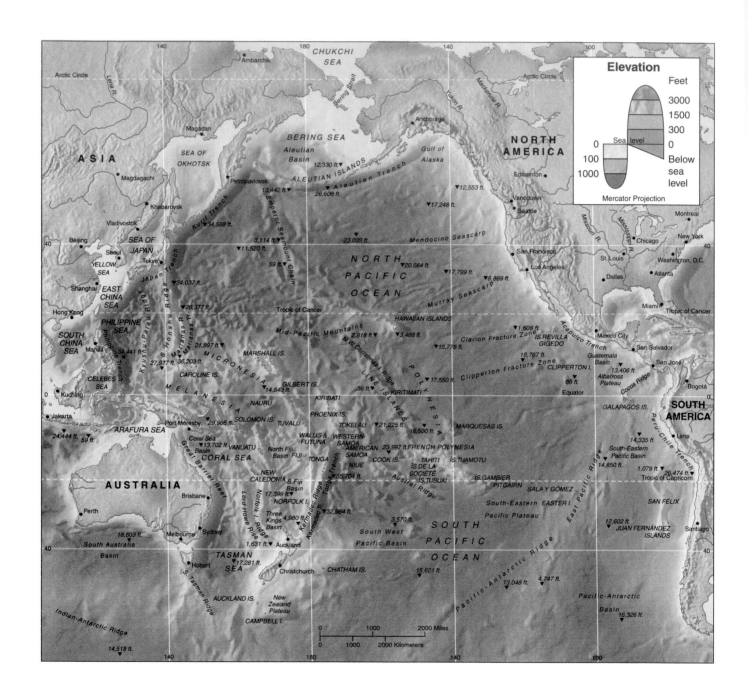

Arctic Circle

140   180   CHUKCHI
                SEA

Ambarchik

Lena R.   Bering Strait

140

100

Arctic Circle   NORTH
                AMERICA

Mackenzie R.

Yukon R.

Anchorage

Magadan

ASIA

BERING SEA

Aleutian
Basin   12,330 ft.▼

Gulf of
Alaska

Edmonton

SEA OF
OKHOTSK

Magdagachi

Petropavlovsk   13,442 ft.▼   ALEUTIAN ISLANDS
                                26,606 ft.   ▼ Aleutian Trench

▼12,553 ft.

Vancouver

Seattle

Khabarovsk

Kuril Trench ▼34,588 ft.

▼17,248 ft.

Montreal

Vladivostok

SEA OF
JAPAN

3,114 ft.▼   ▼23,039 ft.   Mendocino Seascarp

Missouri R.

Mississippi R.

Chicago

New York

40

Beijing   40
Seoul   YELLOW
         SEA
Tokyo

▼11,520 ft.

59 ft.▼

NORTH
PACIFIC
OCEAN

▼20,664 ft.

San Francisco

St. Louis

Washington, D.C.

Shanghai   EAST
            CHINA
            SEA

Japan Trench ▼34,037 ft.

Emperor Seamount Chain

▼17,799 ft.   ▼9,869 ft.

Los Angeles

Dallas

Atlanta

Murray Seascarp

▼28,377 ft.

Tropic of Cancer

HAWAIIAN ISLANDS

Miami   Tropic of Cancer

Hong Kong

PHILIPPINE
SEA

S. Honshu Ridge

Mid-Pacific Mountains   2,818 ft.▼   ▼3,468 ft.

▼1,608 ft.

Clarion Fracture Zone   IS.REVILLA
                         GIGEDO

Acapulco Trench

Mexico City

SOUTH
CHINA
SEA   Manila   Philippine Trench 34,441 ft.▼

Kyushu-Palau Ridge

Mariana Trench
Marianas R.   21,897 ft.▼

MICRONESIA   MARSHALL IS.

N.W. Christmas Ridge

▼15,778 ft.

Clipperton Fracture Zone   CLIPPERTON I.

18,767 ft.
▼

San Salvador

Guatemala
Basin   13,406 ft.
         ▼

San José

CELEBES
SEA

27,977 ft. 36,203 ft.   CAROLINE IS.

GILBERT IS.   36 ft.▼   KIRITIMATI

POLYNESIA

▼17,550 ft.

66 ft.
▼

Albatross
Plateau

Bogotá

Kuching   0

MELANESIA

14,640 ft.

LINE ISLANDS

Equator

Cocos Ridge

SOUTH
AMERICA

Jakarta

ARAFURA SEA

NAURU

KIRIBATI

GALAPAGOS IS.

24,444 ft. 59 ft.▼

Port Moresby. 29,988 ft.▼

SOLOMON IS.   TUVALU

PHOENIX IS.

TOKELAU   ▼21,225 ft.

18,500 ft.
▼

MARQUESAS IS.

14,335 ft.
▼   Lima

Peru-Chile Trench

Coral Sea
Basin ▼13,702 ft.   VANUATU

WALLIS &
FUTUNA   WESTERN
         SAMOA

AMERICAN   23,997 ft. FRENCH POLYNESIA
SAMOA

South-Eastern
Pacific Basin

14,850 ft.
▼

1,079 ft.
▼

26,474 ft.▼

Tropic of Capricorn

CORAL SEA   North Fiji
             Basin FIJI   TONGA   NIUE   COOK IS.

TAHITI
IS.DE LA   IS.TUAMOTU
SOCIÉTÉ

AUSTRALIA

NEW
CALEDONIA

Lord Howe Rise

Norfolk I. Ridge

Kermadec-Tonga Trench

Tonga Trench ▼35,704 ft.

Austral Ridge   IS.TUBUAI

IS.GAMBIER

PITCAIRN   SALA Y GÓMEZ

East Pacific Ridge

SAN FÉLIX

S.Fiji
Basin
17,399 ft.▼

Brisbane

NORFOLK I.

Three
Kings 4,980 ft.▼
Basin   ▼32,984 ft.

3,570 ft.
▼

SOUTH
PACIFIC

South-Eastern   EASTER I.
Pacific Plateau

12,602 ft.
▼

JUAN FERNÁNDEZ
ISLANDS

Santiago

Perth

18,803 ft.
▼

South Australia
Basin   40

Melbourne   Sydney

1,631 ft.▼ ▼ Auckland

South West
Pacific Basin

OCEAN

Pacific-Antarctic Ridge

40

TASMAN
SEA   ▼17,281 ft.

Hobart

Christchurch   CHATHAM IS.

15,601 ft.
▼

13,048 ft.   4,747 ft.
▼   ▼

Pacific-Antarctic
Basin

AUCKLAND IS.   New
                Zealand
                Plateau

1000   2000 Miles

1000   2000 Kilometers

16,326 ft.
▼

Indian-Antarctic Ridge

CAMPBELL I.

14,518 ft.
▼

140   180   140   100

Elevation

Feet

3000
1500
300

0   Sea level   0

100   Below
1000   sea
        level

Mercator Projection

Map **117** The Atlantic Ocean

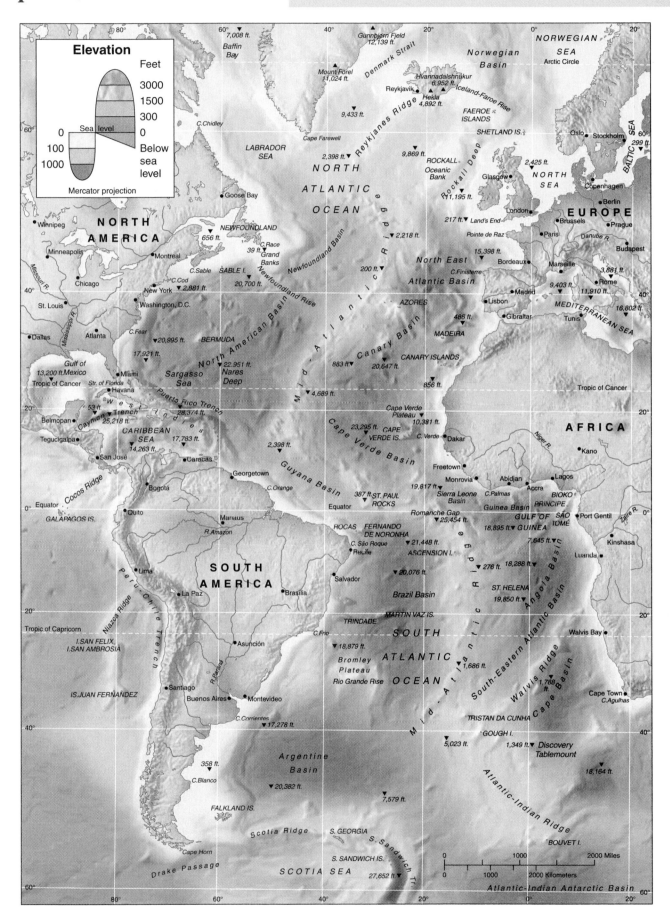

Map **118** The Indian Ocean

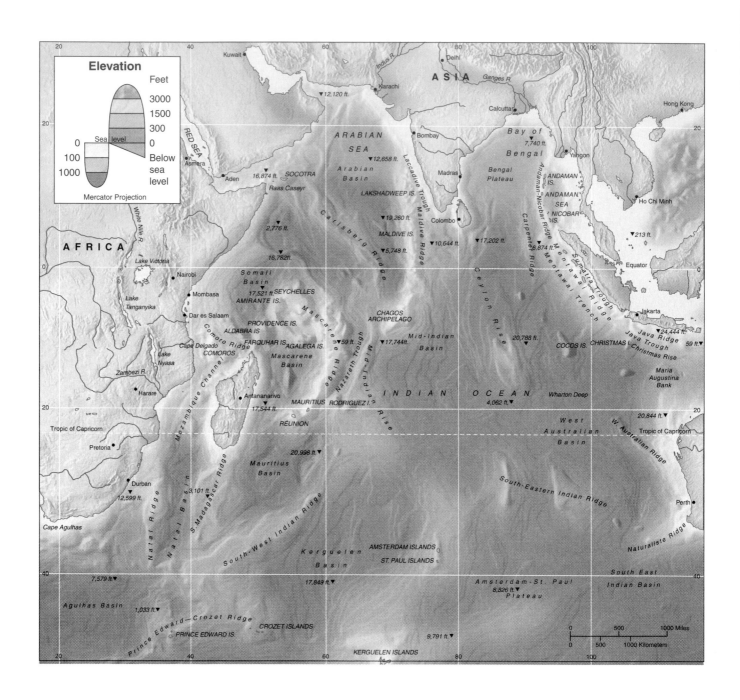

Map 120 Antarctica

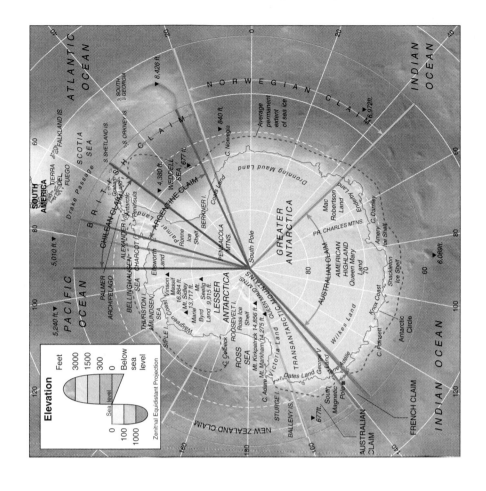

Map 119 The Arctic

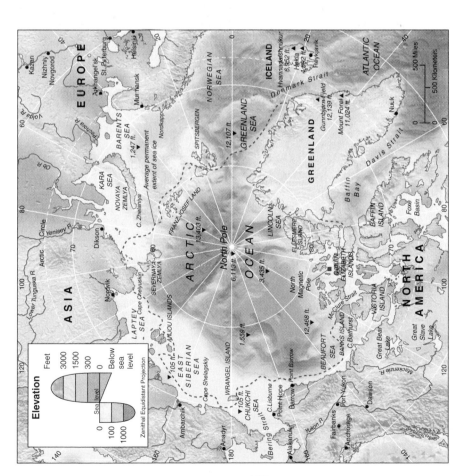

Part VIII

Tables

Table A
World Countries: Area, Population, and Population Density, 2003

COUNTRY	AREA		POPULATION (in thousands)	DENSITY	
	(Mi^2)	(Km^2)	(2003)[a]	(Pop/Mi^2)	(Pop/Km^2)
Afghanistan	251,826	652,229	23,897	95	37
Albania	11,100	28,749	3,166	285	110
Algeria	919,595	2,381,750	31,800	35	13
Andorra	175	453	68	389	150
Angola	481,354	1,246,706	13,625	28	11
Antigua and Barbuda	171	443	67	392	151
Argentina	1,073,400	2,780,104	38,428	36	14
Armenia	11,506	29,801	3,061	266	103
Australia	2,966,155	7,682,337	19,731	7	3
Austria	32,377	83,856	8,116	251	97
Azerbaijan	33,436	86,599	8,370	250	97
Bahamas	5,382	13,939	314	58	23
Bahrain	267	692	724	2,712	1,046
Bangladesh	55,598	143,999	146,736	2,639	1,019
Barbados	166	430	270	1,627	628
Belarus	80,155	207,601	9,895	123	48
Belgium	11,783	30,518	10,318	876	338
Belize	8,866	22,963	256	29	11
Benin	43,475	112,600	6,736	155	60
Bhutan	18,200	47,138	2,257	124	48
Bolivia	424,165	1,098,587	9	0	0
Bosnia and Herzegovina	19,776	51,233	4,161	210	81
Botswana	231,803	600,362	1,785	8	3
Brazil	3,286,488	8,511,999	178,470	54	21
Brunei Darussalam	2,228	5,770	358	161	62
Bulgaria	42,823	110,912	7,897	184	71
Burkina Faso	105,869	274,201	13,002	123	47
Burundi	10,745	27,830	6,825	635	245
Cambodia	69,898	181,036	14,144	202	78
Cameroon	183,569	475,443	16,018	87	34
Canada	3,849,674	9,970,650	31,510	8	3
Cape Verde	1,557	4,033	463	297	115
Central African Republic	240,535	622,985	3,865	16	6
Chad	495,755	1,284,005	8,598	17	7
Chile	292,259	756,950	15,805	54	21
China	3,705,392	9,596,960	1,304,196	352	136
Hong Kong, China	422	1,092	7,049	16,704	6,455
Colombia	439,734	1,138,910	44,222	101	39
Comoros	838	2,170	768	916	354
Democratic Republic of the Congo (formerly Zaire)	905,564	2,345,410	52,771	58	22
Congo Republic	132,047	342,002	3,724	28	11
Costa Rica	19,730	51,101	4,173	212	82
Côte d'Ivoire	124,502	322,460	16,631	134	52

COUNTRY	AREA		POPULATION (in thousands)	DENSITY	
	(Mi^2)	(Km^2)	$(2003)^a$	(Pop/Mi^2)	(Pop/Km^2)
Croatia	21,824	56,538	4,428	203	78
Cuba	42,804	110,862	11,300	264	102
Cyprus	3,571	9,250	802	225	87
Czech Republic	30,387	78,703	10,236	337	130
Denmark	16,629	43,070	5,364	323	125
Djibouti	8,494	22,000	703	83	32
Dominica	290	750	70	241	93
Dominican Republic	18,815	48,730	8,745	465	179
Ecuador	109,484	283,563	13,003	119	46
Egypt	386,662	1,001,454	72	0	0
El Salvador	8,124	21,041	6,515	802	310
Equatorial Guinea	10,831	28,052	494	46	18
Eritrea	46,842	121,320	4,141	88	34
Estonia	17,413	45,100	1,323	76	29
Ethiopia	435,184	1,127,127	70,678	162	63
Fiji	7,054	18,270	839	119	46
Finland	130,127	337,030	5,207	40	15
France	176,460	547,030	60,144	341	110
Gabon	103,347	267,669	1,329	13	5
Gambia	4,363	11,300	1,426	327	126
Georgia	26,911	69,699	5,126	190	74
Germany	137,803	356,910	82,476	599	231
Ghana	92,098	238,534	20,922	227	88
Greece	50,942	131,940	10,976	215	83
Grenada	131	340	89	679	262
Guatemala	42,042	108,889	12,347	294	113
Guinea	94,926	245,858	8,480	89	34
Guinea-Bissau	13,948	36,125	1,493	107	41
Guyana	83,000	214,970	765	9	4
Haiti	10,714	27,749	8,326	777	300
Honduras	43,277	112,087	7	0	0
Hungary	35,920	93,033	9,877	275	106
Iceland	39,768	103,000	290	7	3
India	1,269,340	3,287,590	1,065,462	839	324
Indonesia	741,097	1,919,440	219,883	297	115
Iran	636,294	1,648,000	68,920	108	42
Iraq	168,754	437,072	25,175	149	58
Ireland	27,137	70,285	3,956	146	56
Israel[b]	8,019	20,769	6,433	802	310
Italy	116,305	301,230	57,423	494	191
Jamaica	4,244	10,992	2,651	625	241
Japan	145,882	377,835	127,654	875	338
Jordan	35,445	89,213	5,473	154	61
Kazakhstan	1,049,156	2,717,313	15,433	15	6

World Countries: Area, Population, and Population Density, 2003

COUNTRY	AREA		POPULATION (in thousands)	DENSITY	
	(Mi^2)	(Km^2)	(2003)[a]	(Pop/Mi^2)	(Pop/Km^2)
Kenya	224,961	582,649	31,987	142	55
Kiribati	277	717	96	347	134
Korea, North	46,540	120,539	22,664	487	188
Korea, South	38,023	98,480	47,700	1,255	484
Kuwait	6,880	17,819	2,521	366	141
Kyrgyzstan	76,641	198,500	5,138	67	26
Laos	91,429	236,801	5,657	62	24
Latvia	24,749	64,100	2,307	93	36
Lebanon	4,015	10,399	3,653	910	351
Lesotho	11,720	30,355	1,802	154	59
Liberia	43,000	111,370	3,367	78	30
Libya	679,362	1,759,547	5,551	8	3
Liechtenstein	62	161	33	532	205
Lithuania	25,174	65,201	3,444	137	53
Luxembourg	998	2.585	453	454	175
Macedonia	9,781	25,333	2,056	210	81
Madagascar	226,658	587,044	17,404	77	30
Malawi	45,747	118,485	12,105	265	102
Malaysia	127,317	329,750	24,425	192	74
Maldives	115	298	318	2,765	1,067
Mali	478,767	1,240,006	13,007	27	10
Malta	124	320	318	2,565	994
Marshall Islands	70	181	74	1,057	409
Mauritania	397,954	1,030,700	2,893	7	3
Mauritius	718	1,860	1,221	1,701	656
Mexico	761,603	1,972,550	103,457	136	52
Micronesia	271	702	109	402	155
Moldova	13,012	33,701	4,435	341	132
Monaco	1.21	1.95	32	26,446	16,410
Mongolia	604,427	1,565,000	2,594	4	2
Morocco	172,413	446,550	30,566	177	68
Mozambique	309,494	801,590	18,863	61	24
Myanmar (Burma)	261,969	678,500	49,485	189	73
Namibia	318,259	824,290	1,987	6	2
Nauru	8	21	12	1,500	571
Nepal	54,363	140,800	25,164	463	179
Netherlands	14,413	37,330	16,149	1,120	433
New Zealand	103,738	268,680	3,875	37	14
Nicaragua	49,998	129,494	5,466	109	42
Niger	489,191	1,267,004	11,972	24	9
Nigeria	356,669	923,772	124,009	348	134
Norway	125,182	324,220	4,533	36	14
Oman	82,030	212,458	2,851	35	13
Pakistan	310,402	803,940	153,578	495	191

COUNTRY	AREA		POPULATION (in thousands)	DENSITY	
	(Mi^2)	(Km^2)	$(2003)^a$	(Pop/Mi^2)	(Pop/Km^2)
Palau	177	458	19	107	41
Panama	30,193	78,200	3,120	103	40
Papua New Guinea	178,259	461,690	5,711	32	12
Paraguay	157,048	406,754	5,878	37	14
Peru	496,225	1,285,222	27,167	55	21
Philippines	115,831	300,002	79,999	691	267
Poland	120,728	312,685	38,537	319	123
Portugal	35,552	92,080	10,062	283	109
Qatar	4,247	11,000	610	144	55
Romania	91,699	237,500	22,334	244	94
Russian Federation	6,592,745	17,075,200	143,246	22	8
Rwanda	10,169	26,338	8,387	825	318
St. Kitts and Nevis	104	269	39	375	145
St. Lucia	238	616	149	626	242
St. Vincent/Grenadines	131	340	120	916	353
Samoa	1,104	2,860	178	161	62
San Marino	23	60	28	1,217	467
São Tomé and Principe	372	963	161	433	167
Saudi Arabia	756,982	1,960,582	24,217	32	12
Senegal	75,749	196,190	10,095	133	51
Serbia-Montenegro	39,517	102,350	10,527	266	103
Seychelles	175	453	80	457	177
Sierra Leone	27,699	71,740	4,971	179	69
Singapore	244	633	4,253	17,430	6,719
Slovakia	18,859	48,845	5,042	267	103
Slovenia	7,836	20,296	1,984	253	98
Solomon Islands	10,985	28,450	477	43	17
Somalia	246,201	637,660	9,890	40	16
South Africa	471,444	1,221,040	45,026	96	37
Spain	194,885	504,752	41,060	211	81
Sri Lanka	25,332	65,610	19,065	753	291
Sudan	967,500	2,505,824	33,610	35	13
Suriname	63,039	163,270	436	7	3
Swaziland	6,704	17,363	1,077	161	62
Sweden	173,732	449,966	8,876	51	20
Switzerland	15,943	41,292	7,169	450	174
Syria	71,498	185,180	17,800	249	96
Taiwan	13,892	35,980	22,548	1,623	627
Tajikistan	55,251	143,100	6,245	113	44
Tanzania	364,900	945,090	36,977	101	39
Thailand	198,456	514,000	62,833	317	122
Togo	21,925	56,786	4,909	224	86
Tonga	290	751	104	359	138
Trinidad and Tobago	1,980	5,128	1,303	658	254

Table A (Continued)
World Countries: Area, Population, and Population Density, 2003

COUNTRY	AREA		POPULATION (in thousands)	DENSITY	
	(Mi^2)	(Km^2)	(2003)[a]	(Pop/Mi^2)	(Pop/Km^2)
Tunisia	63,170	163,610	9,832	156	60
Turkey	301,382	780,580	71,325	237	91
Turkmenistan	188,456	488,101	4,867	26	10
Tuvalu	10	26	11	1,100	423
Uganda	93,135	236,040	25,827	277	109
Ukraine	233,090	603,703	48,523	208	80
United Arab Emirates	31,969	82,880	2,995	94	36
United Kingdom	94,525	244,820	59,251	627	242
United States	3,717,797	9,629,091	294,043	79	31
Uruguay	68,039	176,220	3,415	50	19
Uzbekistan	172,742	447,402	26,093	151	58
Vanuatu	5,699	14,760	212	37	14
Venezuela	352,145	912,055	25,699	73	28
Vietnam	127,243	329,560	81,377	640	247
Yemen	203,850	527,970	20,010	98	38
Zambia	290,586	752,617	10,812	37	14
Zimbabwe	150,803	390,580	12,891	85	33

[a]Primary source for population figures: United Nations Population Division.

[b]The figures for Israel do not include the West Bank and Gaza. These territories combined have a population of 3,557,000 (2003), and an area of approximately 2,400 square miles. The West Bank has an estimated population density of 956 persons per square mile and Gaza an estimated population density of 8,820 persons per square mile.

Sources: World Development Indicators 2003 (The World Bank); *World Almanac and Book of Facts; World Population Prospects: The 2000 Revision* (United Nations Population Information Network, 2001).

Table B
World Countries: Form of Government, Capital City, Major Languages

Notes: Unless indicated otherwise, republics are multi-party. Theocratic normally refers to fundamentalist Islamic rule. Transitional governments are those still in the process of change from a previous form (eg. Single-party communist state to multi-party republic).

COUNTRY	GOVERNMENT	CAPITAL	MAJOR LANGUAGES
Afghanistan	Transitional	Kabul	Dari, Pashtu, Uzbek, Turkmen
Albania	Multi-party democracy	Tiranë	Albanian, Greek
Algeria	Republic	Algiers	Arabic, Berber, dialects, French
Andorra	Parliamentary democracy	Andorra	Cataln, French, Spanish, Portuguese
Angola	Multi-party republic	Luanda	Portugese; Bantu and other African
Antigua and Barbuda	Parliamentary democracy	St. John's	English, local dialects
Argentina	Federal republic	Buenos Aires	Spanish, English, Italian, German, other
Armenia	Republic	Yerevan	Armenian, Russian, other
Australia	Federal parliamentary democracy	Canberra	English, indigenous
Austria	Federal republic	Vienna	German
Azerbaijan	Republic	Baku	Azerbaijani, Russian, Armenian, other
Bahamas	Parliamentary democracy; independent Commonwealth	Nassau	English, Creole
Bahrain	Constitutional hereditary monarchy	Al Manamah	Arabic, English, Farsi, Urdu
Bangladesh	Parliamentary democracy	Dhaka	Bangla, English
Barbados	Parliamentary democracy	Bridgetown	English
Belarus	Republic	Minsk	Byelorussian, Russian, other
Belgium	Constitutional monarchy	Brussels	Dutch (Flemish), French, German
Belize	Parliamentary democracy	Belmopan	English, Spanish, Garifuna, Mayan
Benin	Multi-party republic	Porto-Novo	French, Fon, Yoruba
Bhutan	Monarchy; special treaty relationship with India	Timphu	Dzongkha, Tibetan, Nepalese
Bolivia	Republic	La Paz, Sucre	Spanish, Quechua, Aymara
Bosnia-Herzegovina	Emerging federal democratic republic	Sarajevo	Croatian, Serbian, Bosnian
Botswana	Parliamentary republic	Gaborone	English, Setswana
Brazil	Federal republic	Brasilia	Portugese, Spanish, English, French
Brunei	Constitutional monarchy	Bandar Seri Begawan	Malay, English, Chinese
Bulgaria	Parliamentary democracy	Sofia	Bulgarian
Burkina Faso	Parliamentary republic	Ouagadougou	French, indigenous
Burundi	Republic	Bujumbura	French, Kirundi, Swahili
Cambodia	Multi-party democracy (under UN supervision)	Phnom Penh	Khmer, French, English
Cameroon	Multi-party republic	Yaoundé	English, French, indigenous
Canada	Federal parliamentary	Ottawa	English, French, other
Cape Verde	Republic	Cidade de Praia	Portugese, Crioulu
Central African Republic	Republic	Bangui	French, Sangho
Chad	Republic	N'Djamena	French, Arabic, Sara, other indigenous
Chile	Republic	Santiago	Spanish
China	Single-party communist state	Beijing	Various Chinese dialects
Colombia	Republic	Bogotá	Spanish
Comoros	Republic	Moroni	Arabic, French, Shikomoro
Democratic Republic of the Congo (formerly Zaire)	Republic/transitional from military dictatorship	Kinshasa	French, Lingala, Kingwana, Kikongo, Tshilluba
Congo Republic	Multi-party republic	Brazzaville	French, Lingala, Monokutuba, Kikongo
Costa Rica	Democratic republic	San José	Spanish

Notes: Unless indicated otherwise, republics are multi-party. Theocratic normally refers to fundamentalist Islamic rule. Transitional governments are those still in the process of change from a previous form (eg. Single-party communist state to multi-party republic).

COUNTRY	GOVERNMENT	CAPITAL	MAJOR LANGUAGES
Côte d'Ivoire	Multi-party republic	Abidjan, Yamoussoukro	French, indigenous
Croatia	Parliamentary democracy	Zagreb	Croatian
Cuba	Single-party communist state	Havana	Spanish
Cyprus	Republic	Nicosia	Greek, Turkish, English
Czech Republic	Parliamentary democracy	Prague	Czech
Denmark	Constitutional monarchy	Copenhagen	Danish, Faroese, Greenlandic, German
Djibouti	Republic	Djibouti	French, Somali, Afar, Arabic
Dominica	Parliamentary democracy, republic within Commonwealth	Roseau	English, French
Dominican Republic	Republic	Santo Domingo	Spanish
East Timor	Republic	Dili	Tetum, Portuguese, Indonesian, English
Ecuador	Republic	Quito	Spanish, Quechua, indigenous
Egypt	Republic	Cairo	Arabic
El Salvador	Republic	San Salvador	Spanish, Nahua
Equatorial Guinea	Republic	Malabo	Spanish, French, indigenous, English
Eritrea	Transitional government	Asmara	Afar, Amharic, Arabic, Tigre, other indigenous
Estonia	Parliamentary republic	Tallinn	Estonian, Russian, Ukranian, Finnish
Ethiopia	Federal republic	Addis Ababa	Amharic, Tigrinya, Orominga, Somali, Arabic, English
Fiji	Republic	Suva	English, Fijian, Hindustani
Finland	Republic	Helsinki	Finnish, Swedish
France	Republic	Paris	French
Gabon	Multi-party republic	Libreville	French, Fang, indigenous
The Gambia	Multi-party democratic republic	Banjul	English, Mandinka, Wolof, Fula
Georgia	Republic	Tbilisi	Georgian, Russian, Armenian
Germany	Federal republic	Berlin	German
Ghana	Parliamentary democracy	Accra	English, indigenous
Greece	Parliamentary republic	Athens	Greek
Grenada	Parliamentary democracy	St. George's	English, French
Guatemala	Republic	Guatemala City	Spanish, Quiche, Cakchiquel, other indigenous
Guinea	Republic	Conakry	French, indigenous
Guinea-Bissau	Multi-party republic	Bissau	Portugese, Crioulo, indigenous
Guyana	Republic within Commonwealth	Georgetown	English, Creole, Hindi, Urdu, indigenous
Haiti	Republic	Port-au-Prince	Creole, French
Honduras	Republic	Tegucigalpa	Spanish, indigenous
Hungary	Parliamentary democracy	Budapest	Hungarian
Iceland	Republic	Reykjavk	Icelandic
India	Federal republic	New Delhi	English, Hindi, 14 other official
Indonesia	Republic	Jakarta	Bahasa Indonesian, English, Dutch, Javanese
Iran	Theocratic republic	Tehran	Farsi, Turkish, Kurdish
Iraq	In transition following US-led invasion	Baghdad	Arabic, Kurdish, Assyrian, Armenian
Ireland	Republic	Dublin	English, Irish Gaelic
Israel	Parliamentary democracy	Jerusalem	Hebrew, Arabic, English
Italy	Republic	Rome	Italian
Jamaica	Parliamentary democracy	Kingston	English, Creole

Notes: Unless indicated otherwise, republics are multi-party. Theocratic normally refers to fundamentalist Islamic rule. Transitional governments are those still in the process of change from a previous form (eg. Single-party communist state to multi-party republic).

COUNTRY	GOVERNMENT	CAPITAL	MAJOR LANGUAGES
Japan	Constitutional monarchy	Tokyo	Japanese
Jordan	Constitutional monarchy	Amman	Arabic
Kazakhstan	Republic	Astana	Kazakh, Russian
Kenya	Republic	Nairobi	English, Swahili, indigenous
Kiribati	Republic	Tarawa	English, I-Kiribati
Korea, North	Single-party communist state	Pyongyang	Korean
Korea, South	Republic	Seoul	Korean
Kuwait	Constitutional monarchy	Kuwait	Arabic, English
Kyrgyzstan	Republic	Bishkek	Kirghiz, Russian
Laos	Single-party communist state	Vientiane	Lao, French, English, indigenous
Latvia	Parliamentary democracy	Riga	Latvian, Lithuanian, Russian
Lebanon	Republic	Beirut	Arabic, French, Armenian, English
Lesotho	Constitutional monarchy	Maseru	English, Sesotho, Zulu, Xhosa
Liberia	Republic	Monrovia	English, indigenous
Libya	Single party/military dictatorship	Tripoli	Arabic
Liechtenstein	Constitutional monarchy	Vaduz	German
Lithuania	Parliamentary democracy	Vilnius	Lithuanian, Russian, Polish
Luxembourg	Constitutional monarchy	Luxembourg	French, Luxembourgian, German
Macedonia	Parliamentary democracy	Skopje	Macedonian, Albanian, Turkish, Serbo-Croatian
Madagascar	Republic	Antananarivo	Malagasy, French
Malawi	Multi-party democracy	Lilongwe	Chichewa, English, Tombuka
Malaysia	Constitutional monarchy	Kuala Lumpur	Malay, Chinese, English, indigenous
Maldives	Republic	Male	Dhivehi
Mali	Republic	Bamako	French, Bambara, indigenous
Malta	Parliamentary democracy	Valletta	English, Maltese
Marshall Islands	Constitutional government (free association with U.S.)	Majuro	English, Polynesian dialects, Japanese
Mauritania	Republic	Nouakchott	Arabic, Wolof, Pular, French, Solinke
Mauritius	Parliamentary democracy	Port Louis	English, Creole, French, Hindi, Urdu, Bojpoori, Hakka
Mexico	Federal republic	Mexico City	Spanish, indigenous
Micronesia	Constitutional government (free association with U.S.)	Palikir	English, Trukese, Pohnpeian, Yapese, others
Moldova	Republic	Chisinau	Moldavian, Russian, Gagauz
Monaco	Constitutional monarchy	Monaco	French, English, Italian, Monegasque
Mongolia	Republic	Ulaanbaatar	Khalkha Mongol, Turkic, Russian
Morocco	Constitutional monarchy	Rabat	Arabic, Berber dialects, French
Mozambique	Republic	Maputo	Portuguese, indigenous
Myanmar (Burma)	Military regime	Rangoon	Burmese, indigenous
Namibia	Republic	Windhoek	Afrikaans, English, German, indigenous
Nauru	Republic	Yaren district	Nauruan, English
Nepal	Constitutional monarchy	Kathmandu	Nepali, indigenous
Netherlands	Constitutional monarchy	Amsterdam	Dutch
New Zealand	Parliamentary democracy	Wellington	English, Maori
Nicaragua	Republic	Managua	Spanish, English, indigenous

World Countries: Form of Government, Capital City, Major Languages

Notes: Unless indicated otherwise, republics are multi-party. Theocratic normally refers to fundamentalist Islamic rule. Transitional governments are those still in the process of change from a previous form (eg. Single-party communist state to multi-party republic).

COUNTRY	GOVERNMENT	CAPITAL	MAJOR LANGUAGES
Niger	Provisional military	Niamey	French, Hausa, Djerma
Nigeria	Military/transitional	Abuja	English, Hausa, Fulani, Yorbua, Ibo
Norway	Constitutional monarchy	Oslo	Norwegian, Sami
Oman	Monarchy	Muscat	Arabic, English, Baluchi, Urdu
Pakistan	Federal republic	Islamabad	Punjabi, Sindhi, Siraiki, Pashtu, Urbu, English, others
Palau	Constitutional government (free association with U.S.)	Koror	English, Palauan, Sonsolorese, Tobi, Angaur, Japanese
Panama	Constitutional democracy	Panama	Spanish, English
Papua New Guinea	Parliamentary democracy	Port Moresby	Various indigenous, English, Motu
Paraguay	Constitutional republic	Asunción	Spanish, Guarani
Peru	Republic	Lima	Quechua, Spanish, Aymara
Philippines	Republic	Manila	English, Filipino, 8 major dialects
Poland	Republic	Warsaw	Polish
Portugal	Parliamentary democracy	Lisbon	Portuguese
Qatar	Traditional monarchy	Doha	Arabic, English
Romania	Republic	Bucharest	Romanian, Hungarian, German
Russia	Federation	Moscow	Russian, numerous other
Rwanda	Republic	Kigali	French, Kinyarwanda, English, Kiswahili
St. Kitts and Nevis	Constitutional monarchy	Basseterre	English
St. Lucia	Parliamentary democracy	Castries	English, French
St. Vincent/Grenadines	Parliamentary monarchy independent within Commonwealth	Kingstown	English, French
Samoa	Constitutional monarchy	Apia	Samoan, English
San Marino	Republic	San Marino	Italian
São Tomé and Principe	Republic	São Tome	Portuguese
Saudi Arabia	Monarchy	Riyadh	Arabic
Senegal	Republic	Dakar	French, Wolof, indigenous
Serbia and Montenegro	Republic	Belgrade	Serbian, Albanian
Seychelles	Republic	Victoria	English, French, Creole
Sierra Leone	Constitutional democracy	Freetown	English, Krio, Mende, Temne
Singapore	Parliamentary republic	Singapore	Chinese, English, Malay, Tamil
Slovakia	Parliamentary democracy	Bratislava	Slovak, Hungarian
Slovenia	Republic	Ljubljana	Slovenian, Serbo-Croatian, other
Solomon Islands	Parliamentary democracy	Honiara	English, indigenous
Somalia	No permanent national government	Mogadishu	Arabic, Somali, English, Italian
South Africa	Republic	Pretoria	Afrikaans, English, Zulu, Xhosa, other
Spain	Parliamentary monarchy	Madrid	Castilian Spanish, Catalan, Galician, Basque
Sri Lanka	Republic	Colombo	English, Sinhala, Tamil
Sudan	Provisional military	Khartoum	Arabic, Nubian, others
Suriname	Constitutional democracy	Paramaribo	Dutch, Sranang Tongo, English, Hindustani, Javanese
Swaziland	Monarchy within Commonwealth	Mbabane	English, siSwati
Sweden	Constitutional monarchy	Stockholm	Swedish
Switzerland	Federal republic	Bern	German, French, Italian, Romansch
Syria	Republic (under military regime)	Damascus	Arabic, Kurdish, Armenian, Aramaic

Notes: Unless indicated otherwise, republics are multi-party. Theocratic normally refers to fundamentalist Islamic rule. Transitional governments are those still in the process of change from a previous form (eg. Single-party communist state to multi-party republic).

COUNTRY	GOVERNMENT	CAPITAL	MAJOR LANGUAGES
Taiwan	Multi-party democracy	Taipei	Mandarin Chinese, Taiwanese, Hakka
Tajikistan	Republic	Dushanbe	Tajik, Russian
Tanzania	Republic	Dar es Salaam	Kiswahili, English, Arabic, indigenous
Thailand	Constitutional monarchy	Bangkok	Thai, English
Togo	Republic/transitional	Lomé	French, indigenous
Tonga	Constitutional monarchy	Nuku'alofa	Tongan, English
Trinidad and Tobago	Parliamentary democracy	Port-of-Spain	English, Hindi, French, Spanish, Chinese
Tunisia	Republic	Tunis	Arabic, French
Turkey	Parliamentary republic	Ankara	Turkish, Kurdish, Arabic, Armenian, Greek
Turkmenistan	Republic	Ashkhabad	Turkmen, Russian, Uzbek, other
Tuvalu	Constitutional monarchy	Funafuti	Tuvaluan, English, Samoan, Kiribati
Uganda	Republic	Kampala	English, Luganda, Swahili, Arabic, indigenous
Ukraine	Republic	Kiev	Ukranian, Russian, Romanian, Polish, Hungarian
United Arab Emirates	Federated monarchy	Abu Dhabi	Arabic, English, Farsi, Hindi, Urdu
United Kingdom	Constitutional monarchy	London	English, Welsh, Scottish Gaelic
United States	Federal republic	Washington	English, Spanish
Uruguay	Republic	Montevideo	Spanish, Portunol/Brazilero
Uzbekistan	Republic	Tashkent	Uzbek, Russian, Kazakh, Tajik, other
Vanuatu	Republic	Port-Vila	English, French, Bislama (pidgin)
Venezuela	Federal republic	Caracas	Spanish, indigenous
Vietnam	Single-party communist state	Hanoi	Vietnamese, French, Chinese, English, Khmer
Yemen	Republic	San`aa	Arabic
Zambia	Republic	Lusaka	English, Tonga, Lozi, other indigenous
Zimbabwe	Parliamentary democracy	Harare	English, Shona, Sindebele, other

Source: *The World Factbook 2002* (CIA, Washington, DC).

Table C
World Countries: Basic Economic Indicators

COUNTRY	GROSS NATIONAL INCOME (GNI) 2001[a]		PURCHASING POWER PARITY GNI 2001[b]			AVERAGE ANNUAL % GROWTH IN GDP			STRUCTURE OF ECONOMIC OUTPUT (GDP) 2001 (value added in % of GDP)			
	Total ($U.S. billions)	Per Capita ($U.S.)	Total ($U.S. billions)	Per Capita ($U.S.)	Rank	1980–1990	1990–1999	2000–01	Agriculture	Industry	Manufacturing	Services
Afghanistan	–	–	–	–	–	–	–	–	–	–	–	–
Albania	4.2	1,340	12	3,810	130	1.5	3.2	6.5	50	23	13	26
Algeria	51.0	1,650	182[c]	5,910[c]	99	2.7	1.6	2.1	10	55	8	36
Angola	6.7	500	23[c]	1,690[c]	171	3.7	0.4	3.2	8	67	4	25
Argentina	260.3	7	412	10,980	63	–0.4	4.9	–4.5	5	27	17	69
Armenia	2.2	570	10	2,730	145	–	–3.2	9.6	28	34	22	38
Australia	385.9	19,900	478	24,630	24	3.4	4.1	3.9	4	26	13	70
Austria	194.7	23,940	215	26,380	17	2.2	1.9	1.0	2	33	22	65
Azerbaijan	5.3	650	23	2,890	141	–	–9.6	9.9	17	46	6	36
Bangladesh	48.6	360	213	1,600	173	4.3	4.7	5.3	23	25	15	52
Belarus	12.9	1,290	76	7,630	83	–	–3.0	4.1	11	39	33	50
Belgium	245.3	23,850	269	26,150	18	2.0	1.7	1.0	2	27	20	71
Benin	2.4	380	6	970	190	2.9	4.7	5.0	36	14	9	50
Bolivia	8.1	950	19	2,240	155	–0.2	4.2	1.2	16	29	15	56
Bosnia–Herzegovina	5.0	1,240	25	6,250	92	–	35.2	6.0	15	31	16	55
Botswana	5.3	3,100	13	7,410	84	10.3	4.3	6.3	2	47	4	51
Brazil	528.9	3,070	1,219	7,070	86	2.7	3.0	1.5	9	34	21	57
Bulgaria	13.2	1,650	54	6,740	89	3.4	–2.7	4.0	14	29	18	57
Burkina Faso	2.5	220	13[c]	1,120[c]	185	3.6	3.8	5.6	38	21	15	41
Burundi	0.7	100	5[c]	680[c]	203	4.4	–2.9	3.2	50	19	9	31
Cambodia	3.3	270	22	1,790	168	–	4.8	6.3	37	22	–	41
Cameroon	8.7	580	24	1,580	174	3.4	1.3	5.3	43	20	11	38
Canada	681.6	21,930	825[c]	26,530[c]	15	3.3	2.7	1.5	–	–	–	–
Central African Republic	1.0	260	5[c]	1,300[c]	181	1.4	1.8	1.5	55	21	9	24
Chad	1.6	200	8	1,060	187	3.7	2.1	8.5	39	14	10	48
Chile	70.6	4,590	136	8,840	76	4.2	7.2	2.8	9	34	16	57
China	1,131.2	890	5,027	3,950	127	10.2	10.7	7.3	15	51	35	34
Hong Kong, China	170.3	25,30	172	25,560	19	6.9	3.9	0.1	0	14	6	86
Colombia	81.6	1,890	292	6,790	88	3.6	3.3	1.4	13	30	16	57
Democratic Republic of the Congo (formerly Zaire)	4.2	80	33	630	205	1.6	–5.1	–4.5	56	19	4	25
Congo Republic	2.0	640	2	680	203	3.3	–0.5	2.9	6	66	4	28
Costa Rica	15.7	4,060	36	9,260	74	3.0	5.1	0.9	9	29	21	62
Cote d'Ivoire	10.3	630	23	1,400	179	0.7	3.7	–0.9	24	22	19	54
Croatia	19.9	4,550	39	8,930	75	–	0.2	4.1	9	33	23	58
Cuba	–	–[g]	–	–	–	–	–	–	7	46	37	47
Czech Republic	54.3	5,310	146	14,320	55	1.7	0.8	3.3	4	41	–	55
Denmark	164.0	30,600	153	23,490	9	2.3	2.4	1.0	33	26	17	71
Dominican Republic	19.0	2,230	57	6,650	90	3.1	5.8	2.7	11	33	16	55
Ecuador	14.0	1,080	38	2,960	140	2.0	2.2	5.6	11	33	18	56
Egypt	99.6	1,530	232	3,560	131	5.4	4.4	2.9	17	33	19	50
El Salvador	13.0	2,040	33	5,160	107	0.2	5.0	1.8	9	30	23	61
Eritrea	0.7	160	4	1,030	189	–	5.0	9.7	19	22	11	59
Estonia	5.3	3,870	13	9,650	71	2.2	–1.3	5.0	6	29	19	65
Ethiopia	6.7	100	53	800	198	2.3	4.6	7.7	52	11	7	37
Finland	123.4	23,780	125	24,030	28	3.3	2.4	0.4	3	33	26	63
France	1,380.7[h]	22,730[h]	1,425	24,080	27	2.3	1.5	1.8	3	26	18	72

–171–

Table C (Continued)
World Countries: Basic Economic Indicators

COUNTRY	GROSS NATIONAL INCOME (GNI) 2001[a]		PURCHASING POWER PARITY GNI 2001[b]			AVERAGE ANNUAL % GROWTH IN GDP			STRUCTURE OF ECONOMIC OUTPUT (GDP) 2001 (value added in % of GDP)			
	Total ($U.S. billions)	Per Capita ($U.S.)	Total ($U.S. billions)	Per Capita ($U.S.)	Rank	1980–1990	1990–1999	2000–01	Agriculture	Industry	Manufacturing	Services
Gabon	4.0	3,160	7	5,190	105	0.9	3.2	2.5	8	51	5	42
The Gambia	0.4	320	3[c]	2,010[c]	160	3.6	2.8	6.0	40	14	5	46
Georgia	3.1	590	14	2,580	148	0.4	–	4.5	21	23	–	57
Germany	1,939.6	24	2,078	25,240	21	2.2	1.3	0.6	1	31	24	68
Ghana	5.7	290	43[c]	2,170[c]	157	3.0	4.3	4.0	36	25	9	39
Greece	121.0	11,430	186	17,520	47	1.8	2.2	4.1	8	21	12	71
Guatemala	19.6	1,680	51	4,380	120	0.8	4.2	2.1	23	19	13	58
Guinea	3.1	410	14	2	164	–	4.2	3.6	24	38	4	38
Guinea–Bissau	0.2	160	1	890	193	4.0	0.3	0.2	56	13	10	31
Haiti	3.9	480	15[c]	1,870[c]	166	−0.2	−1.3	−1.7	–	–	–	–
Honduras	5.9	900	18	2,760	144	2.7	3.3	2.6	14	32	20	55
Hungary	49.2	4,830	122	11,990	59	1.3	1.0	3.8	–	–	–	–
India	477.4	460	2,913	2,820	143	5.8	6.0	5.4	25	26	16	48
Indonesia	144.7	690	591	2,830	142	6.1	4.7	3.3	16	47	26	37
Iran	108.7	1,680	383	5,940	98	1.7	3.6	4.8	19	33	16	48
Iraq	–	_e	–	–	–	−6.8	–	–	–	–	–	–
Ireland	87.7	22,850	104	27,170	14	3.2	6.9	5.8	4	42	33	55
Israel	106.6	17	125	19,630	40	3.5	5.2	−0.9	–	–	–	–
Italy	1,128.8	19,390	1,422	24,530	25	2.4	1.4	1.8	3	29	21	68
Jamaica	7.3	2,800	9	3	133	2.0	0.3	1.7	6	31	13	63
Japan	4,523.3	36	3,246	25,550	20	4.0	1.3	−0.6	1	32	22	67
Jordan	8.8	1,750	20	3,880	128	2.5	5.3	4.2	2	25	15	73
Kazakhstan	20.1	1,350	92	6,150	94	–	−5.9	13.2	9	39	16	52
Kenya	10.7	350	30	970	190	4.2	2.2	1.1	19	18	13	63
Korea, North	–	_d	–	–	–	–	–	–	–	–	–	–
Korea, South	447.6	9,460	713	15	54	9.4	5.7	3.0	4	41	30	54
Kuwait	37.4	18,270	44	21,530	35	1.3	–	−1.0	–	–	–	–
Kyrgyzstan	1.4	280	13	2,630	147	–	−5.4	5.3	38	27	8	35
Laos	1.6	300	8[c]	1,540[c]	175	–	6.6	5.7	51	23	18	26
Latvia	7.6	3,230	18	7,760	82	3.5	−4.8	7.6	5	26	15	69
Lebanon	17.6	4,010	19	4,400	119	–	7.7	1.3	12	22	10	66
Lesotho	1.1	530	6[c]	2,980[c]	139	4.4	4.4	4.0	16	42	14	42
Liberia	0.5	140	–	–	196	–	–	5.3	–	–	–	–
Libya	–	_g	–	–	–	−5.7	–		–	–	–	–
Lithuania	11.7	3,350	29	8,350	78	–	−4.0	5.9	7	35	23	58
Luxembourg	18.8	41,550						1	1	20		79
Macedonia	3.5	1,690	12	6,040	97	–	−0.8	−4.1	11	31	20	58
Madagascar	4.2	260	13	820	197	1.1	1.7	6.0	30	14	12	56
Malawi	1.7	151	6	560	206	2.5	3.6	−1.5	34	18	13	48
Malaysia	79.3	3,330	188	7,910	81	5.3	7.3	0.4	9	49	31	42
Mali	2.5	230	9	770	200	2.8	3.6	1.4	38	26	4	36
Mauritania	1.0	360	5	1,940	162	1.8	4.2	4.6	21	29	8	50
Mauritius	4.6	3,830	12	9,860	70	6.2	5.1	7.2	6	31	23	62
Mexico	550.2	5,530	820	8	80	0.7	2.7	−0.3	4	27	19	69
Moldova	1.5	400	10	2,300	154	3.0	−11.0	6.1	26	24	18	50
Mongolia	1.0	400	4	1,710	170	5.4	0.7	1.4	30	17	5	53
Morocco	34.7	1,190	102	3,500	132	4.2	2.3	6.5	16	18	50	53
Mozambique	3.8	210	19[c]	1,050[c]	188	−0.1	6.2	13.9	22	26	12	52
Myanmar (Burma)	–	_d	–	–	–	0.6	6.3	–	57	10	7	33

Table C (Continued)
World Countries: Basic Economic Indicators

COUNTRY	GROSS NATIONAL INCOME (GNI) 2001[a]		PURCHASING POWER PARITY GNI 2001[b]			AVERAGE ANNUAL % GROWTH IN GDP			STRUCTURE OF ECONOMIC OUTPUT (GDP) 2001 (value added in % of GDP)			
	Total ($U.S. billions)	Per Capita ($U.S.)	Total ($U.S. billions)	Per Capita ($U.S.)	Rank	1980–1990	1990–1999	2000–01	Agriculture	Industry	Manufacturing	Services
Namibia	3.5	1,960	13[c]	7,410[c]	85	0.9	3.4	2.7	11	33	11	56
Nepal	5.8	250	32	1,360	180	4.6	4.9	4.8	39	22	9	39
Netherlands	390.3	24,330	439	27,390	13	2.3	2.7	1.1	3	27	17	70
New Zealand	51.0	13,250	70	18	43	1.8	3.1	3.2	–	–	–	–
Nicaragua	–	–	–	–	158	–2.0	3.2	–	–	–	–	–
Niger	2.0	180	10[c]	880[c]	194	–0.1	2.4	7.6	40	17	7	43
Nigeria	37.1	290	102	790	199	1.6	2.4	3.9	30	46	4	25
Norway	160.8	36	132	29,340	7	2.8	3.8	1.4	2	43	–	55
Oman	–	_g		–	–	8.4	5.9	–	–	–	–	–
Pakistan	60.0	420	263	1,860	167	6.3	3.8	2.7	25	23	16	52
Panama	9.5	3,260	16[c]	5,440[c]	104	0.5	4.2	0.3	7	9	7	77
Papua New Guinea	3.0	580	13[c]	2,450[c]	149	1.9	4.7	–3.5	26	42	8	32
Paraguay	7.6	1,350	29[c]	5,180[c]	106	2.5	2.4	2.7	20	26	13	54
Peru	52.2	1,980	118	4,470	117	–0.3	5.0	0.2	9	30	15	62
Philippines	80.8	1,030	319	4	125	1.0	3.2	3.4	15	31	22	54
Poland	163.6	4,230	362	9,370	73	1.8	4.5	1.0	4	37	20	59
Portugal	109.3	10,900	178	17,710	46	3.1	2.5	1.7	4	30	19	66
Puerto Rico	42.1	10,950	69	18,090	44	4.0	3.1	5.6	1	43	40	56
Romania	38.6	1,720	130	5,780	101	0.5	–0.8	5.3	15	35	–	50
Russian Federation	253.4	1,750	995	6,880	87	–	–6.1	5.0	7	37	–	56
Rwanda	1.9	220	11	1,240	183	2.2	–1.5	6.7	40	22	10	38
Saudi Arabia	181.1	8	284	13,290	54	0.0	1.6	1.2	–	–	–	–
Senegal	4.7	490	14	1,480	176	3.1	3.3	5.7	18	27	18	55
Serbia–Montenegro	9.9	930	–	–	–	–	–	–	15	32	–	53
Sierra Leone	0.7	140	2	460	208	0.3	–4.7	5.4	50	30	5	20
Singapore	88.8	21,500	94	22,850	32	6.6	8.0	–2.0	0	34	23	68
Slovak Republic	20.3	4	64	11,780	60	2.0	1.8	3.3	4	29	21	67
Slovenia	19.4	9,760	34	17,060	49	–	2.4	3.0	3	38	28	58
Somalia	–	d	–	–	–	–	–	–	–	–	–	–
South Africa	121.9	2,820	472[c]	10,910[c]	64	1.2	1.9	2.2	3	31	19	66
Spain	588.0	14,800	816	19,860	39	3.0	2.2	2.8	4	30	19	66
Sri Lanka	16.4	880	61	3,260	134	4.0	5.3	–1.4	19	27	16	54
Sudan	10.7	340	56	1,750	169	0.4	8.2	6.9	39	19	10	42
Swaziland	1.4	1,300	5	4,430[c]	118	–	–	1.6	17	44	36	39
Sweden	225.9	25,400	212	23,800	29	2.3	1.6	1.2	2	32	–	71
Switzerland	277.2	38,330	224	30,970	5	2.0	0.6	1.3	–	–	–	–
Syria	17.3	1,040	52	3,160	136	1.5	5.7	2.8	22	28	20	50
Tajikistan	1.1	180	7	1,140	184	–	–	10.2	29	29	25	41
Tanzania	9.4[i]	270[i]	18	520	207	[i]	2.8	5.7	45	16	7	39
Thailand	118.5	1,940	381	6,230	93	7.6	4.7	1.8	10	40	32	49
Togo	1.3	270	8	1,620	172	1.7	2.4	2.7	39	21	10	39
Trinidad and Tobago	7.8	5,960	11	8,620	77	–0.8	2.7	5.0	2	44	8	55
Tunisia	20.0	2,070	59	6,090	96	3.3	4.6	4.9	12	29	18	60
Turkey	167.3	2,530	386	5,830[c]	100	5.4	3.8	–7.4	14	26	15	61
Turkmenistan	5.1	950	23	4,240	124	–	–6.8	20.5	29	51	–	20
Uganda	5.9	260	33[c]	1,460[c]	177	2.9	7.2	4.6	36	21	10	43
Ukraine	35.2	720	210	4,270	131	–	–10.7	9.1	17	39	23	44

Table C *(Continued)*
World Countries: Basic Economic Indicators

COUNTRY	GROSS NATIONAL INCOME (GNI) 2001[a]		PURCHASING POWER PARITY GNI 2001[b]			AVERAGE ANNUAL % GROWTH IN GDP			STRUCTURE OF ECONOMIC OUTPUT (GDP) 2001 (value added in % of GDP)			
	Total ($U.S. billions)	Per Capita ($U.S.)	Total ($U.S. billions)	Per Capita ($U.S.)	Rank	1980–1990	1990–1999	2000–01	Agriculture	Industry	Manufacturing	Services
United Arab Emirates	–	_f	–	–	–	–3.5	2.9	–	–	–	–	–
United Kingdom	1,476.8	25,120	1,431	24,340	26	3.2	2.5	2.2	1	27	19	72
United States	9,780.8	34,280	10	34,280	3	3.0	3.3	0.3	2	25	17	73
Uruguay	19.2	5,710	28	8,250	79	0.4	3.8	–3.1	6	28	56	67
Uzbekistan	13.8	550	60	2,410	152	–	–1.2	4.5	34	23	9	43
Venezuela	117.2	4,760	138	5,590	102	1.1	1.7	2.7	5	50	20	45
Vietnam	32.8	410	164	2,070	159	4.6	8.1	6.8	24	38	20	39
West Bank and Gaza	4.2	1,350	–	–	–	–	3.7	–11.9	8	27	15	66
Yemen, Rep.	8.2	450	13	730	202	–	3.2	3.1	16	50	7	35
Zambia	3.3	320	8	750	201	1.0	0.2	4.9	22	26	11	52
Zimbabwe	6.2	480	28	2,220	156	3.6	2.8	–8.4	18	24	14	58

a. Gross National Income (GNI) has replaced GNP in the World Bank Atlas Method's estimate of national income
b. Calculated using the World Bank Atlas method.
c. The estimate is based on regression; others are extrapolated from the latest International Comparison Programme benchmark estimates.
d. Estimated to be low income ($745 or less).
e. Estimated to be lower middle income ($746 to $2,975).
f. Estimated to be high income ($9,206 or more).
g. Estimated to be upper middle income ($2,976–to $9,205).
h.GNI and GNI per capita estimates include the French overseas departments of French Guiana, Guadeloupe, Martinique, and Reunion.
i. Data refer to mainland Tanzania only.

Sources: *World Development Indicators, 2003* (World Bank)

Table D
World Countries: Population Growth, 1950-2025

COUNTRY	POPULATION (thousands)			AVERAGE ANNUAL POPULATION CHANGE (percent)		AVERAGE ANNUAL INCREMENT TO THE POPULATION (mid-year population, in thousands)		
	1950	2000[a]	2025[a]	1975-1980	2001-2015[a]	1985-1990	1995-2000	2005-2010
WORLD	2,518,629.0	6,070,581.0	7,851,455.0	1.7	2.5	85,831.0		
AFRICA								
Algeria	8,753.0	30,245.0	42,429.0	3.1	1.5	631.8	566.0	539.3
Angola	4,131.0	12,386.0	19,268.0	2.7	2.6	130.2	187.2	267.5
Benin	2,046.0	6,222.0	11,120.0	2.5	2.4	135.6	184.8	204.6
Botswana	419.0	1,725.0	1,614.0	3.5	0.5	43.3	21.5	-15.4
Burkina Faso	3,960.0	11,905.0	24,527.0	2.5	2.1	233.8	307.0	360.3
Burundi	2,456.0	6,267.0	12,328.0	2.3	1.7	95.2	123.0	166.5
Cameroon	4,466.0	15,127.0	20,831.0	2.8	1.7	326.5	370.9	374.8
Central African Republic	1,314.0	3,715.0	5,193.0	2.3	1.5	57.5	62.0	60.0
Chad	2,658.0	7,861.0	15,770.0	2.1	2.9	171.2	259.8	340.4
Congo Democratic Republic	12,184.0	48,571.0	95,448.0	3.0	2.6	1,146.2	1,234.4	1,907.8
Congo Republic	808.0	3,447.0	6,750.0	2.9	2.7	56.3	62.4	67.2
Côte d'Ivoire	2,775.0	15,827.0	22,140.0	3.9	1.6	411.1	350.2	394.2
Egypt	21,834.0	67,784.0	103,165.0	2.4	1.5	1,318.5	1,199.9	1,125.0
Equatorial Guinea	226.0	456.0	812.0	-0.7	–	8.7	11.1	13.6
Eritrea	1,140.0	3,712.0	7,261.0	2.6	2.3	35.4	134.9	153.2
Ethiopia	18,434.0	65,590.0	116,006.0	2.4	2.1	1,530.7	1,670.9	1,848.4
Gabon	469.0	1,258.0	1,915.0	3.1	2.2	11.6	13.9	8.6
Gambia	294.0	1,312.0	2,177.0	3.1	2.0	33.0	42.3	48.0
Ghana	4,900.0	19,593.0	30,618.0	1.9	1.6	435.4	380.2	285.2
Guinea	2,550.0	8,117.0	13,704.0	1.5	1.9	173.7	63.1	193.9
Guinea-Bissau	505.0	1,367.0	2,774.0	4.7	2.2	22.1	28.4	35.1
Kenya	6,265.0	30,549.0	39,917.0	3.8	1.4	723.5	604.9	206.8
Lesotho	734.0	1,785.0	1,608.0	2.5	0.8	41.5	39.6	11.5
Liberia	824.0	2,943.0	6,081.0	3.1	2.3	-2.9	236.3	107.3
Libya	1,029.0	5,237.0	7,785.0	4.4	1.9	92.8	92.2	136.3
Madagascar	4,230.0	15,970.0	30,249.0	2.5	2.5	308.1	433.2	590.5
Malawi	2,881.0	11,370.0	18,245.0	3.3	1.8	416.6	169.9	103.8
Mali	3,520.0	11,904.0	25,679.0	2.1	2.1	164.0	305.1	390.7
Mauritania	825.0	2,645.0	4,973.0	2.5	2.3	47.5	65.2	94.9
Morocco	8,953.0	29,108.0	40,721.0	2.3	1.4	565.7	535.1	515.0
Mozambique	6,442.0	17,861.0	25,350.0	2.8	1.6	78.7	359.2	75.1
Namibia	511.0	1,894.0	2,350.0	2.7	1.2	58.6	33.5	7.5
Niger	2,500.0	10,742.0	25,722.0	3.2	2.8	207.6	259.7	322.0
Nigeria	29,790.0	114,746.0	192,115.0	2.8	1.9	2,530.7	3,225.5	3,161.6
Rwanda	2,162.0	7,724.0	12,509.0	3.3	1.6	187.9	311.1	49.1
Senegal	2,500.0	9,393.0	15,663.0	2.8	2.0	191.6	278.9	336.2
Sierra Leone	1,944.0	4,415.0	7,593.0	2.0	1.9	106.3	143.8	162.5
Somalia	2,264.0	8,720.0	20,978.0	7.0	3.1	45.8	192.4	266.1
South Africa	13,683.0	44,000.0	42,962.0	2.2	0.4	943.4	383.4	-419.6
Sudan	9,190.0	31,437.0	47,536.0	3.1	2.0	634.6	902.5	1,059.5
Tanzania	7,886.0	34,837.0	53,435.0	3.1	1.7	799.7	832.1	970.9
Togo	1,329.0	4,562.0	7,551.0	2.7	1.9	122.0	160.9	115.3
Tunisia	3,530.0	9,586.0	12,843.0	2.6	1.3	168.9	124.3	104.8

COUNTRY	POPULATION (thousands)			AVERAGE ANNUAL POPULATION CHANGE (percent)		AVERAGE ANNUAL INCREMENT TO THE POPULATION (mid-year population, in thousands)		
	1950	2000[a]	2025[a]	1975-1980	2001-2015[a]	1985-1990	1995-2000	2005-2010
Uganda	5,310.0	23,487.0	54,883.0	3.2	2.4	590.9	598.7	877.5
Zambia	2,440.0	10,419.0	14,401.0	3.4	1.2	210.9	174.1	190.9
Zimbabwe	2,744.0	12,650.0	12,857.0	3.0	0.6	308.9	78.1	-58.6
NORTH AND CENTRAL AMERICA						353.9		
Belize	69.0	240.0	356.0	1.7	–	5.0	6.3	7.2
Canada	13,737.0	30,769.0	36,128.0	1.2	0.6	369.8	331.8	289.5
Costa Rica	966.0	3,929.0	5,621.0	3.0	1.4	76.6	65.4	58.0
Cuba	5,850.0	11,202.0	11,479.0	0.9	0.3	93.2	48.4	37.3
Dominican Republic	2,353.0	8,353.0	10,955.0	2.4	1.3	141.1	136.4	147.1
El Salvador	1,951.0	6,209.0	8,418.0	2.1	1.6	87.1	110.8	117.7
Guatemala	2,969.0	11,428.0	19,456.0	2.5	2.4	255.9	317.6	366.3
Haiti	3,261.0	8,005.0	10,670.0	2.1	1.7	111.8	89.1	114.6
Honduras	1,380.0	6,457.0	10,115.0	3.4	2.1	117.2	151.1	134.9
Jamaica	1,403.0	2,580.0	3,263.0	1.2	1.1	18.3	16.8	23.9
Mexico	27,737.0	98,933.0	129,866.0	2.7	1.4	1,594.3	1,572.4	1,425.0
Nicaragua	1,134.0	5,073.0	8,318.0	3.1	2.1	91.0	107.6	100.9
Panama	860.0	2,950.0	4,290.0	2.5	1.3	44.8	39.8	32.5
Trinidad and Tobago	636.0	1,289.0	1,340.0	1.3	0.8	6.5	-4.9	-6.1
United States	157,813.0	285,003.0	358,030.0	0.9	0.8	2,296.3	2,503.8	2,429.2
SOUTH AMERICA								
Argentina	17,150.0	37,074.0	47,043.0	1.5	1.0	445.4	427.4	401.1
Bolivia	2,714.0	8,317.0	12,495.0	2.4	1.8	127.7	155.2	128.3
Brazil	53,975.0	171,796.0	216,372.0	2.4	1.1	2,756.3	1,975.6	1,285.4
Chile	6,082.0	15,224.0	19,651.0	1.5	1.0	212.2	189.7	148.3
Colombia	12,568.0	4,120.0	58,157.0	2.3	1.3	636.0	681.0	630.9
Ecuador	3,387.0	12,420.0	16,704.0	2.8	1.5	262.0	264.1	257.2
Guyana	423.0	759.0	724.0	0.7	–	-3.2	-3.5	4.1
Paraguay	1,488.0	5,470.0	9,173.0	3.2	2.1	113.5	141.6	162.8
Peru	7,632.0	25,952.0	35,622.0	2.7	1.3	472.9	491.4	432.4
Suriname	215.0	425.0	486.0	-0.5	–	3.9	3.3	1.4
Uruguay	2,239.0	3,342.0	3,875.0	0.6	0.6	19.4	23.7	26.7
Venezuela	5,094.0	24,277.0	31,189.0	3.4	1.5	465.5	397.3	351.7
ASIA								
Afghanistan	8,151.0	21,391.0	44,940.0	0.9	2.5	170.3	879.9	724.7
Armenia	1,354.0	3,112.0	2,866.0	1.8	0.3	-0.7	-13.8	7.6
Azerbaijan	2,896.0	8,157.0	10,222.0	1.6	0.7	103.6	23.6	61.8
Bangladesh	41,783.0	137,952.0	208,268.0	2.8	1.6	2,028.8	2,001.0	2,119.6
Bhutan	734.0	2,063.0	3,701.0	2.3	–	34.7	42.4	48.8
Cambodia	4,346.0	13,147.0	21,899.0	-1.8	1.5	313.1	271.6	314.6
China[b]	556,924.0	1,282,472.0	1,454,141.0	1.5	0.6	16,833.4	11,408.3	8,726.8
Georgia	3,527.0	5,262.0	4,429.0	0.7	-0.7	49.9	-53.5	-14.4
India	357,561.0	1,016,938.0	1,369,284.0	2.1	1.2	16,448.0	16,317.4	15,140.5
Indonesia	79,538.0	211,559.0	270,113.0	2.1	1.1	3,283.4	3,702.8	3,388.7
Iran	16,913.0	664,423.0	90,927.0	3.3	1.6	1,632.8	818.3	1,000.3

COUNTRY	POPULATION (thousands)			AVERAGE ANNUAL POPULATION CHANGE (percent)		AVERAGE ANNUAL INCREMENT TO THE POPULATION (mid-year population, in thousands)		
	1950	2000[a]	2025[a]	1975-1980	2001-2015[a]	1985-1990	1995-2000	2005-2010
Iraq	5,158.0	23,224.0	41,707.0	3.3	1.9	488.2	623.7	719.5
Israel	1,258.0	6,042.0	8,598.0	2.3	1.5	87.4	107.5	73.6
Japan	83,625.0	127,034.0	123,444.0	0.9	-0.2	556.6	252.5	-30.4
Jordan	472.0	5,035.0	8,116.0	2.3	2.2	126.9	159.4	145.2
Kazakhstan	6,703.0	15,640.0	15,388.0	1.1	0.1	148.5	-42.0	85.9
Korea, North	10,815.0	22,268.0	24,665.0	1.6	0.6	307.4	27.2	4,294.4
Korea, South	18,859.0	46,835.0	50,165.0	1.6	0.4	943.5	459.2	322.0
Kuwait	152.0	2,247.0	3,930.0	6.2	2.1	81.8	70.6	90.4
Kyrgyzstan	1,740.0	4,921.0	6,484.0	1.9	1.1	76.8	30.0	80.9
Laos	1,755.0	5,279.0	8,635.0	1.2	2.2	110.7	130.3	155.3
Lebanon	1,443.0	3,478.0	4,554.0	-0.7	1.2	11.8	48.7	46.0
Malaysia	6,110.0	23,001.0	33,479.0	2.3	1.5	391.7	436.4	438.2
Mongolia	761.0	2,500.0	3,368.0	2.8	1.3	62.1	37.4	44.4
Myanmar (Burma)	17,832.0	47,544.0	59,760.0	2.1	1.0	452.8	317.4	162.3
Nepal	8,643.0	23,518.0	37,831.0	2.5	2.0	457.5	559.0	616.3
Oman	456.0	2,609.0	4,785.0	5.0	2.2	58.3	80.5	104.3
Pakistan	39,659.0	142,654.0	249,766.0	2.6	2.2	2,984.4	2,984.8	2,936.8
Philippines	1,996.0	75,711.0	108,589.0	2.3	1.6	1,450.5	1,658.1	1,666.1
Saudi Arabia	3,201.0	22,147.0	39,751.0	5.6	2.9	527.8	678.3	922.2
Singapore	1,022.0	4,016.0	4,905.0	1.3	1.1	56.1	134.2	169.1
Sri Lanka	7,483.0	18,595.0	21,464.0	1.7	1.1	234.4	186.9	153.5
Syria	3,495.0	16,560.0	26,979.0	3.1	2.1	391.1	399.2	431.5
Tajikistan	1,532.0	6,089.0	8,193.0	2.8	1.5	149.0	115.3	168.7
Thailand	19,626.0	60,925.0	73,869.0	2.4	0.6	755.5	598.8	468.5
Turkey	21,484.0	68,281.0	88,995.0	2.1	1.1	1,083.1	895.5	732.4
Turkmenistan	1,211.0	4,643.0	6,549.0	2.5	1.1	85.4	83.3	95.8
United Arab Emirates	70.0	2,820.0	3,944.0	14.0	1.8	76.1	38.6	40.0
Uzbekistan	6,314.0	24,913.0	33,774.0	2.6	1.3	473.1	381.7	485.8
Vietnam	27,369.0	78,137.0	104,649.0	2.2	1.2	1,321.7	1,178.3	1,123.4
Yemen	4,316.0	18,017.0	43,204.0	3.2	3.0	436.3	524.0	782.1
EUROPE								
Albania	1,215.0	3,113.0	3,629.0	1.9	1.0	60.3	50.7	34.4
Austria	6,935.0	8,102.0	7,979.0	-0.1	-0.1	32.1	17.8	11.3
Belarus	7,745.0	10,034.0	8,950.0	0.6	-0.5	46.7	-7.5	-1.4
Belgium	8,639.0	10,251.0	10,516.0	0.1	0.0	22.2	20.9	5.3
Bosnia-Herzegovina	2,661.0	3,977.0	4,183.0	0.9	0.5	29.7	96.0	15.5
Bulgaria	7,251.0	8,099.0	6,609.0	0.3	-0.7	-9.9	-95.1	-74.3
Croatia	3,850.0	4,446.0	4,088.0	0.5	-0.3	10.1	-34.6	11.2
Czech Republic	8,925.0	10,269.0	9,806.0	0.6	-0.2	-0.1	-10.6	-14.7
Denmark	4,271.0	5,322.0	5,469.0	0.2	0.1	5.5	20.8	12.0
Estonia	1,101.0	1,369.0	1,017.0	0.6	-0.5	7.0	-10.5	-4.9
Finland	4,009.0	5,177.0	5,289.0	0.3	0.1	16.9	12.3	4.8
France	41,829.0	59,296.0	64,165.0	0.4	0.3	312.8	236.0	142.8
Germany	68,376.0	82,282.0	81,959.0	-0.1	-0.2	339.1	229.8	152.4
Greece	7,566.0	10,903.0	10,707.0	1.3	0.1	44.5	22.4	11.0
Hungary	9,338.0	10,012.0	8,865.0	0.3	-0.6	-55.4	-31.4	-31.1

COUNTRY	POPULATION (thousands)			AVERAGE ANNUAL POPULATION CHANGE (percent)		AVERAGE ANNUAL INCREMENT TO THE POPULATION (mid-year population, in thousands)		
	1950	2000[a]	2025[a]	1975-1980	2001-2015[a]	1985-1990	1995-2000	2005-2010
Iceland	143.0	282.0	325.0	0.9	-	2.7	1.8	1.1
Ireland	2,969.0	3,819.0	4,668.0	1.4	0.8	-6.4	37.2	31.9
Italy	47,104.0	57,536.0	52,939.0	0.4	-0.4	4.0	74.2	-67.5
Latvia	1,949.0	2,373.0	1,857.0	0.4	-0.7	12.3	-23.5	-12.9
Lithuania	2,567.0	3,501.0	3,035.0	0.7	-0.2	22.2	-10.4	-3.6
Macedonia	1,230.0	2,024.0	2,199.0	1.4	0.4	6.9	11.0	7.2
Moldova	2,341.0	4,283.0	4,096.0	0.9	-0.2	49.9	-5.8	16.0
Netherlands	10,114.0	15,898.0	17,123.0	0.7	0.4	92.0	86.6	62.5
Norway	3,265.0	4,473.0	4,859.0	0.4	0.4	17.9	24.4	18.2
Poland	24,824.0	38,671.0	37,337.0	0.9	0.0	178.7	8.5	11.2
Portugal	8,405.0	10,016.0	9,834.0	1.4	-0.1	5.1	15.9	9.6
Romania	16,311.0	22,480.0	20,806.0	0.9	-0.3	69.0	-56.3	-49.8
Russian Federation	102,192.0	145,612.0	124,428.0	0.6	-0.5	820.8	-422.7	-281.7
Serbia-Montenegro	7,131.0	10,555.0	10,230.0	0.9	0.1	21.0	-5.2	-4.6
Slovak Republic	3,463.0	5,391.0	5,397.0	1.0	0.0	23.6	9.3	6.1
Slovenia	1,473.0	1,990.0	1,859.0	1.0	-0.2	4.6	3.6	1.2
Spain	28,009.0	40,752.0	40,369.0	1.1	0.0	163.2	49.1	-2.8
Sweden	7,014.0	8,856.0	9,055.0	0.3	0.0	40.5	9.5	0.5
Switzerland	4,694.0	7,173.0	6,801.0	-0.1	-0.1	54.8	19.2	7.9
Ukraine	37,298.0	49,688.0	40,775.0	0.4	-0.7	142.7	-432.6	-264.4
United Kingdom	49,816.0	58,689.0	63,275.0	0.0	0.0	189.0	178.9	94.6
OCEANIA								
Australia	8,219.0	19,153.0	23,205.0	0.9	0.7	246.8	209.7	167.0
Fiji	289.0	814.0	965.0	1.9	–	7.8	11.3	12.8
New Zealand	1,908.0	3,784.0	4,379.0	0.2	0.5	12.3	50.8	38.5
Papua New Guinea	1,798.0	5,334.0	8,443.0	2.5	1.9	89.3	116.4	125.1
Solomon Islands	90.0	437.0	783.0	3.5	–	11.0	13.8	14.3

a Data include projections based on 1990 base year population data
b Includes Hong Kong and Macao

Source: United Nations Population Division and International Labour Organisation. *World Resources 2000-2001;* (World Resources Institute), U.S. Bureau of the Census International Data Base (2000).

Table E
World Countries: Basic Demographic Data, 1975-2001

COUNTRY	CRUDE BIRTH RATE (births per 1,000 population)		LIFE EXPECTANCY AT BIRTH (years)		LIFE EXPECTANCY OF FEMALES AS A PERCENTAGE OF MALES (years)		TOTAL FERTILITY RATE		PERCENTAGE OF POPULATION IN SPECIFIC AGE GROUPS					
									1980			2001		
	1975-1980	2001	1975-80	2001	1975-80	1995-00	1975-80	2001	<15	15-65	>65	<15	15-65	>65
WORLD	28.3	21	59.7	67	106.0	107.9	3.9	2.6	35.2	58.9	5.9	29.6	63.4	7.0
AFRICA	46.0	39	47.9	46	106.7	105.6	6.5	–	44.7	52.2	3.1			
Algeria	45.0	23	57.5	71	103.5	102.9	7.2	2.9	46.5	49.6	3.9	35.4	60.7	3.9
Angola	50.2	47	40.0	47	108.1	106.6	6.8	6.6	44.7	52.4	2.9	47.4	49.7	2.9
Benin	51.4	39	47.0	53	108.2	105.7	7.1	5.4	45.1	50.8	4.1	45.9	51.4	2.7
Botswana	46.6	31	56.5	39	106.6	104.3	6.4	3.9	48.7	49.3	2.0	42.0	56.7	2.2
Burkina Faso	50.8	44	42.9	44	106.0	102.2	7.8	6.4	47.4	49.8	2.8	47.1	50.2	2.7
Burundi	44.7	39	46.0	42	107.2	107.3	6.8	5.9	44.7	51.8	3.4	46.0	51.4	2.6
Cameroon	45.5	36	48.5	49	106.4	105.6	6.5	4.7	44.4	52.0	3.6	41.6	54.8	3.7
Central African Republic	44.1	36	44.5	44	111.9	109.3	5.9	4.7	41.7	54.4	4.0	42.3	54.2	3.5
Chad	44.1	45	41.0	48	108.1	106.5	5.9	6.3	41.9	54.5	3.6	49.6	47.4	3.0
Dem. Republic of the Congo (formerly Zaire)	47.8	45	48.0	45	107.1	106.1	6.5	6.1	46.0	51.1	2.8	47.6	49.8	2.6
Congo Republic	45.8	42	48.7	51	111.3	110.8	6.3	5.9	45.1	51.5	3.4	46.3	50.5	3.2
Côte d'Ivoire	50.7	37	47.9	46	107.1	102.0	7.4	4.7	46.6	51.0	2.5	42.1	55.3	2.6
Egypt	38.9	25	54.1	68	104.5	104.6	5.3	3.2	39.5	56.5	4.0	34.7	61.1	4.2
Equatorial Guinea	42.7	–	42.0	–	107.9	108.3	5.7	–	41.0	54.8	4.1	–	–	–
Eritrea	45.1	38	45.3	51	106.8	106.1	6.1	5.4	44.2	53.1	2.6	45.2	52.2	2.6
Ethiopia	49.0	43	42.0	42	107.9	104.7	6.8	5.6	46.1	51.2	2.7	46.0	51.2	2.8
Gabon	32.9	35	47.0	53	107.3	105.8	4.4	4.1	34.4	59.5	6.1	40.3	54.1	5.7
The Gambia	48.8	38	39.0	53	108.3	108.8	6.5	4.9	42.6	54.4	2.8	40.3	56.5	3.2
Ghana	45.1	29	51.0	56	106.9	106.9	6.5	4.1	44.9	52.3	2.8	43.3	52.2	4.6
Guinea	51.6	38	38.8	46	102.6	102.1	7.0	5.1	45.8	51.6	2.6	44.3	53.1	2.6
Guinea-Bissau	42.4	40	37.5	45	108.6	106.9	5.6	5.7	39.0	57.0	4.0	43.6	52.9	3.6
Kenya	53.6	35	53.4	46	107.8	103.9	8.1	4.3	50.1	46.5	3.4	43.0	54.3	2.7
Lesotho	41.9	32	51.8	43	107.0	103.6	5.7	4.3	41.9	53.9	4.2	39.7	56.1	4.3
Liberia	47.4	44	49.5	47	106.3	106.5	6.8	5.9	44.3	52.0	3.7	44.5	52.8	2.8
Libya	47.3	27	55.8	72	106.3	105.8	7.4	3.4	46.7	51.1	2.2	33.6	62.9	3.5
Madagascar	46.9	39	49.5	55	106.3	105.3	6.6	5.3	45.9	51.4	2.7	44.8	52.2	3.0
Malawi	57.2	45	43.1	38	103.3	102.5	7.6	6.2	47.5	50.3	2.3	44.4	52.0	3.6
Mali	50.7	46	40.0	41	108.1	105.8	7.1	6.2	46.8	50.7	2.5	47.1	49.9	3.0
Mauritania	44.7	41	45.5	51	107.3	105.8	6.5	4.5	43.7	53.3	3.0	43.9	52.9	3.2
Mauritius	26.7	–	64.9	–	108.3	–	3.1	–	35.6	60.8	3.6	25.5	68.3	6.2
Morocco	39.4	22	55.8	68	106.3	106.1	5.9	2.8	43.2	52.7	4.1	34.1	61.7	4.2
Mozambique	45.4	40	43.5	42	107.6	106.8	6.5	5.1	43.4	53.4	3.2	42.8	53.5	3.7
Namibia	41.9	35	51.3	44	105.0	101.9	6.0	4.9	43.1	53.4	3.5	41.7	54.6	3.8
Niger	59.7	50	40.5	46	108.2	106.4	8.1	7.2	46.8	50.8	2.5	49.0	48.7	2.4
Nigeria	46.6	39	45.0	46	107.4	106.1	6.5	5.2	44.3	3.1	2.6	43.9	53.5	2.6
Rwanda	52.8	44	45.0	40	107.4	107.7	8.5	5.8	48.8	48.8	2.4	47.6	49.3	3.1
Senegal	49.3	36	42.8	52	104.8	105.8	7.0	5.0	45.3	51.8	2.8	44.4	52.9	2.7
Sierra Leone	48.8	44	35.2	37	108.9	108.3	6.5	5.7	43.0	53.9	3.1	44.6	52.9	2.6
Somalia	50.4	50	42.0	47	107.9	108.8	7.0	7.0	46.0	51.0	3.0	47.9	49.7	2.4
South Africa	37.3	25	55.9	47	111.3	111.5	5.1	2.8	40.3	55.9	3.9	32.3	63.1	4.6
Sudan	47.1	34	46.7	58	106.2	103.7	6.7	4.5	44.9	52.4	2.7	39.9	56.6	3.5
Swaziland	46.2	35	49.9	45	109.9	–	6.5	4.3	45.9	51.3	2.9	42.2	55.0	2.8
Tanzania	47.5	39	49.0	44	107.2	104.2	6.8	5.2	47.6	50.1	2.3	45.2	52.4	2.4
Togo	45.2	35	48.0	49	107.1	104.1	6.6	5.0	44.6	52.3	3.2	43.9	53.0	3.2
Tunisia	36.4	17	60.0	72	101.7	104.4	5.7	2.1	41.6	54.6	3.8	28.9	65.1	5.9

COUNTRY	CRUDE BIRTH RATE (births per 1,000 population)		LIFE EXPECTANCY AT BIRTH (years)		LIFE EXPECTANCY OF FEMALES AS A PERCENTAGE OF MALES (years)		TOTAL FERTILITY RATE		PERCENTAGE OF POPULATION IN SPECIFIC AGE GROUPS					
									1980			2001		
	1975-1980	2001	1975-80	2001	1975-80	1995-00	1975-80	2001	<15	15-65	>65	<15	15-65	>65
Uganda	50.3	45	47.0	43	107.0	102.5	6.9	6.1	47.8	49.7	2.5	49.0	49.1	1.9
Zambia	51.6	39	49.3	37	106.9	102.5	7.2	5.2	49.4	48.2	2.4	45.1	52.7	2.2
Zimbabwe	44.2	29	53.8	39	106.9	102.3	6.6	3.7	47.9	49.5	2.6	44.6	52.2	3.2
NORTH AMERICA			73.3		111.2	108.7	1.8		22.5	66.4	11.0			
Canada	15.4	11	74.2	79	110.8	107.8	1.8	1.5	22.7	67.9	9.4	18.8	68.6	12.6
United States	15.1	15	73.2	78	111.2	109.6	1.8	2.1	22.5	66.3	11.2	21.2	66.2	12.6
CENTRAL AMERICA			63.7		110.2	106.7	5.4		45.1	51.2	3.6			
Belize	40.9	–	69.7	–	102.5	104.1	6.2	–	47.2	48.6	4.8	–	–	–
Costa Rica	31.7	20	71.0	78	106.4	106.7	3.9	2.4	38.9	57.5	3.6	31.2	63.1	5.7
Cuba	17.2	12	73.0	77	104.8	105.4	2.1	1.6	32.9	60.5	7.6	21.1	68.8	10.1
Dominican Republic	34.9	23	62.0	67	106.1	105.8	4.7	2.7	42.3	54.6	3.1	33.0	62.7	4.4
El Salvador	41.5	26	57.2	70	119.5	108.9	5.7	3.0	46.1	50.8	3.1	35.3	59.7	5.0
Guatemala	44.3	33	56.4	65	107.2	109.8	6.4	4.4	45.9	51.3	2.8	43.2	53.3	3.5
Haiti	36.8	32	50.7	52	106.3	109.8	5.4	4.3	40.7	54.8	4.4	40.2	56.3	3.5
Honduras	44.9	30	57.7	66	107.7	105.9	6.6	4.1	47.2	50.1	2.7	41.1	55.2	3.4
Jamaica	28.8	21	70.1	76	106.3	105.5	4.0	2.4	40.2	53.0	6.8	30.6	62.4	7.0
Mexico	37.1	24	65.3	73	110.3	107.1	5.3	2.5	45.1	51.1	3.8	33.6	61.4	5.0
Nicaragua	45.7	30	57.6	69	108.5	107.6	6.4	3.5	47.7	49.8	2.5	42.1	54.9	3.1
Panama	31.0	21	69.0	75	105.8	105.5	4.1	2.5	40.5	55.0	4.5	30.9	63.5	5.6
Trinidad and Tobago	29.3	15	68.3	72	107.6	105.5	3.4	1.8	34.2	60.2	5.5	24.9	68.8	6.3
SOUTH AMERICA			62.9		108.8	109.0	4.3		37.8	57.6	4.6			
Argentina	25.7	19	68.7	74	110.4	110.0	3.4	2.5	30.5	61.4	8.1	27.5	62.8	9.7
Bolivia	41.0	30	50.1	63	108.8	105.0	5.8	3.8	42.6	53.9	3.5	39.1	56.5	4.4
Brazil	32.6	19	61.8	68	108.1	112.7	4.3	2.2	38.1	57.8	4.2	28.4	66.4	5.2
Chile	24.0	17	67.2	76	110.5	108.3	3.0	2.1	33.5	60.9	5.6	27.8	65.0	7.2
Colombia	31.7	22	64.0	72	107.3	110.4	4.1	2.5	40.0	56.2	3.7	32.4	62.9	4.7
Ecuador	38.2	24	61.4	70	105.9	107.5	5.4	2.9	42.8	53.2	4.0	33.4	61.9	4.7
Guyana	31.5	–	60.7	–	108.4	111.4	3.9	–	40.8	55.2	4.0	–	–	–
Paraguay	35.9	30	66.6	71	106.7	107.4	5.2	3.9	42.2	53.3	4.5	39.1	57.3	3.5
Peru	38.0	24	58.5	70	106.7	107.6	5.4	2.7	41.9	54.5	3.6	32.9	62.2	4.9
Suriname	29.5	–	65.1	–	107.8	107.3	4.2	–	39.7	55.8	4.5	–	–	–
Uruguay	20.3	16	69.6	74	110.2	111.4	2.9	2.2	26.9	62.5	10.5	24.7	62.7	12.6
Venezuela	34.2	23	67.6	74	109.1	108.6	4.5	2.8	40.7	56.1	3.2	33.5	62.1	4.4
ASIA			58.5		102.6	106.2	4.2		37.6	58.0	4.4			
Afghanistan	50.8	48	40.0	43	100.0	102.2	7.2	6.8	43.0	54.5	2.5	43.7	53.5	2.8
Armenia	20.9	11	72.3	74	109.0	110.0	2.5	1.4	30.4	63.6	6.0	23.0	67.7	9.3
Azerbaijan	24.7	16	68.5	65	112.1	110.4	3.6	2.1	34.5	60.0	5.4	28.4	64.4	7.2
Bangladesh	47.2	28	46.6	62	97.9	100.0	6.7	3.0	46.0	50.5	3.4	37.0	59.7	3.3
Bhutan	42.8	–	42.6	–	104.8	103.3	5.9	–	41.3	55.6	3.1	–	–	–
Cambodia	30.0	29	31.2	54	108.3	107.8	4.1	3.9	39.2	57.9	2.9	43.0	54.2	2.8
China	21.5	15	65.3	70	103.0	105.8	3.3	1.9	35.5	59.7	4.7	24.8	68.1	7.1
Georgia	18.1	8	70.7	73	111.9	111.6	2.4	1.1	25.7	65.1	9.1	20.1	66.8	13.1
India	34.7	25	52.9	63	96.3	101.6	4.8	3.0	38.5	57.4	4.0	33.1	61.9	5.0
Indonesia	35.4	21	52.8	66	104.9	106.3	4.7	2.4	41.0	55.6	3.3	30.2	65.1	4.7

COUNTRY	CRUDE BIRTH RATE (births per 1,000 population)		LIFE EXPECTANCY AT BIRTH (years)		LIFE EXPECTANCY OF FEMALES AS A PERCENTAGE OF MALES (years)		TOTAL FERTILITY RATE		PERCENTAGE OF POPULATION IN SPECIFIC AGE GROUPS					
									1980			2001		
	1975-1980	2001	1975-80	2001	1975-80	1995-00	1975-80	2001	<15	15-65	>65	<15	15-65	>65
Iran	44.7	22	58.6	69	101.4	101.4	6.5	2.6	44.9	51.8	3.3	32.6	62.7	4.7
Iraq	41.9	30	61.4	62	103.0	104.9	6.6	4.2	46.0	51.3	2.7	40.9	56.2	2.9
Israel	26.0	21	73.1	79	104.9	105.2	3.4	2.8	33.2	58.2	8.6	27.6	62.6	9.7
Japan	15.2	9	75.5	81	107.4	107.8	1.8	1.4	23.6	67.4	9.0	14.5	67.9	17.6
Jordan	45.0	29	61.2	72	106.1	104.3	7.4	3.6	49.4	47.5	3.1	38.2	58.7	3.1
Kazakhstan	24.9	15	65.4	63	117.1	114.2	3.1	1.8	32.4	61.5	6.1	26.1	66.4	7.5
Korea, North	"22.2	18	65.8	61	110.3	108.7	3.3	2.1	39.5	57.0	3.5	27.1	67.1	5.8
Korea, South	"23.9	13	64.8	74	111.6	110.1	2.9	1.4	34.0	62.2	3.8	21.3	71.8	7.0
Kuwait	40.1	20	69.6	77	106.2	105.4	5.9	2.6	40.2	58.3	1.4	32.2	65.5	2.3
Kyrgyzstan	29.9	20	64.2	66	114.2	114.2	4.1	2.5	37.1	57.1	5.8	33.4	60.6	6.0
Laos	45.1	36	43.5	54	106.9	105.7	6.7	4.9	42.0	55.2	2.8	42.4	54.1	3.5
Lebanon	30.1	20	65.0	71	106.2	105.8	4.3	2.3	40.1	54.5	5.4	31.5	62.6	5.9
Malaysia	30.4	22	65.3	55	105.7	105.7	4.2	2.9	39.3	57.0	3.7	33.7	62.1	4.2
Mongolia	39.2	22	56.3	65	104.5	104.6	6.7	2.5	43.2	53.9	2.9	33.2	62.7	4.0
Myanmar (Burma)	37.5	24	51.2	57	106.4	105.0	5.3	2.9	39.6	56.4	4.0	32.7	62.7	4.6
Nepal	43.4	32	46.2	59	96.6	98.2	6.2	4.2	42.9	54.1	3.0	40.7	55.5	3.8
Oman	46.1	27	54.9	74	104.3	105.8	7.2	4.1	44.6	52.8	2.6	43.2	54.2	2.5
Pakistan	47.3	33	53.4	63	101.3	103.1	7.0	4.6	44.4	52.7	2.9	41.2	55.5	3.3
Philippines	35.9	26	59.9	70	105.5	104.4	5.0	3.3	41.9	55.3	2.8	36.9	59.2	3.9
Saudi Arabia	45.9	33	58.8	73	104.0	104.2	7.3	5.9	44.3	52.9	2.8	40.9	56.2	2.9
Singapore	17.2	–	70.8	78	106.6	105.3	1.9	1.4	27.0	68.2	4.7	21.5	71.2	7.3
Sri Lanka	28.5	18	66.8	73	105.4	105.6	3.8	2.1	35.3	60.4	4.3	26.0	67.6	6.4
Syria	46.0	29	60.1	70	106.2	105.9	7.4	3.6	48.5	48.3	3.2	40.0	56.9	3.1
Tajikistan	37.2	21	64.5	67	108.3	109.3	5.9	3.0	42.9	52.5	4.6	38.6	57.0	4.4
Thailand	31.6	15	61.2	69	106.6	109.0	4.3	1.8	40.0	56.5	3.5	23.6	70.1	6.3
Turkey	32.0	20	60.3	70	107.8	107.4	4.5	2.3	39.2	56.0	4.7	28.3	65.9	5.8
Turkmenistan	35.3	20	61.6	65	11.2	111.2	5.3	2.3	41.3	54.4	4.3	36.6	59.1	4.3
United Arab Emirates	30.5	17	66.8	75	106.5	102.7	5.7	3.1	28.6	70.1	1.3	26.1	71.2	2.7
Uzbekistan	34.4	21	65.1	67	111.0	110.9	5.1	2.5	40.9	54.0	5.1	36.5	59.0	4.5
Vietnam	38.3	19	55.8	69	108.2	107.7	5.6	2.2	42.5	52.7	4.8	32.4	62.3	5.3
Yemen	53.7	41	44.1	57	101.1	101.7	7.6	6.1	50.2	47.2	2.6	46.2	51.0	2.8
EUROPE			71.3		111.4	110.6	2.0		22.2	65.5	12.3			
Albania	30.3	17	68.9	74	106.6	108.6	4.2	2.2	35.9	58.9	5.2	28.6	64.5	6.9
Austria	11.5	9	72.0	78	110.4	108.1	1.7	1.3	20.4	64.2	15.4	16.5	67.8	15.7
Belarus	16.1	9	71.1	68	114.7	119.3	2.1	1.3	22.9	66.4	10.7	18.0	68.3	13.6
Belgium	12.4	11	72.3	78	109.6	109.4	1.7	1.6	20.2	65.5	14.3	17.2	66.1	16.6
Bosnia-Herzegovina	19.6	12	69.9	74	107.0	107.0	2.2	1.6	27.3	66.7	6.1	18.6	71.3	10.2
Bulgaria	15.8	9	71.3	72	107.9	110.2	2.2	1.3	22.1	66.0	11.9	15.2	68.6	16.2
Croatia	15.6	9	70.6	74	110.4	111.6	2.0	1.4	21.2	67.2	11.7	16.7	68.1	15.2
Czech Republic	17.4	9	70.6	75	110.4	111.0	2.3	1.6	23.5	63.2	13.4	16.1	70.1	13.8
Denmark	12.3	12	74.2	77	108.4	106.8	1.7	1.8	20.8	64.7	14.4	18.4	66.7	14.9
Estonia	15.0	9	69.7	71	115.3	119.0	2.1	1.2	21.7	65.8	12.5	17.1	68.0	14.9
Finland	13.6	11	72.7	78	112.6	110.9	1.6	1.7	20.3	67.7	12.0	17.9	67.0	15.0
France	14.0	13	73.7	79	111.6	110.8	1.9	1.9	22.3	63.8	14.0	18.7	65.1	16.1
Germany	10.4	9	72.5	78	109.4	108.1	1.5	1.4	18.5	65.9	15.6	15.3	68.3	16.4
Greece	15.6	11	73.7	78	105.7	106.5	2.3	1.3	22.8	64.0	13.1	14.9	67.0	18.1
Hungary	16.2	10	69.4	72	109.8	111.9	2.1	1.3	21.9	64.6	13.4	16.8	68.7	14.5
Iceland	19.2	–	76.3	–	108.0	105.1	2.3	–	27.6	62.7	10.1	–	–	–

COUNTRY	CRUDE BIRTH RATE (births per 1,000 population)		LIFE EXPECTANCY AT BIRTH (years)		LIFE EXPECTANCY OF FEMALES AS A PERCENTAGE OF MALES (years)		TOTAL FERTILITY RATE		PERCENTAGE OF POPULATION IN SPECIFIC AGE GROUPS					
									1980			2001		
	1975-1980	2001	1975-80	2001	1975-80	1995-00	1975-80	2001	<15	15-65	>65	<15	15-65	>65
Ireland	21.2	15	72.0	77	107.2	106.7	3.5	1.9	30.6	58.7	10.7	21.5	67.2	11.2
Italy	13.0	9	73.6	79	109.2	108.0	1.9	1.2	22.3	64.6	13.1	14.2	67.4	18.4
Latvia	14.0	8	69.2	70	115.6	119.3	2.0	1.2	20.4	66.5	13.0	16.5	68.6	14.9
Lithuania	15.4	9	70.8	73	114.2	118.7	2.1	1.3	23.6	65.1	11.3	18.9	67.4	13.7
Macedonia	22.6	13	69.6	73	105.2	105.6	2.7	1.8	28.6	64.5	6.9	22.2	67.5	10.2
Moldova	19.6	9	64.8	67	111.2	112.5	2.4	1.4	26.7	65.5	7.8	21.8	67.1	11.0
Netherlands	12.7	13	75.3	78	109.0	108.0	1.6	1.7	22.3	66.2	11.5	18.5	67.9	13.7
Norway	12.9	13	75.3	79	108.9	108.0	1.8	1.8	22.2	63.1	14.8	20.0	65.0	15.1
Poland	19.2	10	70.9	74	111.9	113.2	2.3	1.3	24.3	65.6	10.1	18.8	69.0	12.2
Portugal	18.2	11	70.2	76	110.6	109.7	2.4	1.5	25.9	63.6	10.5	17.1	67.6	15.2
Romania	19.1	10	69.6	70	107.0	112.1	2.6	1.3	26.7	63.1	10.3	17.7	68.8	13.5
Russian Federation	15.8	9	67.4	66	118.1	119.6	1.9	1.2	21.6	68.1	10.2	17.5	69.9	12.5
Serbia-Montenegro	18.4	12	70.3	73	106.5	107.1	2.4	1.7	24.1	66.1	9.8	20.1	66.2	13.7
Slovak Republic	20.6	10	70.4	73	110.9	111.5	2.5	1.3	26.1	63.5	10.4	19.3	69.3	11.3
Slovenia	16.6	9	71.0	76	111.6	109.8	2.2	1.2	23.4	65.3	11.4	15.7	70.2	14.2
Spain	17.4	10	74.3	78	108.4	109.3	2.6	1.2	26.6	62.7	10.7	15.0	68.1	16.9
Sweden	11.7	10	75.2	80	109.3	106.6	1.7	1.6	19.6	64.1	16.3	17.9	64.6	17.5
Switzerland	11.6	10	75.2	80	109.2	109.3	1.5	1.4	19.7	66.4	13.8	16.8	67.8	15.4
Ukraine	14.9	8	69.3	68	114.8	115.6	2.0	1.2	21.4	66.7	11.9	17.1	68.6	14.3
United Kingdom	12.4	11	72.8	77	109.0	106.6	1.7	1.7	20.9	64.0	15.1	18.6	65.4	16.1
OCEANIA			68.2		108.7	106.7	2.8		29.3	62.7	8.0			
Australia	16.0	13	73.4	79	109.8	108.0	2.1	1.8	25.3	65.1	9.6	20.5	67.5	12.0
Fiji	33.2	–	67.1	–	105.3	105.6	4.0	–	39.1	58.0	2.8	–	–	–
New Zealand	17.2	15	72.4	78	109.2	108.1	2.2	2.0	26.7	63.3	10.0	22.3	65.9	11.8
Papua New Guinea	39.6	32	49.7	57	101.0	103.5	5.9	4.4	43.0	55.5	1.6	39.9	57.6	2.5
Solomon Islands	45.0	–	65.3	–	106.2	105.7	7.1	–	47.6	49.3	3.1	–	–	–
DEVELOPING COUNTRIES			56.7		103.8	104.8	4.7		39.3	56.6	4.1			
DEVELOPED COUNTRIES			72.2		110.8	111.2	1.9		22.5	65.9	11.6			

Sources: United Nations Population Division; *World Resources 2000-2001* (World Resources Institute); *World Development Indicators,* 2003 (World Bank).

Table F
World Countries: Mortality, Health, and Nutrition

COUNTRY	MORTALITY						HEALTH			NUTRITION	
	Infant Mortality		Adult Mortality								
	Infant Mortality (per 1,000 live births)	Under-five Mortality (per 1,000)	Male (per 1,000)		Female (per 1,000)		Health Expenditure per Capita ($US) 1997-2000[a]	Physicians per 1,000 People 1995-2000[a]	Hospital Beds per 1,000 people 1995-2000[a]	Average daily per capita supply of calories (kilocalories)[c] 1987	Average daily per capita supply of calories (kilocalories)[c] 1997
	2001	2001	1980	2000-2001	1980	2000-2001					
Afghanistan	–	257	–	437	–	376	8	0.1	–	–	–
Albania	23	25	140	209	82	95	41	1.3	3.2	2,556	2,961
Algeria	39	49	226	155	197	119	64	1.0	2.1	2,757	2,853
Angola	154	260	569	492	458	386	24	0.1	–	2,063	2,183
Argentina	16	19	205	184	102	92	658	2.7	3.3	3,096	3,093
Armenia	31	35	158	223	85	106	38	3.2	0.7	–	2,371
Australia	6	6	178	100	53	55	1,698	2.5	7.9	3,159	3,224
Austria	5	5	197	124	92	59	1,872	3.1	8.6	3,419	3,536
Azerbaijan	77	96	262	261	127	150	8	3.6	9.7	–	2,236
Bangladesh	51	77	383	262	388	252	14	0.2	–	2,062	2,086
Belarus	17	20	255	381	95	133	57	4.4	12.2	–	3,226
Belgium	5	6	173	127	90	66	1,936	3.9	7.3	3,454	3,619
Benin	94	158	486	384	397	328	11	0.1	–	1,961	2,487
Bolivia	60	77	357	264	273	219	67	1.3	1.7	2,153	2,174
Bosnia- Herzegovina	15	18	181	200	108	93	50	1.4	1.8	–	2,266
Botswana	80	110	341	703	278	669	191	–	–	2,337	2,183
Brazil	31	36	221	259	161	136	267	1.3	3.1	2,745	2,974
Bulgaria	14	16	190	239	106	109	59	3.4	7.4	3,699	2,686
Burkina Faso	104	197	467	559	362	507	8	0.0[b]	1.4	2,181	2,121
Burundi	114	190	489	648	400	603	3	–	–	2,023	1,685
Cambodia	97	138	473	373	355	264	19	0.3	–	1,868	2,048
Cameroon	96	155	489	488	415	440	24	0.1	–	2,178	2,111
Canada	5	7	161	101	85	57	2,058	6.8	3.9	3,105	3,119
Central African Republic	115	180	540	620	424	573	8	0.0b	0.9	1,914	2,016
Chad	117	200	556	449	449	361	6	–	–	1,596	2,032
Chile	10	12	218	151	120	67	336	1.1	2.7	2,518	2,796
China	31	39	185	161	148	110	45	1.7	2.4	2,608[d]	2,897[d]
Hong Kong, China	3	–	150	99	87	51	–	1.3	–	–	–
Colombia	19	23	237	238	162	115	186	1.2	1.5	2,341	2,597
Democratic Republic of the Congo (formerly Zaire)	129	205	–	571	–	493	9	0.1	–	2,132	1,755
Congo, Rep.	81	108	408	475	298	406	22	0.3	–	2,326	2,144
Costa Rica	9	11	159	131	100	78	273	0.9	1.7	2,717	2,649
Côte d'Ivoire	102	175	421	553	346	494	16	0.1	–	2,677	2,610
Croatia	7	8	233	152	106	114	434	2.3	–	–	2,445
Cuba	7	9	135	143	94	94	169	5.3	5.1	3,125	2,480
Czech Republic	4	5	225	167	102	75	358	3.1	8.8	–	3,244
Denmark	8	4	163	129	102	81	2,512	3.4	4.5	3,211	3,407
Dominican Republic	41	47	183	234	138	146	151	2.2	1.5	2,330	2,288
Ecuador	24	30	229	199	176	120	26	1.7	1.6	2,430	2,679
Egypt	35	41	257	210	204	147	51	1.6	2.1	3,120	3,287
El Salvador	33	39	410	250	178	148	184	1.1	1.6	2,312	2,562
Eritrea	72	111	–	493	–	441	9	0.0[b]	–	–	1,622
Estonia	11	12	291	316	110	114	218	3.0	7.4	–	2,849
Ethiopia	116	172	491	594	401	535	5	–	–	1,677	1,858

–183–

COUNTRY	MORTALITY						HEALTH			NUTRITION	
	Infant Mortality		Adult Mortality								
	Infant Mortality (per 1,000 live births)	Under-five Mortality (per 1,000)	Male (per 1,000)		Female (per 1,000)		Health Expenditure per Capita ($US) 1997-2000[a]	Physicians per 1,000 People 1995-2000[a]	Hospital Beds per 1,000 people 1995-2000[a]	Average daily per capita supply of calories (kilocalories)[c] 1987	Average daily per capita supply of calories (kilocalories)[c] 1997
	2001	2001	1980	2000-2001	1980	2000-2001					
Finland	4	5	206	144	74	61	1,559	3.1	7.5	2,941	3,100
France	4	6	190	137	85	59	2,057	3.0	8.2	3,543	3,518
Gabon	60	90	474	380	387	330	120	–	–	2,460	2,556
The Gambia	91	126	584	373	466	320	10	0.0[b]	–	2,498	2,350
Georgia	24	29	210	250	94	133	41	4.4	4.8	–	2,614
Germany	4	5	177	126	90	60	2,422	3.6	9.1	3,478	3,382
Ghana	57	100	400	379	334	326	11	0.1	–	1,979	2,611
Greece	5	5	134	114	86	47	884	4.4	4.9	3,481	3,649
Guatemala	43	58	336	286	266	182	79	0.9	1.0	2,384	2,339
Guinea	109	169	589	432	507	366	13	0.1	–	2,060	2,232
Guinea-Bissau	130	211	535	495	517	427	9	0.2	–	2,378	2,430
Haiti	79	123	348	524	275	373	21	0.2	0.7	1,848	1,869
Honduras	31	38	306	221	237	157	62	0.8	1.1	2,206	2,403
Hungary	8	9	270	295	130	123	315	3.2	8.2	3,768	3,313
India	67	93	261	250	279	191	23	–	–	2,228	2,496
Indonesia	33	45	368	230	308	178	19	–	–	2,458	2,886
Iran	35	42	221	170	190	139	258	0.9	1.6	2,659	2,836
Iraq	107	133	207	258	191	208	375	0.6	1.5	3,418	2,619
Ireland	6	6	175	108	103	62	1,692	2.3	9.7	3,623	3,565
Israel	6	6	138	99	85	56	2,021	3.8	6.0	3,075	3,278
Italy	4	6	163	110	80	53	1,498	6.0	4.9	3,512	3,507
Jamaica	17	20	186	169	121	127	165	1.4	2.1	2,630	2,553
Japan	3	5	129	98	70	44	2,908	1.9	16.5	2,870	2,932
Jordan	27	33	–	199	–	144	137	1.7	1.8	2,780	3,014
Kazakhstan	81	99	312	366	140	201	44	3.5	8.5	–	3,085
Kenya	78	122	417	578	339	529	28	0.1	–	2,025	1,977
Korea, North	42	55	270	238	156	192	18	3.0	–	2,509	1,837
Korea, South	5	5	270	186	156	71	584	1.3	6.1	3,110	3,155
Kuwait	9	10	172	100	116	68	586	1.9	2.8	3,021	3,096
Kyrgyzstan	52	61	296	335	131	299	12	3.0	9.5	–	2,447
Laos	87	100	531	355	439	299	11	0.2	–	2,102	2,108
Latvia	17	21	281	328	106	122	174	2.8	10.3	–	2,864
Lebanon	28	32	241	192	181	136	499	2.1	2.7	3,040	3,277
Lesotho	91	132	371	667	279	630	28	0.1	–	2,216	2,244
Liberia	157	235	–	448	–	385	2	0.0	–	–	–
Libya	16	19	276	210	218	157	246	1.3	4.3	3,308	3,289
Lithuania	8	9	243	286	92	106	185	4.0	9.2	–	3,261
Macedonia	22	26	–	160	–	89	106	2.2	4.9	–	2,664
Madagascar	84	136	353	385	278	322	9	0.1	–	2,292	2,022
Malawi	114	183	429	701	349	653	11	–	1.3	2,027	2,043
Malaysia	8	8	230	202	149	113	101	0.7	2.0	2,616	2,977
Mali	141	231	454	518	362	446	10	0.1	0.2	1,967	2,030
Mauritania	120	183	505	357	416	302	14	0.1		2,509	2,622
Mauritius	17	19	277	228	181	109	134	0.9	–	–	–
Mexico	24	29	205	180	121	101	311	1.8	1.1	3,022	3,097
Moldova	27	32	289	325	173	165	11	3.5	12.1	–	2,567

Table F (Continued)
World Countries: Mortality, Health, and Nutrition

COUNTRY	MORTALITY						HEALTH			NUTRITION	
	Infant Mortality		Adult Mortality								
	Infant Mortality (per 1,000 live births)	Under-five Mortality (per 1,000)	Male (per 1,000)		Female (per 1,000)		Health Expenditure per Capita ($US) 1997-2000[a]	Physicians per 1,000 People 1995-2000[a]	Hospital Beds per 1,000 people 1995-2000[a]	Average daily per capita supply of calories (kilocalories)[c] 1987	Average daily per capita supply of calories (kilocalories)[c] 1997
	2001	2001	1980	2000-2001	1980	2000-2001					
Mongolia	61	76	320	280	273	199	23	2.4	–	2,034	1,917
Morocco	39	44	264	174	207	113	50	0.5	1.0	3,047	3,078
Mozambique	125	197	468	674	361	612	9	–	–	1,785	1,832
Myanmar (Burma)	77	109	284	343	313	245	153	0.3	–	2,697	2,862
Namibia	55	67	427	695	366	661	136	0.3	–	2,199	2,183
Nepal	66	91	376	314	395	314	–	0.0[b]	0.2	2,144	2,366
Netherlands	5	6	133	95	74	64	1,900	3.2	10.8	3,076	3,284
New Zealand	6	6	177	108	91	69	1,062	2.2	6.2	3,201	3,395
Nicaragua	36	43	277	225	189	161	43	0.9	1.5	2,330	2,186
Niger	156	265	562	473	453	308	5	0.0[b]	0.1	2,033	2,097
Nigeria	110	183	535	443	453	393	8	–	–	2,103	2,735
Norway	4	4	144	106	71	60	2,832	2.9	14.6	3,304	3,357
Oman	12	13	389	187	326	135	295	1.3	2.2	–	–
Pakistan	84	109	283	221	291	198	18	0.6	–	2,224	2,476
Panama	19	25		145		93	268	1.7	2.2	2,302	2,430
Papua New Guinea	70	94	514	359	478	329	31	0.1	–	2,137	2,224
Paraguay	26	30	198	173	144	129	112	1.1	1.3	2,564	2,566
Peru	30	39	287	190	229	139	100	0.9	1.5	2,276	2,302
Philippines	29	38	323	249	259	142	33	1.2	–	2,244	2,366
Poland	8	9	254	226	105	88	246	2.2	4.9	3,441	3,366
Portugal	5	6	199	164	95	66	862	3.2	4.0	3,400	3,667
Puerto Rico	10	–	159	149	78	56	–	1.8	3.3	–	–
Romania	19	21	216	260	119	117	48	1.8	7.6	2,944	3,253
Russian Federation	18	21	341	424	120	153	92	4.2	12.1	–	2,904
Rwanda	96	183	503	667	409	599	12	–	–	2,042	2,057
Saudi Arabia	23	28	283	181	241	116	448	1.7	2.3	2,488	2,783
Senegal	79	138	586	355	516	303	22	0.1	0.4	2,104	2,418
Serbia-Montenegro	17	19	164	180	106	100	50	2.0	5.3	–	3,031
Sierra Leone	182	316	540	587	527	531	6	0.1	–	2,126	2,035
Singapore	3	4	199	114	115	61	814	1.6	–	–	–
Slovak Republic	8	9	226	210	105	83	210	3.5	7.1	–	2,984
Slovenia	4	5	250	170	105	76	788	2.3	5.7	–	3,101
Somalia	47	225	–	516	–	452	19	0.0	–	–	–
South Africa	56	71	–	594	–	543	255	0.6	–	2,976	2,990
Spain	4	6	144	122	69	49	1,073	3.3	4.1	3,150	3,310
Sri Lanka	17	19	210	244	152	124	31	0.4	–	2,253	2,302
Sudan	65	107	537	341	462	291	13	0.1	–	2,208	2,395
Swaziland	106	149	–	635	–	595	56	0.2	–	–	–
Sweden	3	3	142	89	76	56	2,179	2.9	3.6	2,898	3,194
Switzerland	5	6	145	99	70	58	3,573	3.5	17.9	3,358	3,223
Syria	23	28	–	170	–	132	30	1.3	1.4	3,166	3,352
Tajikistan	91	116	190	293	129	204	6	2.0	–	–	2,001
Tanzania	104	165	451	569	370	520	12	0.0[b]	–	2,288	1,995
Thailand	24	28	280	245	210	150	71	0.4	2.0	2,133	2,360
Togo	79	141	457	460	375	406	8	0.1	–	1,946	2,469
Trinidad and Tobago	17	20	234	209	166	133	268	0.8	5.1	2,975	2,661

World Countries: Mortality, Health, and Nutrition

COUNTRY	MORTALITY						HEALTH			NUTRITION	
	Infant Mortality		Adult Mortality								
	Infant Mortality (per 1,000 live births)	Under-five Mortality (per 1,000)	Male (per 1,000)		Female (per 1,000)		Health Expenditure per Capita ($US) 1997-2000[a]	Physicians per 1,000 People 1995-2000[a]	Hospital Beds per 1,000 people 1995-2000[a]	Average daily per capita supply of calories (kilocalories)[c] 1987	Average daily per capita supply of calories (kilocalories)[c] 1997
	2001	2001	1980	2000-2001	1980	2000-2001					
Tunisia	21	27	227	169	224	99	110	0.7	1.7	3,067	3,283
Turkey	36	43	–	218	–	120	150	1.2	2.6	3,496	3,525
Turkmenistan	69	87	263	280	154	157	52	3.0	–	–	2,306
Uganda	79	124	463	617	395	567	10	–	–	2,113	2,085
Ukraine	17	20	282	365	112	135	26	3.0	11.8	–	2,795
United Arab Emirates	8	9	153	143	106	93	767	1.8	2.6	3,038	3,390
United Kingdom	6	7	160	109	96	66	1,747	1.8	4.1	3,215	3,276
United States	7	8	194	141	102	82	4,499	2.8	3.6	3,430	3,699
Uruguay	14	16	176	185	91	89	653	3.7	4.4	2,613	2,816
Uzbekistan	52	68	219	282	116	176	29	3.3	8.3	–	2,433
Venezuela	19	22	219	178	123	99	233	2.4	1.5	2,602	2,321
Vietnam	30	38	262	203	204	139	21	0.5	1.7	2,193	2,484
West Bank and Gaza	21	25	–	157	–	100	–	0.5	1.2	–	–
Yemen	79	107	382	278	304	226	20	0.2	0.6	2,126	2,051
Zambia	112	202	482	725	413	687	18	0.1	–	2,017	1,970
Zimbabwe	76	123	389	650	321	612	43	0.1	–	2,112	2,145

Sources: World Development Indicators, 2003 (World Bank); *World Resources 2000–2001* (World Resources Institute)

a. Data are for the most recent year available
b. Less than 0.05
c. has replaced "calories available as a percentage of need" as a benchmark for nutrition
d. data for China include Taiwan

Table G
World Countries: Education and Literacy, 1990–2002

COUNTRY	EDUCATIONAL INPUTS					OUTCOMES							
	Public Expenditure on Education 2000[b] (% of government expenditure)	Primary Pupil-Teacher Ratio 2000 (pupils per teacher)	Primary School Enrollment 2000 (% of age group)	Secondary School Enrollment 2000 (% of age group)	Tertiary School Enrollment 2000 (% of age group)	ADULT ILLITERACY (% above age 15)				YOUTH ILLITERACY RATE (% aged 15–24)			
						Male		Female		Male		Female	
						1990	2001	1990	2001	1990	2001	1990	2001
Afghanistan	–	43	15	–	–	–	–	–	–	–	–	–	–
Albania	–	22	107	78	15	14	8	32	22	3	1	7	3
Algeria	–	28	112	71	15	32	23	59	42	13	6	32	15
Angola	–	35	74	15	1	–	–	–	–	–	–	–	–
Argentina	11.8	22	120	97	48	4	3	4	3	2	2	2	1
Armenia	–	–	78	73	20	1	1	4	2	0[a]	0[a]	1	0[a]
Australia	–	–	102	161[d]	63	–	–	–	–	–	–	–	–
Austria	12.4	13	104	99	58	–	–	–	–	–	–	–	–
Azerbaijan	24.4	19	98	80	22	–	–	–	–	–	–	–	–
Bangladesh	15.7	57	100	46	7	54	50	77	69	45	43	68	60
Belarus	–	17	109	84	56	0[a]	0[a]	1	0[a]	0[a]	0[a]	0[a]	0[a]
Belgium	11.6	12	105	–	57	–	–	–	–	–	–	–	–
Benin	–	54	95	22	4	59	47	84	75	37	28	74	63
Bolivia	23.1	24	116	80	36	13	8	30	20	4	2	11	6
Bosnia-Herzegovina	–	–	–	–	–	–	–	–	–	–	–	–	–
Botswana	–	27	108	93	5	34	25	30	19	21	15	13	8
Brazil	12.9	26	162	108	17	18	13	20	13	12	6	9	3
Bulgaria	–	18	103	94	41	2	1	4	2	1	0[a]	1	0[a]
Burkina Faso	–	47	44	10	–	75	65	92	85	64	53	86	75
Burundi	–	50	65	10	I	50	43	73	58	42	33	55	36
Cambodia	10.1	53	110	19	3	49	20	86	42	34	16	73	25
Cameroon	12.5	63	108	20	5	28	20	46	35	10	8	15	11
Canada	–	15	99	103	60	–	–	–	–	–	–	–	–
Central African Republic	–	74	75	–	2	53	39	79	63	34	23	61	39
Chad	–	71	73	11	1	63	47	81	64	42	25	62	38
Chile	17.5	25	103	75	38	6	4	6	4	2	1	2	1
China	–	22	106	63	7	14	7	33	21	3	1	8	3
Hong Kong, China	–	–	–	–	–	5	3	16	10	2	1	1	0a
Colombia	–	26	112	70	23	11	8	12	8	6	4	4	2
Democratic Republic of the Congo (formerly Zaire)	–	26	47	18	1	38	26	66	48	19	11	42	24
Congo Republic	12.6	51	97	42	5	23	12	42	24	5	2	10	3
Costa Rica	–	25	107	60	16	6	4	6	4	3	2	2	1
Côte d'Ivoire	21.5	48	81	23	7	56	40	77	62	40	29	59	46
Croatia	10.4	18	–	–	–	1	1	5	3	0[a]	0[a]	0[a]	0[a]
Cuba	15.1	11	102	85	24	5	3	5	3	1	0[a]	1	0[a]
Czech Republic	9.5	18	104	95	30	–	–	–	–	–	–	–	–
Denmark	15.3	10	102	128	59	–	–	–	–	–	–	–	–
Dominican Republic	15.7	40	124	59	–	20	16	21	16	13	9	12	8
Ecuador	8.0	23	115	57	–	10	7	15	10	4	2	5	3
Egypt	–	22	100	86	39	40	33	66	55	29	23	49	36
El Salvador	13.4	26	109	54	18	24	18	31	23	15	11	17	12
Eritrea	–	45	59	28	2	42	32	72	54	27	19	54	39
Estonia	–	14	103	92	58	0[a]	0[a]	0[a]	0[a]	–	0[a]	–	0[a]
Ethiopia	13.8	55	64	18	2	64	52	80	68	52	38	64	50

COUNTRY	EDUCATIONAL INPUTS					OUTCOMES							
	Public Expenditure on Education 2000[b] (% of government expenditure)	Primary Pupil-Teacher Ratio 2000 (pupils per teacher)	Primary School Enrollment 2000 (% of age group)	Secondary School Enrollment 2000 (% of age group)	Tertiary School Enrollment 2000 (% of age group)	ADULT ILLITERACY (% above age 15)				YOUTH ILLITERACY RATE (% aged 15–24)			
						Male		Female		Male		Female	
						1990	2001	1990	2001	1990	2001	1990	2001
Finland	12.5	16	102	126	–	–	–	–	–	–	–	–	–
France	11.5	19	105	108	54	–	–	–	–	–	–	–	–
Gabon	–	49	144	60	8	–	–	–	–	–	–	–	–
The Gambia	14.2	37	82	36	–	68	55	80	69	49	33	66	49
Georgia	–	16	95	73	35	–	–	–	–	–	–	–	–
Germany	9.7	15	104	99	46	–	–	–	–	–	–	–	–
Ghana	–	33	80	36	3	30	19	53	35	12	6	25	11
Greece	7.0	13	99	98	–	2	1	8	4	1	0[a]	0[a]	0[a]
Guatemala	11.4	33	102	37	–	31	23	47	38	20	14	34	27
Guinea	25.6	44	67	14	–	–	–	–	–	–	–	–	–
Guinea-Bissau	4.8	44	83	20	0	54	45	89	75	30	26	79	54
Haiti	10.9	–	–	–	–	57	47	63	51	44	35	46	34
Honduras	–	34	106	–	15	31	25	32	24	23	16	20	13
Hungary	14.1	11	102	99	40	1	1	1	1	0[a]	0[a]	0[a]	0[a]
India	12.7	40	102	49	10	38	31	64	54	27	20	46	34
Indonesia	–	22	110	57	15	13	8	27	17	3	2	7	3
Iran	20.4	25	86	78	10	27	16	45	30	8	4	18	8
Iraq	–	21	102	38	14	43	45	67	76	29	40	48	70
Ireland	13.2	20	119	–	48	–	–	–	–	–	–	–	–
Israel	–	12	114	93	53	3	3	9	7	1	0[a]	2	1
Italy	9.5	11	101	96	50	2	1	3	2	0[a]	0[a]	0[a]	0[a]
Jamaica	11.1	36	100	83	16	22	17	14	9	13	9	5	2
Japan	9.3	20	101	102	48	–	–	–	–	–	–	–	–
Jordan	5.0	–	101	88	29	10	5	28	15	2	1	4	1
Kazakhstan	–	19	99	88	31	1	0[a]	2	1	–	0[a]	–	0[a]
Kenya	22.5	30	94	31	3	19	11	39	23	7	4	13	5
Korea, North	–	–	–	–	–	–	–	–	–	–	–	–	–
Korea, South	17.4	32	101	94	78	2	1	7	3	0[a]	0[a]	0[a]	0[a]
Kuwait	–	14	85	56	21	20	16	27	20	12	8	13	6
Kyrgyzstan	–	24	101	86	41	–	–	–	–	–	–	–	–
Laos	8.8	30	113	38	3	47	23	80	46	28	15	62	28
Latvia	–	15	100	91	63	0[a]	0[a]	0[a]	0[a]	0[a]	0[a]	0[a]	0[a]
Lebanon	11.1	17	99	76	42	12	8	27	19	5	3	11	7
Lesotho	18.5	48	115	33	3	35	27	11	6	23	17	3	1
Liberia	–	36	118	38	–	45	29	77	62	25	14	61	46
Libya	–	8	116	90	–	17	9	49	31	1	0[a]	17	6
Lithuania	–	16	101	95	52	1	0[a]	1	0[a]	0[a]	0[a]	0[a]	0[a]
Macedonia	–	22	99	84	24	–	–	–	–	–	–	–	–
Madagascar	10.2	50	103	14	2	34	26	50	39	22	16	33	23
Malawi	24.6	56	137	36	0	31	25	64	52	24	19	49	38
Malaysia	26.7	18	99	70	28	13	8	25	16	5	2	6	2
Mali	–	63	61	15	2	67	63	81	83	62	52	83	74
Mauritania	18.9	42	83	21	4	53	49	74	69	44	43	65	59
Mauritius	12.1	26	109	77	11	15	12	25	18	9	6	9	5
Mexico	22.6	27	113	75	21	10	7	15	11	4	2	6	3
Moldova	15.0	20	84	71	28	1	0[a]	4	2	0[a]	0[a]	0[a]	0[a]
Mongolia	2.2	32	99	61	33	2	1	3	2	1	1	1	1

Table G
World Countries: Education and Literacy, 1990–2002

COUNTRY	EDUCATIONAL INPUTS					OUTCOMES							
	Public Expenditure on Education 2000[b] (% of government expenditure)	Primary Pupil-Teacher Ratio 2000 (pupils per teacher)	Primary School Enrollment 2000 (% of age group)	Secondary School Enrollment 2000 (% of age group)	Tertiary School Enrollment 2000 (% of age group)	ADULT ILLITERACY (% above age 15)				YOUTH ILLITERACY RATE (% aged 15–24)			
						Male		Female		Male		Female	
						1990	2001	1990	2001	1990	2001	1990	2001
Morocco	26.1	28	94	39	10	47	37	75	63	32	23	58	40
Mozambique	12.3	64	92	12	1	51	39	82	70	34	24	68	52
Myanmar (Burma)	9.0	32	89	39	12	13	11	26	19	10	9	14	9
Namibia	–	32	112	62	6	23	17	28	18	14	10	11	6
Nepal	14.1	37	118	51	5	53	39	86	75	34	23	73	56
Netherlands	10.7	10	108	124[d]	55	–	–	–	–	–	–	–	–
New Zealand	–	16	100	112	69	–	–	–	–	–	–	–	–
Nicaragua	13.8	36	104	54	–	36	33	34	33	32	29	28	27
Niger	–	42	35	6	1	82	76	95	91	75	67	91	86
Nigeria	–	–	–	–	–	41	27	62	42	19	10	34	15
Norway	16.2	–	101	115	70	–	–	–	–	–	–	–	–
Oman	–	24	72	68	8	33	19	62	36	5	0[a]	25	3
Pakistan	7.8	44	75	–	–	50	42	79	71	36	28	67	57
Panama	–	25	112	69	35	10	7	12	9	4	3	5	4
Papua New Guinea	17.5	36	84	21	2	34	29	52	42	25	20	38	28
Paraguay	11.2	20	111	60	–	8	5	12	8	4	3	5	3
Peru	21.1	25	128	81	29	8	5	21	14	3	2	8	5
Philippines	20.6	35	113	77	31	7	5	8	5	3	1	3	1
Poland	11.4	11	100	101	56	0[a]	0[a]	1	0[a]	0[a]	0[a]	0[a]	0[a]
Portugal	13.1	13	121	114[d]	50	9	5	16	10	1	0[a]	0[a]	0[a]
Puerto Rico	–	–	–	–	–	9	6	9	6	5	3	3	2
Romania	–	20	99	82	27	1	1	5	3	1	0[a]	1	0[a]
Russian Federation	–	17	–	83	64	0[a]	0[a]	1	1	0[a]	0[a]	0[a]	0[a]
Rwanda	–	51	119	12	2	37	26	56	38	22	14	33	17
Saudi Arabia	–	12	68	68	22	22	16	49	32	9	5	21	9
Senegal	–	51	75	18	4	62	52	81	71	50	40	70	57
Serbia and Montenegro	–	20	–	–	–	–	–	–	–	–	–	–	–
Sierra Leone	–	44	93	26	2	–	–	–	–	–	–	–	–
Singapore	23.6	–	–	–	–	6	4	17	11	1	0[a]	1	0[a]
Slovak Republic	13.8	19	103	87	30	–	–	–	–	–	–	–	–
Slovenia	–	14	100	–	61	0[a]	0[a]	0[a]	0[a]	0[a]	0[a]	0[a]	0[a]
Somalia	–	–	–	–	–	–	–	–	–	–	–	–	–
South Africa	25.8	–	111	87	15	18	14	20	15	11	8	12	9
Spain	11.3	33	105	116	59	2	1	5	3	0[a]	0[a]	0[a]	0[a]
Sri Lanka	–	14	106	72	–	7	5	15	11	4	3	6	3
Sudan	–	–	55	29	7	39	30	68	52	24	17	46	27
Swaziland	–	33	125	60	5	26	19	30	21	15	10	15	8
Sweden	13.4	11	110	149[d]	70	–	–	–	–	–	–	–	–
Switzerland	15.2	14	107	100	42	–	–	–	–	–	–	–	–
Syria	11.1	24	109	43	6	18	11	53	38	8	4	33	20
Tajikistan	11.8	22	104	79	14	1	0[a]	3	1	0[a]	0[a]	0[a]	0[a]
Tanzania	–	40	63	6	1	23	15	49	32	10	1	22	11
Thailand	31.0	21	95	82	35	5	3	11	6	1	1	2	2
Togo	23.2	34	124	39	4	36	27	71	56	19	12	55	35
Trinidad and Tobago	16.7	20	100	81	6	6	1	11	2	0[a]	0[a]	0[a]	0[a]

Table G
World Countries: Education and Literacy, 1990–2002

COUNTRY	EDUCATIONAL INPUTS					OUTCOMES							
	Public Expenditure on Education 2000[b] (% of government expenditure)	Primary Pupil-Teacher Ratio 2000 (pupils per teacher)	Primary School Enrollment 2000 (% of age group)	Secondary School Enrollment 2000 (% of age group)	Tertiary School Enrollment 2000 (% of age group)	ADULT ILLITERACY (% above age 15)				YOUTH ILLITERACY RATE (% aged 15–24)			
						Male		Female		Male		Female	
						1990	2001	1990	2001	1990	2001	1990	2001
Tunisia	17.4	23	117	78	22	28	18	54	38	7	2	25	10
Turkey	–	–	101	58	15	11	6	33	23	3	1	12	6
Turkmenistan	–	–	–	–	–	–	–	–	–	–	–	–	–
Uganda	–	59	136	19	3	31	22	57	42	20	14	39	27
Ukraine	15.7	20	78	105	43	0[a]	0[a]	1	0[a]	0[a]	0[a]	0[a]	0[a]
United Arab Emirates	–	16	99	75	12	29	25	30	20	19	12	11	5
United Kingdom	11.4	18	99	156[d]	60	–	–	–	–	–	–	–	–
United States	–	15	101	95	73	–	–	–	–	–	–	–	–
Uruguay	–	21	109	98	36	4	3	3	2	1	1	1	1
Uzbekistan	–	–	–	–	–	1	0[a]	2	1	0[a]	0[a]	0[a]	0[a]
Venezuela	–	–	102	59	28	10	7	12	8	5	3	3	1
Vietnam	–	28	106	67	10	6	5	13	9	5	5	5	4
West Bank and Gaza	–	–	–	–	–	–	–	–	–	–	–	–	–
Yemen	32.8	30	79	48	11	45	32	87	73	26	16	75	51
Zambia	17.6	45	78	24	2	22	14	41	27	14	9	24	14
Zimbabwe	–	37	95	44	4	13	7	25	15	3	1	9	4

a. Less than 0.5
b. Data are for the most recent year available
c. Gross National Income (GNI) has replaced GNP in the World Bank Atlas Method's estimate of national income
d. Includes training for the unemployed

Source: World Development Indicators 2003 (World Bank).

Table H
World Countries: Agricultural Operations, 1996–2002

COUNTRY			AGRICULTURAL INPUTS				AGRICULTURAL OUTPUT AND PRODUCTIVITY			
			Agricultural Machinery							
	Arable Land (hectares per capita)	Irrigated Land (% of cropland)	Land under Cereal Production (thousand hectares)	Fertilizer Consumption (hundreds of grams per hectare of arable land)	Tractors Per Thousand Agricultural Workers	Tractors Per Hundred Hectares of Agricultural Land	Crop Production Index (1989–91 = 100)	Food Production Index (1989–91 = 100)	Livestock Production Index (1989–91 = 100)	Agricultural Productivity (agricultural value added per worker in 1995$)
	1998–2000	1998–2000	1999–2001	1998–2000	1998–2000	1998–2000	1990–2001	1990–2001	1999–2001	1999–2001
Afghanistan	0.31	29.6	2,400	7	0	1	–	–	–	–
Albania	0.19	48.6	213	261	11	142	–	–	–	2,101
Algeria	0.26	6.8	1,965	121	38	121	127.9	133.1	128.3	1,939
Angola	0.24	2.2	904	9	2	34	151.2	146.3	136.8	131
Argentina	0.68	5.7	11,004	323	191	112	163.9	142.9	112.1	10,351
Armenia	0.13	51.3	178	153	69	354	98.5	73.9	60.2	5,435
Australia	2.67	4.6	17,514	455	700	62	171.2	145.7	114.8	33,225
Austria	0.17	0.3	820	1,665	1,728	2,512	102.6	107.6	106.3	31,091
Azerbaijan	0.21	74.3	635	62	34	192	57.4	78.0	79.6	768
Bangladesh	0.06	47.6	11,736	1,593	0	7	132.4	135.6	140.9	311
Belarus	0.61	1.8	2,462	1,403	107	132	86.7	59.5	56.8	2,180
Belgium	0.08	4.2	324	3,687	1,234	1,298	146.1	113.6	109.7	29,098
Benin	0.31	0.6	843	228	0	1	181.9	157.6	110.8	608
Bolivia	0.24	5.9	775	25	4	29	163.0	144.8	126.8	748
Bosnia-Herzegovina	0.13	0.4	367	872	284	580	–	–	–	7,811
Botswana	0.21	0.3	159	127	20	175	80.4	95.5	97.4	575
Brazil	0.32	4.4	17,629	1,200	59	152	129.1	145.9	162.2	4,798
Bulgaria	0.53	17.6	1,867	333	77	58	59.8	66.0	60.7	8,277
Burkina Faso	0.34	0.7	2,948	114	0	5	137.8	135.4	141.4	183
Burundi	0.13	5.9	202	41	0	2	92.3	91.8	77.3	150
Cambodia	0.31	7.1	2,050	7	0	5	149.0	153.3	166.1	363
Cameroon	0.41	0.5	734	76	0	1	132.9	130.8	122.0	1,189
Canada	1.49	1.6	18,016	572	1,761	156	123.7	126.7	129.6	43,428
Central African Republic	0.53	–	151	3	0	0	133.3	139.4	137.3	490
Chad	0.47	0.6	2,156	49	0	0	144.5	138.0	119.0	213
Chile	0.13	78.4	586	2,416	55	273	131.2	137.8	147.4	6,040
China	0.10	39.6	86,688	2,871	2	62	146.1	175.9	217.9	334
Hong Kong, China	–	–	–	–	–	–	59.3	58.0	57.0	–
Colombia	0.06	19.6	1,119	2,342	6	83	104.1	120.2	122.9	3,590
Democratic Republic of Congo (formerly Zaire)	0.14	0.1	2,066	2	0	4	81.9	85.3	101.5	218
Congo Republic	0.06	0.5	10	287	1	40	126.0	128.3	133.4	471
Costa Rica	0.06	21.2	79	8,572	21	311	149.9	148.0	132.7	5,272
Côte d'Ivoire	0.19	1.0	1,439	264	1	13	135.8	138.0	122.9	1,057
Croatia	0.33	0.2	665	1,354	12	17	86.6	67.3	50.0	9,449
Cuba	0.33	19.5	208	423	98	215	58.4	62.2	68.3	–
Czech Republic	0.30	0.7	1,624	936	173	273	92.0	77.1	66.9	6,235
Denmark	0.43	19.4	1,523	1,667	1,104	557	91.9	104.2	116.8	57,896
Dominican Republic	0.13	17.2	160	860	4	22	87.3	110.1	141.0	3,179
Ecuador	0.13	28.8	897	1,062	7	57	157.6	156.1	153.1	1,716
Egypt	0.05	100.0	2,716	4,284	10	303	151.2	155.9	161.4	1,324
El Salvador	0.09	4.9	380	1,497	4	61	104.0	117.0	120.2	1,710
Eritrea	0.12	4.3	333	189	0	12	148.9	126.8	111.0	85
Estonia	0.81	0.4	326	268	556	454	69.5	43.4	37.2	4,265
Ethiopia	0.16	1.8	7,562	163	0	3	153.2	141.1	118.4	141
Finland	0.42	2.9	1,154	1,419	1,268	893	94.7	89.8	91.7	40,463

COUNTRY			AGRICULTURAL INPUTS				AGRICULTURAL OUTPUT AND PRODUCTIVITY			
			Agricultural Machinery							
	Arable Land (hectares per capita)	Irrigated Land (% of cropland)	Land under Cereal Production (thousand hectares)	Fertilizer Consumption (hundreds of grams per hectare of arable land)	Tractors Per Thousand Agricultural Workers	Tractors Per Hundred Hectares of Agricultural Land	Crop Production Index (1989–91 = 100)	Food Production Index (1989–91 = 100)	Livestock Production Index (1989–91 = 100)	Agricultural Productivity (agricultural value added per worker in 1995$)
	1998–2000	1998–2000	1999–2001	1998–2000	1998–2000	1998–2000	1990–2001	1990–2001	1999–2001	1999–2001
France	0.31	10.8	8,963	2,491	1,326	692	107.9	105.3	104.6	58,177
Gabon	0.27	3.0	16	11	7	46	119.9	115.9	119.0	2,047
The Gambia	0.17	0.9	133	70	0	2	145.0	139.8	106.1	298
Georgia	0.15	44.2	352	480	19	131	52.5	78.7	94.9	–
Germany	0.14	4.0	6,911	2,460	965	887	119.4	98.2	87.8	32,814
Ghana	0.19	0.2	1,304	39	1	10	178.6	169.6	117.7	569
Greece	0.26	37.2	1,254	1,688	302	877	111.6	103.4	96.1	14,079
Guatemala	0.12	6.8	655	1,511	2	32	131.8	134.7	134.1	2,115
Guinea	0.12	6.4	803	36	0	6	153.8	155.3	174.4	274
Guinea-Bissau	0.26	4.9	123	40	0	1	144.7	140.0	126.3	322
Haiti	0.07	8.2	460	203	0	3	87.6	99.8	145.6	–
Honduras	0.22	4.6	446	1,242	7	35	105.1	112.1	149.5	990
Hungary	0.47	4.2	2,758	897	171	192	80.1	76.3	70.1	5,159
India	0.16	32.2	100,967	1,063	6	93	124.6	129.9	144.1	402
Indonesia	0.10	14.7	15,270	1,326	1	36	119.4	118.3	116.4	744
Iran	0.25	42.5	7,234	833	37	139	135.9	139.0	145.1	3,698
Iraq	0.23	63.6	2,646	732	76	95	73.6	73.3	68.1	–
Ireland	0.29	–	282	6,252	1,035	1,590	111.0	113.5	110.6	–
Israel	0.05	46.4	80	3,406	343	734	100.3	115.0	122.5	–
Italy	0.14	24.3	4,185	2,116	1154	1,979	104.7	104.9	106.6	26,690
Jamaica	0.07	9.1	2	1,320	12	177	118.1	118.8	122.3	1,540
Japan	0.04	54.6	2,026	3,192	707	4,691	88.0	92.5	94.1	30,828
Jordan	0.05	19.1	34	882	28	197	114.9	134.1	176.6	825
Kazakhstan	1.47	10.4	12,121	12	44	28	84.1	70.3	45.3	1,649
Kenya	0.14	1.5	1,945	349	1	36	110.4	107.7	106.5	216
Korea, North	0.08	73	1,287	1,530	21	441	–	–	–	–
Korea, South	0.04	60.4	1,163	4,794	66	981	113.0	131.0	161.6	13,782
Kuwait	0.00	81.0	1	1,341	10	132	166.2	208.1	210.0	–
Kyrgyzstan	0.28	74.8	607	210	47	190	143.5	121.4	78.1	1,636
Laos	0.17	18.1	765	88	1	12	153.7	163.4	185.7	614
Latvia	0.77	1.1	422	278	324	293	74.3	41.9	31.7	2,671
Lebanon	0.04	32.3	39	3,416	113	298	138.3	143.7	164.7	28,322
Lesotho	0.16	–	233	170	6	62	164.6	116.9	87.5	553
Liberia	0.12	0.5	147	0	0	9	–	–	–	525
Libya	0.35	21.9	337	356	306	187	133.8	157.5	162.1	–
Lithuania	0.83	0.2	947	509	394	337	74.6	59.8	49.4	3,131
Macedonia	0.29	8.8	220	744	420	919	105.2	94.6	86.1	4,155
Madagascar	0.19	31.3	1,409	30	1	12	103.3	109.5	110.0	155
Malawi	0.20	1.3	1,622	217	0	7	149.4	160.8	113.2	123
Malaysia	0.08	4.8	719	7,843	24	238	119.7	143.2	155.1	6,843
Mali	0.44	3.0	2,411	107	1	6	135.0	119.5	111.8	290
Mauritania	0.19	9.8	246	29	1	8	133.9	109.4	106.0	500
Mauritius	0.09	18.9	0	3,463	6	37	90.4	101.7	141.3	5,580
Mexico	0.26	23.8	10,269	726	20	69	121.9	133.6	145.1	1,801
Moldova	0.42	14.1	854	28	82	239	56.8	45.9	34.5	1,661
Mongolia	0.52	6.8	222	28	22	56	30.9	103.4	109.2	1,428
Morocco	0.31	13.2	5,258	393	10	48	93.1	103.8	121.9	1,512
Mozambique	0.23	2.6	1,793	26	1	14	140.8	128.5	103.2	138

COUNTRY			AGRICULTURAL INPUTS				AGRICULTURAL OUTPUT AND PRODUCTIVITY			
			Agricultural Machinery							
	Arable Land (hectares per capita)	Irrigated Land (% of cropland)	Land under Cereal Production (thousand hectares)	Fertilizer Consumption (hundreds of grams per hectare of arable land)	Tractors Per Thousand Agricultural Workers	Tractors Per Hundred Hectares of Agricultural Land	Crop Production Index (1989–91 = 100)	Food Production Index (1989–91 = 100)	Livestock Production Index (1989–91 = 100)	Agricultural Productivity (agricultural value added per worker in 1995$)
	1998–2000	1998–2000	1999–2001	1998–2000	1998–2000	1998–2000	1990–2001	1990–2001	1999–2001	1999–2001
Myanmar (Burma)	0.21	17.8	6,919	186	1	10	167.9	163.6	157.9	–
Namibia	0.47	0.9	308	3	11	39	121.5	118.0	117.7	1,618
Nepal	0.13	38.2	3,289	325	0	16	127.8	127.6	125.5	200
Netherlands	0.06	59.8	209	5,082	590	1,673	115.4	102.9	100.5	58,280
New Zealand	0.41	8.7	138	4,743	447	489	142.7	128.0	117.9	28,791
Nicaragua	0.50	3.2	398	141	7	11	138.4	144.1	139.5	–
Niger	0.43	1.5	7,569	7	0	0	144.5	136.0	125.2	201
Nigeria	0.23	0.8	18,995	63	2	11	159.7	156.2	127.4	714
Norway	0.20	–	328	2,230	1,266	1,549	79.2	91.9	99.1	34,535
Oman	0.01	77.2	2	3,586	1	77	166.1	162.5	133.6	–
Pakistan	0.16	82.1	12,443	1,312	13	151	126.0	145.2	155.1	712
Panama	0.18	5.3	129	716	20	100	94.0	106.7	126.1	2,738
Papua New Guinea	0.04	–	3	480	1	58	122.2	122.8	140.9	815
Paraguay	0.42	2.9	552	299	24	74	114.6	137.5	132.6	3,389
Peru	0.14	28.5	1,210	619	5	36	173.0	171.8	160.0	1,834
Philippines	0.07	15.4	6,581	1,269	1	21	119.9	131.1	162.7	1,428
Poland	0.36	0.7	8,778	1,092	293	930	85.3	85.9	83.4	1,601
Portugal	0.20	24.3	566	1,224	242	844	93.8	102.9	119.0	7,552
Puerto Rico	0.01	49.4	4	–	–	–	67.9	83.8	89.2	–
Romania	0.42	27.8	5,732	343	96	178	92.4	96.3	88.2	3,193
Russian Federation	0.86	3.6	37,953	108	96	65	78.7	63.7	51.5	2,648
Rwanda	0.10	0.4	266	3	0	1	100.4	103.8	116.8	251
Saudi Arabia	0.18	42.8	629	1,002	15	26	88.1	83.3	144.6	–
Senegal	0.25	3.0	1,247	162	0	2	129.5	136.6	147.4	334
Serbia and Montenegro	–	–	–	–	–	–	–	–	–	–
Sierra Leone	0.10	5.4	220	4	0	2	72.9	81.9	126.0	359
Singapore	0.00	–	–	36,137	19	650	48.2	39.5	39.5	44,907
Slovak Republic	0.27	11.0	–	711	–	167	–	–	–	–
Slovenia	0.09	1.1	100	4,468	4,476	6,108	94.4	108.9	104.7	4,912
Somalia	0.12	18.7	507	5	1	18	–	–	–	–
South Africa	0.35	8.9	4,741	523	46	55	106.9	107.7	103.0	3,837
Spain	0.34	19.9	6,637	1,688	636	642	112.3	114.8	128.4	22,088
Sri Lanka	0.05	34.7	888	2,791	2	88	120.3	122.2	135.0	734
Sudan	0.54	11.8	6,764	23	1	6	159.6	161.7	158.3	–
Swaziland	0.17	36.7	58	320	26	164	87.9	89.0	82.5	1,922
Sweden	0.31	–	1,188	1,049	1083	623	89.2	97.1	102.5	36,365
Switzerland	0.06	5.7	181	2,607	673	2,699	90.1	94.1	93.5	–
Syria	0.29	22.1	3,060	766	68	201	156.3	147.0	134.6	2,618
Tajikistan	0.12	83.5	373	257	37	410	51.6	51.5	39.5	1,322
Tanzania	0.12	3.3	3,005	64	1	20	97.2	103.8	121.8	185
Thailand	0.25	27.1	11,154	1,113	10	148	118.5	118.3	130.1	904
Togo	0.56	0.3	680	73	0	0	140.7	132.9	107.5	531
Trinidad and Tobago	0.06	2.5	4	836	53	360	108.2	114.7	100.8	3,036
Tunisia	0.31	7.6	1,411	388	38	121	122.1	132.4	162.8	3,168
Turkey	0.38	16.7	13,174	890	63	372	113.8	111.5	106.8	1,852
Turkmenistan	0.32	–	779	558	75	307	91.8	134.1	136.8	1,518

COUNTRY			AGRICULTURAL INPUTS				AGRICULTURAL OUTPUT AND PRODUCTIVITY			
			Agricultural Machinery							
	Arable Land (hectares per capita)	Irrigated Land (% of cropland)	Land under Cereal Production (thousand hectares)	Fertilizer Consumption (hundreds of grams per hectare of arable land)	Tractors Per Thousand Agricultural Workers	Tractors Per Hundred Hectares of Agricultural Land	Crop Production Index (1989–91 = 100)	Food Production Index (1989–91 = 100)	Livestock Production Index (1989–91 = 100)	Agricultural Productivity (agricultural value added per worker in 1995$)
	1998–2000	1998–2000	1999–2001	1998–2000	1998–2000	1998–2000	1990–2001	1990–2001	1999–2001	1999–2001
Uganda	0.23	0.1	1,371	8	1	9	135.5	131.8	127.3	342
Ukraine	0.66	7.2	12,801	140	93	110	63.9	49.5	46.4	1,521
United Arab Emirates	0.02	47.2	1	8,034	4	70	284.6	270.7	203.4	–
United Kingdom	0.10	1.8	3,168	3,195	908	822	97.9	95.5	95.7	33,520
United States	0.64	12.5	57,456	1,090	1,542	271	119.8	122.4	122.1	50,777
Uruguay	0.39	13.5	542	974	174	257	150.4	136.6	199.9	8,010
Uzbekistan	0.18	88.3	1,520	1,733	57	380	88.8	118.0	116.3	1,088
Venezuela	0.10	16.9	736	920	60	201	118.2	123.2	118.6	5,304
Vietnam	0.07	40.9	8,322	3,420	5	224	166.8	153.9	154.9	253
West Bank and Gaza	–	–	–	–	–	–	–	–	–	–
Yemen	0.09	29.5	621	111	2	37	129.6	134.1	145.9	406
Zambia	0.53	0.9	735	64	2	11	100.7	106.0	117.5	195
Zimbabwe	0.26	3.5	1,787	544	7	73	121.3	110.0	114.5	361

Source: World Development Indicators 2003 (World Bank).

Table I
Land Use and Deforestation, 1980–2000

COUNTRY	LAND AREA	RURAL POPULATION DENSITY	LAND USE						FOREST AREA	AVERAGE ANNUAL DEFORESTATION
	(thousands of sq. km.)	(people per sq. km. of arable land)	Arable Land (% of land area)		Permanent Cropland (% of land area)		Other (% of land area)		(thousands of sq. km.)	Decline in forest area %
	2000	2000	1980	2000	1980	2000	1980	2000	2000	1990-2000
Afghanistan	652	262	12.1	12.1	0.2	0.2	87.7	87.6	14	0.8
Albania	27	313	21.4	21.1	4.3	4.4	74.4	74.5	10	0.8
Algeria	2,382	170	2.9	3.2	0.3	0.2	96.8	96.6	21	-1.3
Angola	1,247	288	2.3	2.4	0.4	0.2	97.3	97.4	698	0.2
Argentina	2,737	17	9.1	9.1	0.8	0.8	90.1	90.1	346	0.8
Armenia	28	252	–	17.6	–	2.3	–	80.1	4	-1.3
Australia	7,682	4	5.7	6.5	0.0	0.0	94.2	93.4	1,581	0.0
Austria	83	190	18.6	16.9	1.2	0.9	80.2	82.2	39	-0.2
Azerbaijan	87	236	–	19.0	–	3.0	–	78.0	11	-1.3
Bangladesh	130	1,208	68.3	62.5	2.0	2.7	29.6	34.8	13	-1.3
Belarus	207	50	–	29.6	–	0.6	–	69.8	94	-3.2
Belgium[a]	33	34	23.2	24.8	0.4	0.7	76.4	74.5	7	0.2
Benin	111	186	12.2	17.6	4.0	2.4	83.8	80.0	27	2.3
Bolivia	1,084	161	1.7	1.8	0.2	0.2	98.1	98.0	531	0.3
Bosnia-Herzegovina	51	454	–	9.8	–	2.9	–	87.3	23	0.0
Botswana	567	231	0.7	0.7	0.0	0.0	99.3	99.4	124	0.9
Brazil	8,457	60	4.6	6.3	1.2	1.4	94.2	92.3	5,325	0.4
Bulgaria	111	60	34.6	40.0	3.2	1.9	62.2	58.1	37	-0.6
Burkina Faso	274	248	10.0	13.9	0.1	0.2	89.8	85.9	71	0.2
Burundi	26	689	35.8	35.0	10.1	14.0	54.0	50.9	1	9.0
Cambodia	177	270	11.3	21.0	0.4	0.6	88.3	78.4	93	0.6
Cameroon	465	128	12.7	12.8	2.2	2.6	85.1	84.6	239	0.9
Canada	9,221	14	4.9	4.9	0.0	0.0	95.0	95.0	2,446	0.0
Central African Republic	623	113	3.0	3.1	0.1	0.1	96.9	96.8	229	0.1
Chad	1,259	167	2.5	2.8	0.0	0.0	97.5	97.2	127	0.6
Chile	749	109	5.4	2.6	0.3	0.4	94.3	96.9	155	0.1
China[b]	9,327	653	10.4	13.3	0.4	1.2	89.3	85.5	1,635	-1.2
Hong Kong,China	1	–	7.0	–	1.0	–	92.0	–	–	–
Colombia	1,039	376	3.6	2.7	1.4	1.7	95.0	95.6	496	0.4
Congo, Democratic Republic	2,267	–	3.0	3.0	0.3	0.5	96.6	96.5	1,352	0.4
Congo,Rep.	342	597	0.4	0.5	0.1	0.1	99.6	99.4	221	0.1
Costa Rica	51	694	5.5	4.4	4.4	5.5	90.1	90.1	20	0.8
Côte d'Ivoire	318	306	6.1	9.3	7.2	13.8	86.6	76.9	71	3.1
Croatia	56	127	–	26.1	–	2.3	–	71.6	18	-0.1
Cuba	110	76	23.9	33.1	6.4	7.6	69.7	59.3	23	-1.3
Czech Republic	77	85	–	39.9	–	3.1	–	57.1	26	0.0
Denmark	42	35	62.3	53.8	0.3	0.2	37.4	46.1	5	-0.2
Dominican Republic	48	264	22.1	22.7	7.2	10.3	70.6	67.0	14	0.0
Ecuador	277	277	5.6	5.7	3.3	5.2	91.1	89.2	106	1.2
Egypt	995	1,298	2.3	2.8	0.2	0.5	97.5	96.7	1	-3.4
El Salvador	21	445	27.0	27.0	8.0	12.1	65.0	60.9	1	4.6
Eritrea	101	669	–	4.9	–	0.0	–	95.0	16	0.3
Estonia	42	37	–	26.5	–	0.3	–	73.2	21	-0.6
Ethiopia	1,000	543	–	10.0	–	0.7	–	89.3	46	0.8
Finland	305	97	8.4	7.2	–	0.0	–	92.8	219	0.0

-195-

Table I *(Continued)*
Land Use and Deforestation, 1980–2000

COUNTRY	LAND AREA	RURAL POPULATION DENSITY	LAND USE						FOREST AREA	AVERAGE ANNUAL DEFORESTATION
	(thousands of sq. km.)	(people per sq. km. of arable land)	Arable Land (% of land area)		Permanent Cropland (% of land area)		Other (% of land area)		(thousands of sq. km)	Decline in forest area %
	2000	2000	1980	2000	1980	2000	1980	2000	2000	1990-2000
France	550	79	31.8	33.5	2.5	2.1	65.7	64.4	153	-0.4
Gabon	258	70	1.1	1.3	0.6	0.7	98.2	98.1	218	0.0
Gambia	10	393	15.9	23.0	–	0.5	–	76.5	5	-1.0
Georgia	70	303	–	11.4	–	3.9	–	84.7	30	0.0
Germany	349	87	34.4	33.1	1.4	0.6	64.1	66.3	107	0.0
Ghana	228	342	8.4	15.9	7.5	9.7	84.2	74.5	63	1.7
Greece	129	154	22.5	21.3	7.9	8.6	69.6	70.1	36	-0.9
Guatemala	108	505	11.7	12.5	4.4	5.0	83.9	82.4	29	1.7
Guinea	246	607	2.0	3.6	0.9	2.4	97.1	94.0	69	0.5
Guinea-Bissau	28	274	9.1	10.7	1.1	1.8	89.9	87.6	22	0.9
Haiti	28	914	19.8	20.3	12.5	12.7	67.7	67.0	1	5.7
Honduras	112	284	13.9	9.5	1.8	3.2	84.3	87.2	54	1.0
Hungary	92	78	54.4	49.8	3.3	2.2	42.2	48.0	18	-0.4
India	2,973	454	54.8	54.4	1.8	2.7	43.4	42.9	641	-0.1
Indonesia	1,812	594	9.9	11.3	4.4	7.2	85.6	81.5	1,050	1.2
Iran	1,622	160	8.0	8.8	0.5	1.2	91.5	90.0	73	0.0
Iraq	437	145	12.0	11.9	0.4	0.8	87.6	87.3	8	0.0
Ireland	69	148	16.1	15.2	0.0	0.0	83.9	84.7	7	-3.0
Israel	21	156	15.8	16.1	4.3	4.1	80.0	79.7	1	-4.9
Italy	294	239	32.2	27.1	10.0	9.7	57.7	63.2	100	-0.3
Jamaica	11	649	16.6	16.1	5.5	9.2	77.8	74.7	3	1.5
Japan	377	603	11.4	12.3	1.6	1.0	87.0	86.7	241	0.0
Jordan	89	427	3.4	2.7	0.4	1.8	96.2	95.5	1	0.0
Kazakhstan	2,671	31	–	8.0	–	0.1	–	92.0	121	-2.2
Kenya	569	501	6.7	7.0	0.8	0.9	92.5	92.1	171	0.5
Korea,North	120	521	13.4	14.1	2.4	2.5	84.2	83.4	82	0.0
Korea,South	99	496	20.9	17.4	1.4	2.0	77.8	80.6	63	0.1
Kuwait	18	989	0.1	0.4	–	0.1	–	99.4	0	-5.2
Kyrgyzstan	192	236	–	7.1	–	0.3	–	92.5	10	-2.6
Laos	231	486	2.9	3.8	0.1	0.4	97.0	95.8	126	0.4
Latvia	62	51	–	29.7	–	0.5	–	69.8	29	-0.4
Lebanon	10	234	20.5	18.6	8.9	13.9	70.6	67.5	0	0.3
Lesotho	30	451	9.6	10.7	–	–	–	–	0	0.0
Liberia	96	454	3.9	3.9	2.1	2.2	94.0	93.8	35	2.0
Libya	1,760	36	1.0	1.0	0.2	0.2	98.8	98.8	4	-1.4
Lithuania	65	38	–	45.3	–	0.9	–	53.8	20	-0.2
Macedonia	25	149	–	21.8	–	1.7	–	76.4	9	0.0
Madagascar	582	377	4.3	5.0	0.9	1.0	94.8	94.0	117	0.9
Malawi	94	419	13.3	22.3	0.9	1.5	85.8	76.2	26	2.4
Malaysia	329	544	3.0	5.5	11.6	17.6	85.4	76.9	193	1.2
Mali	1,220	163	1.6	3.8	0.0	0.0	98.3	96.2	132	0.7
Mauritania	1,025	231	0.2	0.5	0.0	0.0	99.8	99.5	3	2.7
Mauritius	2	697	49.3	49.3	3.4	3.0	47.3	47.8	0	0.6
Mexico	1,909	101	12.1	13.0	0.8	1.3	87.1	85.7	552	1.1
Moldova	33	137	–	55.3	–	11.2	–	33.5	3	-0.2
Mongolia	1,567	89	0.8	0.8	0.0	0.0	99.2	99.2	106	0.5

Table I (Continued)
Land Use and Deforestation, 1980–2000

COUNTRY	LAND AREA	RURAL POPULATION DENSITY	LAND USE						FOREST AREA	AVERAGE ANNUAL DEFORESTATION
	(thousands of sq. km.)	(people per sq. km. of arable land)	Arable Land (% of land area)		Permanent Cropland (% of land area)		Other (% of land area)		(thousands of sq. km)	Decline in forest area %
	2000	2000	1980	2000	1980	2000	1980	2000	2000	1990-2000
Morocco	446	146	16.6	19.6	1.1	2.2	82.3	78.2	30	0.0
Mozambique	784	308	3.6	5.0	0.3	0.3	96.1	94.7	306	0.2
Myanmar (Burma)	658	349	14.6	15.1	0.7	0.9	84.8	84.0	344	1.4
Namibia	823	149	0.8	1.0	0.0	0.0	99.2	99.0	80	0.9
Nepal	143	701	16.0	20.3	0.2	0.5	83.8	79.2	39	1.8
Netherlands	34	184	23.3	26.8	0.9	1.0	75.8	72.1	4	-0.3
New Zealand	268	35	9.3	5.8	3.7	6.4	86.9	87.8	79	-0.5
Nicaragua	121	91	9.5	20.2	1.5	2.4	89.1	77.4	33	3.0
Niger	1,267	192	2.8	3.5	0.0	0.0	97.2	96.4	13	3.7
Nigeria	911	252	30.6	31.0	2.8	2.9	66.6	66.1	135	2.6
Norway	307	129	2.7	2.9	–	–	–	–	89	-0.4
Oman	212	3,049	0.1	0.1	0.1	0.2	99.8	99.7	0	0.0
Pakistan	771	434	25.9	27.6	0.4	0.9	73.7	71.5	25	1.1
Panama	74	250	5.8	6.7	1.6	2.1	92.5	91.2	29	1.6
Papua New Guinea	453	2,067	0.4	0.5	1.1	1.4	98.5	98.1	306	0.4
Paraguay	397	106	4.1	5.8	0.3	0.2	95.6	94.0	234	0.5
Peru	1,280	191	2.5	2.9	0.3	0.4	97.2	96.7	652	0.4
Philippines	298	572	14.5	18.6	14.8	15.1	70.8	66.3	58	1.4
Poland	304	104	48.0	46.0	1.1	1.1	50.9	52.9	93	-0.1
Portugal	92	179	26.5	21.7	7.8	7.8	65.7	70.4	37	-1.7
Puerto Rico	9	2,701	8.3	3.9	5.6	5.2	88.7	90.9	2	0.2
Romania	230	108	42.7	40.7	2.9	2.2	54.4	57.2	64	-0.2
Russian Federation	16,889	32	–	7.4	–	0.1	–	92.5	8,514	0.0
Rwanda	25	887	30.8	36.5	10.3	10.1	58.9	53.4	3	3.9
Saudi Arabia	2,150	80	0.9	1.7	0.0	0.1	99.1	98.2	15	0.0
Senegal	193	212	12.2	12.3	0.0	0.2	87.8	87.5	62	0.7
Serbia-Montenegro	–	–	–	–	–	–	–	–	29	0.0
Sierra Leone	72	651	6.3	6.8	0.7	0.8	93.0	92.3	11	2.9
Singapore	1	0	3.3	1.6	9.8	0.0	86.9	98.4	0	0.0
Slovak Republic	48	158	–	30.4	–	2.8	–	66.8	20	-0.3
Slovenia	20	584	–	8.6	–	1.5	–	89.9	11	-0.2
Somalia	627	610	1.6	1.7	0.0	0.0	98.4	98.3	75	1.0
South Africa	1,221	125	10.2	12.1	0.7	0.8	89.1	87.1	89	0.1
Spain	499	68	31.1	26.7	9.9	9.8	59.0	63.5	144	-0.6
Sri Lanka	65	1,602	13.2	13.8	15.9	15.8	70.8	70.4	19	1.6
Sudan	2,376	122	5.2	6.8	0.0	0.1	94.8	93.1	616	1.4
Swaziland	17	432	10.8	10.3	0.2	0.7	89.0	89.0	5	-1.2
Sweden	412	55	7.2	6.6	–	–	–	–	271	0.0
Switzerland	40	567	9.9	10.4	0.5	0.6	89.6	89.0	12	-0.4
Syria	184	173	28.5	24.7	2.5	4.4	69.1	70.9	5	0.0
Tajikistan	141	614	–	5.2	–	0.9	–	93.9	4	-0.5
Tanzania	884	571	2.5	4.5	1.0	1.1	96.5	94.4	388	0.2
Thailand	511	331	32.3	28.8	3.5	6.5	64.2	64.8	148	0.7
Togo	54	120	36.8	46.1	6.6	2.2	56.6	51.6	5	3.4
Trinidad and Tobago	5	450	13.6	14.6	9.0	9.2	77.4	76.2	3	0.8
Tunisia	155	113	20.5	18.7	9.7	13.5	69.7	67.7	5	-0.2

Table I (Continued)
Land Use and Deforestation, 1980–2000

COUNTRY	LAND AREA (thousands of sq. km.) 2000	RURAL POPULATION DENSITY (people per sq. km. of arable land) 2000	Arable Land (% of land area) 1980	Arable Land (% of land area) 2000	Permanent Cropland (% of land area) 1980	Permanent Cropland (% of land area) 2000	Other (% of land area) 1980	Other (% of land area) 2000	FOREST AREA (thousands of sq. km) 2000	AVERAGE ANNUAL DEFORESTATION Decline in forest area % 1990-2000
Turkey	770	93	32.9	31.4	4.1	3.3	63.0	65.3	102	-0.2
Turkmenistan	470	179	–	3.5	–	0.1	–	96.4	38	0.0
Uganda	200	377	20.4	25.7	8.0	9.6	71.6	64.7	42	2.0
Ukraine	579	49	–	56.2	–	1.6	–	42.2	96	-0.3
United Arab Emirates	84	644	0.2	0.7	0.1	2.2	99.7	97.0	3	-2.8
United Kingdom	241	105	28.7	24.4	0.3	0.2	71.1	75.4	26	-0.8
United States	9,159	36	20.6	19.3	0.2	0.2	79.2	80.5	2,260	-0.2
Uruguay	175	21	8.0	7.4	0.3	0.2	91.7	92.3	13	-5.0
Uzbekistan	414	350	–	10.8	–	0.9	–	88.3	20	-0.2
Venezuela	882	130	3.2	2.8	0.9	1.1	95.9	96.1	495	0.4
Vietnam	325	1,037	18.2	17.7	1.9	4.9	79.8	77.4	98	-0.5
West Bank and Gaza	–	–	–	–	–	–	–	–	–	–
Yemen	528	853	2.6	2.9	0.2	0.2	97.2	96.8	4	1.8
Zambia	743	116	6.9	7.1	0.0	0.0	93.1	92.9	312	2.4
Zimbabwe	387	254	6.4	8.3	0.3	0.3	93.4	91.3	190	1.5

a. Includes Luxembourg.
b. Includes Taiwan.

Source: World Development Indicators 2003 (World Bank).

Table J
World Countries: Energy Production and Use, 1980–2000

COUNTRY	COMMERCIAL ENERGY PRODUCTION		COMMERCIAL ENERGY USE			COMMERCIAL ENERGY USE PER CAPITA			NET ENERGY IMPORTS[a]	
	Thousand Metric Tons of Oil Equivalent		Thousand Metric Tons of Oil Equivalent		Average Annual % Growth	Kg. of Oil Equivalent		Average Annual % Growth	% of Commercial Energy Use	
	1980	2000	1980	2000	1980-2000	1980	2000	1980-2000	1980	2000
Afghanistan	–	–	–	–	–	–	–	–	–	–
Albania	3,428	814	3,049	1,634	-5.6	1,142	521	-6.3	-12	50
Algeria	66,741	149,629	12,089	29,060	3.5	648	956	1.0	-452	-415
Angola	11,301	43,669	4,437	7,667	2.8	632	584	-0.3	-155	-470
Argentina	38,813	81,221	41,868	61,469	2.2	1,490	1,660	0.8	7	-32
Armenia	1,263	632	1,070	2,061	–	346	542	–	–	69
Australia	86,096	232,552	70,372	110,174	2.4	4,790	5,744	1.0	-22	-111
Austria	7,655	9,686	23,450	28,582	1.6	3,105	3,524	1.1	67	66
Azerbaijan	14,821	18,951	15,001	11,703	–	2,433	1,454	–	1	-62
Bangladesh	9,234	15,053	10,930	18,666	4.1	99	142	1.9	20	19
Belarus	2,566	3,466	2,385	24,330	–	247	2,432	–	-8	86
Belgium	7,986	13,233	46,100	59,217	1.8	4,682	5,776	1.6	84	78
Benin	1,212	1,821	1,363	2,362	2.3	394	377	-0.8	11	23
Bolivia	4,372	5,901	2,287	7,929	3.8	455	592	1.5	-79	-20
Bosnia and Herzegovina	–	3,277	–	4,359	–	–	1096	–	–	25
Botswana	–	–	–	–	–	–	–	–	–	–
Brazil	63,372	142,078	111,471	183,165	2.7	917	1,077	1.0	44	22
Bulgaria	7,737	10,005	28,673	18,784	-2.6	3,235	2,299	-2.2	73	47
Burkina Faso	–	–	–	–	–	–	–	–		–
Burundi	–	–	–	–	–	–	–	–	–	–
Cambodia	–	–	–	–	–	–	–	–	–	–
Cameroon	6,707	12,729	3,676	6,355	2.5	421	427	-0.2	-82	-100
Canada	207,417	374,864	193,000	250,967	1.6	7,848	8,156	0.4	-7	-49
Central African Republic			–	–	–	–	–	–	–	–
Chad	–	–	–	–	–	–	–	–	–	–
Chile	5,801	8,299	9,662	24,403	5.7	867	1,604	4.0	40	66
China	615,475	1,107,636	598,498	1,142,439	3.7	610	905	2.4	-3	3
Hong Kong, China	39	48	5,439	15,453	5.6	1,079	2,319	4.2	99	100
Colombia	18,040	74,584	19,349	28,786	2.5	680	681	0.5	7	-159
Democratic Republic of the Congo (formerly Zaire)	8,697	13,546	8,706	14,888	2.7	324	292	-0.6	0	-4
Congo, Rep.	4,024	14,656	862	895	-1.1	516	296	-4.0	-367	-1,538
Costa Rica	767	1,591	1,527	3,281	4.2	669	861	1.5	50	52
Côte d'Ivoire	2,419	6,097	3,662	6,928	3.3	447	433	-0.1	34	12
Croatia	–	3,582	–	7,775	–	–	1,775	–	–	54
Cuba	4,227	6,051	14,910	13,203	-1.6	1,536	1,180	-2.4	72	54
Czech Republic	41,000	29,869	47,254	40,383	-1.2	4,618	3,931	-1.2	13	26
Denmark	952	27,831	19,783	19,456	0.6	3,862	3,643	0.4	95	-43
Dominican Republic	1,327	1,421	3,491	7,804	4.0	613	932	2.1	62	82
Ecuador	11,744	22,520	5,180	8,187	2.1	651	647	-0.3	-127	-175
Egypt	34,168	57,599	15,970	46,423	4.6	391	726	2.3	-114	-24
El Salvador	1,913	2,157	2,537	4,083	2.2	553	651	0.6	25	47
Eritrea	–	–	–	–	–	–	–	–	–	–
Estonia	6,951	2,917	6,275	4,523	–	4,248	3,303	–	-11	36
Ethiopia	10,575	17,583	11,145	18,732	2.6	295	291	-0.1	5	6
Finland	6,912	15,134	25,413	33,147	1.7	5,316	6,409	1.3	73	54
France	45,544	103,730	187,737	257,128	1.9	3,484	4,306	1.5	76	49
Gabon	9,441	16,800	1,493	1,563	-0.2	2,157	1,271	-3.1	-532	-975
The Gambia	–	–	–	–	–	–	–	–	–	–

COUNTRY	COMMERCIAL ENERGY PRODUCTION		COMMERCIAL ENERGY USE			COMMERCIAL ENERGY USE PER CAPITA			NET ENERGY IMPORTS[a]	
	Thousand Metric Tons of Oil Equivalent		Thousand Metric Tons of Oil Equivalent		Average Annual % Growth	Kg. of Oil Equivalent		Average Annual % Growth	% of Commercial Energy Use	
	1980	2000	1980	2000	1980-2000	1980	2000	1980-2000	1980	2000
Georgia	1,504	737	4,474	2,860	–	882	533	–	66	74
Germany	185,628	134,317	360,441	339,640	-0.2	4,602	4,131	-0.5	48	60
Ghana	3,305	5,883	4,027	7,720	3.6	378	400	0.5	19	24
Greece	3,696	9,987	15,960	27,822	3.0	1,628	2,635	2.5	76	64
Guatemala	2,583	5,241	3,847	7,146	3.5	564	628	0.9	33	27
Guinea	–	–	–	–	–	–	–	–	–	–
Guinea-Bissau	–	–	–	–	–	–	–	–	–	–
Haiti	1,877	1,541	3,847	2,039	0.4	392	256	-1.6	11	24
Honduras	1,315	1,522	1,892	3,012	2.8	530	469	-0.2	31	49
Hungary	14,957	11,090	28,940	24,783	-1.0	2,703	2,448	-0.7	48	55
India	221,322	421,565	241,016	501,894	3.8	351	494	1.8	8	16
Indonesia	128,996	229,478	59,933	145,575	4.8	404	706	3.1	-115	-58
Iran	81,142	242,146	38,987	112,725	5.6	997	1,771	3.1	-108	-115
Iraq	136,643	134,089	12,030	27,678	4.3	925	1,190	1.3	-1,036	-384
Ireland	1,894	2,197	8,485	14,623	2.7	2,495	3,854	2.3	78	85
Israel	153	654	8,563	20,200	5.2	2,208	3,241	2.6	98	97
Italy	19,644	26,858	138,629	171,567	1.3	2,456	2,974	1.2	86	84
Jamaica	224	486	2,378	3,920	3.5	1,115	1,524	2.6	91	88
Japan	43,247	105,505	346,492	524,715	2.6	2,967	4,136	2.2	88	80
Jordan	1	286	1,714	5,185	4.9	786	1,061	0.5	100	94
Kazakhstan	76,799	78,102	76,799	39,093	–	5,163	2,594	–	0	-100
Kenya	7,891	12,260	9,791	15,482	2.2	599	515	-0.7	19	21
Korea, North	29,669	42,576	32,631	46,112	1.9	1,898	2,071	0.5	9	8
Korea, South	9,272	33,615	41,372	193,626	9.1	1,085	4,119	8.0	78	83
Kuwait	91,636	111,469	12,249	20,894	1.0	8,908	10,529	0.2	-648	-434
Kyrgyzstan	2,190	1,443	1,717	2,445	–	473	497	–	-28	41
Laos	–	–	–	–	–	–	–	–	–	–
Latvia	261	1,250	566	3,655	–	222	1,541	–	54	66
Lebanon	178	171	2,524	5,058	4.8	841	1,169	2.8	93	97
Lesotho	–	–	–	–	–	–	–	–	–	–
Liberia	–	–	–	–	–	–	–	–	–	–
Libya	96,550	73,904	7,193	16,438	3.6	2,364	3,107	1.0	-1,242	-350
Lithuania	–	3,212	–	7,124	–	–	2,032	–	–	55
Macedonia	–	–	–	–	–	–	–	–	–	–
Madagascar	–	–	–	–	–	–	–	–	–	–
Malawi	–	–	–	–	–	–	–	–	–	–
Malaysia	18,202	76,759	12,162	49,472	7.7	884	2,126	4.9	-50	-55
Mali	–	–	–	–	–	–	–	–	–	–
Mauritania	–	–	–	–	–	–	–	–	–	–
Mauritius	–	–	–	–	–	–	–	–	–	–
Mexico	149,359	229,653	98,898	155,513	2.1	1,464	1,567	0.2	-51	-50
Moldova	35	60	–	2,871	–	–	671	–	–	98
Mongolia	–	–	–	–	–	–	–	–	–	–
Morocco	877	572	4,778	10,293	4.3	247	359	2.2	82	94
Mozambique	7,413	7,219	8,074	7,126	-0.8	668	403	-2.6	8	-1
Myanmar	9,513	15,144	9,430	12,522	1.3	280	262	-0.4	-1	-21
Namibia	–	292	–	1031	–	–	587	–	–	72
Nepal	4,403	6,872	4,576	7,900	2.7	314	343	0.4	4	13
Netherlands	71,830	57,239	64,984	75,799	1.4	4,593	4,762	0.8	-11	24
New Zealand	5,488	15,379	9,213	18,633	3.8	2,959	4,864	2.7	40	17

COUNTRY	COMMERCIAL ENERGY PRODUCTION		COMMERCIAL ENERGY USE			COMMERCIAL ENERGY USE PER CAPITA			NET ENERGY IMPORTS[a]	
	Thousand Metric Tons of Oil Equivalent		Thousand Metric Tons of Oil Equivalent		Average Annual % Growth	Kg. of Oil Equivalent		Average Annual % Growth	% of Commercial Energy Use	
	1980	2000	1980	2000	1980-2000	1980	2000	1980-2000	1980	2000
Nicaragua	907	1,553	1,553	2,746	2.7	531	542	0.0	42	43
Niger	–	–	–	–	–	–	–	–	–	–
Nigeria	148,479	197,726	52,846	90,169	2.6	743	710	-0.4	-181	-119
Norway	55,716	224,993	18,792	25,617	1.8	4,588	5,704	1.3	-197	-778
Oman	15,090	60,084	996	9,750	11.0	905	4,046	6.6	-1,415	-516
Pakistan	20,997	47,124	25,472	63,951	4.8	308	463	2.2	18	26
Panama	526	732	1,399	2,546	2.7	717	892	0.7	62	71
Papua New Guinea	–	–	–	–	–	–	–	–	–	–
Paraguay	1,605	6,886	2,089	3,930	4.2	671	715	1.2	23	-75
Peru	14,655	9,477	11,752	12,695	0.2	678	489	-1.8	-25	25
Philippines	10,670	20,922	21,212	42,424	3.9	442	554	1.5	50	51
Poland	122,222	78,960	123,031	89,975	-1.4	3,458	2,328	-1.8	1	12
Portugal	1,481	3,129	10,291	24,613	4.7	1,054	2,459	4.7	86	87
Puerto Rico	–	–	–	–	–	–	–	–	–	–
Romania	52,587	28,290	65,123	36,330	-3.1	2,933	1,619	-3.2	19	22
Russian Federation	748,647	966,512	763,707	613,969	–	5,494	4,218	–	2	-57
Rwanda	–	–	–	–	–	–	–	–	–	–
Saudi Arabia	533,071	487,889	31,108	105,303	5.0	3,319	5,081	1.0	-1,614	-363
Senegal	1,046	1,723	1,010	3,086	2.4	346	324	-0.3	45	44
Serbia and Montenegro	–	–	–	–	–	–	–	–	–	–
Sierra Leone	–	–	–	–	–	–	–	–	–	–
Singapore	–	64	6,062	24,591	8.7	2,511	6,120	6.0	–	100
Slovak Republic	3,418	5,994	21,056	17,466	-1.4	4,224	3,234	-1.7	84	66
Slovenia	–	3,098	–	6,540	–	–	3,288	–	–	53
Somalia	–	–	–	–	–	–	–	–	–	–
South Africa	73,169	144,469	65,417	107,595	2.1	2,372	2,514	-0.2	-12	-34
Spain	15,636	31,865	68,576	124,881	3.2	1,834	3,084	2.9	77	74
Sri Lanka	3,209	4,530	4,536	8,063	2.5	311	437	1.4	29	44
Sudan	7,089	23,664	8,406	16,216	3.1	435	521	0.7	16	-46
Swaziland	–	–	–	–	–	–	–	–	–	–
Sweden	16,132	30,681	39,911	47,481	1.0	4,803	5,354	0.6	60	35
Switzerland	7,030	11,782	20,861	26,587	1.4	3,301	3,704	0.7	66	56
Syria	9,502	32,890	5,348	18,407	5.4	614	1,137	2.2	-78	-79
Tajikistan	1,986	1,250	1,650	2,911	–	416	470	–	-20	57
Tanzania	9,502	14,601	10,280	15,386	2.0	553	457	-1.0	8	5
Thailand	11,182	41,118	22,808	73,618	7.4	488	1,212	6.0	51	44
Togo	562	1,036	715	1530	3.8	284	338	0.8	21	32
Trinidad and Tobago	13,141	17,884	3,873	8,655	3.3	3,580	6,660	2.4	-239	-106
Tunisia	6,966	7,003	3,907	7,888	3.7	612	825	1.6	-78	11
Turkey	17,077	26,186	31,452	77,104	4.6	707	1,181	2.7	46	66
Turkmenistan	8,034	45,968	7,948	13,855	–	2,778	2,627	–	-1	-231
Uganda	–	–	–	–	–	–	–	–	–	–
Ukraine	109,708	82,330	97,893	139,592	–	1,956	2,820	–	-12	41
United Arab Emirates	89,716	143,589	6,273	29,559	8.2	6,014	10,175	2.8	-1,330	-386
United Kingdom	196,792	272,338	201,284	232,644	1.0	3,573	3,962	0.8	2	-17
United States	1,553,260	1,675,770	1,811,650	2,299,669	1.5	7,973	8,148	0.4	14	27
Uruguay	766	1,028	2,643	3,079	1.7	907	923	1.0	71	67
Uzbekistan	4,615	55,066	4,821	50,151	–	302	2,027	–	4	-10
Venezuela	140,578	225,470	36,148	29,256	2.5	2,395	2,452	0.1	-289	-280
Vietnam	18,364	46,299	19,573	36,965	3.2	364	471	1.2	6	-25

Table J (Continued)
World Countries: Energy Production and Use, 1980–2000

COUNTRY	COMMERCIAL ENERGY PRODUCTION		COMMERCIAL ENERGY USE			COMMERCIAL ENERGY USE PER CAPITA			NET ENERGY IMPORTS[a]	
	Thousand Metric Tons of Oil Equivalent		Thousand Metric Tons of Oil Equivalent		Average Annual % Growth	Kg. of Oil Equivalent		Average Annual % Growth	% of Commercial Energy Use	
	1980	2000	1980	2000	1980-2000	1980	2000	1980-2000	1980	2000
West Bank and Gaza	–	–	–	–	–	–	–	–	–	–
Yemen	60	22,046	1,424	3,526	4.2	167	201	0.3	96	-525
Zambia	4,179	5,916	4,719	6,244	1.3	822	619	-1.6	11	5
Zimbabwe	5,793	8,708	6,570	10,219	2.6	921	809	-0.3	12	15
World	6,912,364	10,010,145	6,930,291	9886146	2.9	1,623	1649	0.9	–	–
Low income	819,169	1,400,460	674,008	1,287,496	4.7	452	569	2.2	-22	-9
Middle income	33,026,123	4,812,604	2,446,876	3,457,150	4.1	1,252	1,318	2.1	-35	-39
Lower middle income	2,028,987	3,252,169	1,856,387	2,573,688	5.0	1,156	1,206	2.9	-9	-26
Upper middle income	1,273,626	1,560,436	590,489	883,462	2.1	1,694	1,805	0.3	-116	-77
Low & middle income	4,121,782	6,213,064	3,120,884	4,744,646	4.3	906	971	2.0	-32	-31
East Asia & Pacific	842,071	1,579,933	776,249	1,549,127	3.9	578	871	2.4	-8	-2
Europe & Central Asia	1,241,969	1,470,085	1,332,884	1,253,443	7.6	3,348	2,653	–	7	-17
Latin America & Carib.	476,911	847,298	381,002	601,859	2.4	1,074	1,181	0.6	-25	-41
Middle East & N. Africa	981,350	1,268,307	138,565	398,549	4.9	798	1,368	2.2	-610	-219
South Asia	256,676	495,144	284,041	600,474	3.9	321	453	1.8	10	18
Sub-Saharan Africa	322,805	552,297	208,143	341,194	2.3	714	669	-0.5	-55	-62
High income	2,790,581	3,797,081	3,809,407	5,141,500	1.8	4,623	5,430	1.1	27	26
Europe EMU	367,292	434,433	952,761	1,160,702	1.2	3,337	3,824	0.9	61	63

Source: World Development Indicators 2003, (World Bank).

a. A negative value indicates that a country is a net exporter.

Table K
World Countries: Energy Efficiency and Emissions, 1980–1999

COUNTRY	GDP PER UNIT OF ENERGY USE		TRADITIONAL FUEL USE		CARBON DIOXIDE EMISSIONS					
	PPP $ Per Kg. of Oil Equivalent		% of Total Energy Use		Total Million Metric Tons		Per Capita Metric Tons		Kg. Per PPP $ of GDP	
	1980	2000	1980	1997	1980	1999	1980	1999	1980	1999
Afghanistan	–	–	–	–	1.7	1.0	0.1	0	–	–
Albania	–	6.7	13.1	7.3	4.8	1.5	1.8	0.5	–	0.2
Algeria	5.5	6.4	1.9	1.5	66.1	90.8	3.5	3.0	1.0	0.5
Angola	–	3.6	64.9	69.7	5.3	10.3	0.8	0.8	–	0.4
Argentina	4.4	7.2	5.9	4.0	107.5	137.5	3.8	3.8	0.6	0.3
Armenia	–	4.5	–	0.0	–	3.1	–	0.8	–	0.4
Australia	2.0	4.3	3.8	4.4	202.8	344.4	13.8	18.2	1.5	0.8
Austria	3.4	7.5	1.2	4.7	52.4	61.4	6.9	7.6	0.7	0.3
Azerbaijan	–	1.9	–	0.0	–	33.6	–	4.2	–	1.8
Bangladesh	5.4	10.8	81.3	46.0	7.6	25.4	0.1	0.2	0.2	0.1
Belarus	–	3.0	–	0.8	–	57.6	–	5.7	–	0.9
Belgium	2.2	4.4	0.2	1.6	131.3	104.4	13.3	10.2	1.3	0.4
Benin	1.2	2.5	85.4	89.2	0.5	1.3	0.1	0.2	0.3	0.2
Bolivia	3.0	3.9	19.3	14.0	4.5	11.2	0.8	1.4	0.6	0.6
Bosnia-Herzegovina	–	5.2	–	10.1	–	4.8	–	1.2	–	0.2
Botswana	–	–	35.7	–	1.0	3.9	1.1	2.4	0.7	0.4
Brazil	4.2	6.7	35.5	28.7	183.4	300.7	1.5	1.8	0.4	0.3
Bulgaria	1.0	2.8	0.5	1.3	75.3	42.1	8.5	5.1	2.7	0.9
Burkina Faso	–	–	91.3	87.1	0.4	1.0	0.1	0.1	0.1	0.1
Burundi	–	–	97.0	94.2	0.1	0.2	0.0	0.0	0.1	0.1
Cambodia	–	–	100.0	89.3	0.3	0.7	0.0	0.1	–	0.0
Cameroon	2.7	3.8	51.7	69.2	3.9	4.7	0.4	0.3	0.4	0.2
Canada	1.4	3.3	0.4	4.7	420.9	438.6	17.1	14.4	1.5	0.6
Central African Republic	–	–	88.9	87.5	0.1	0.3	0.0	0.1	0.1	0.1
Chad	–	–	95.9	97.6	0.2	0.1	0.0	0.0	0.1	0.0
Chile	3.0	5.6	12.3	11.3	27.5	62.5	2.5	4.2	1.0	0.5
China	0.7	4.1	8.4	5.7	1,476.8	2,825.0	1.5	2.3	3.5	0.7
Hong Kong, China	6.2	10.9	0.9	0.7	16.3	41.2	3.2	6.2	0.5	0.3
Colombia	4.7	10.3	15.9	17.7	39.8	63.6	1.4	1.5	0.4	0.2
Democratic Republic of the Congo (formerly Zaire)	3.8	2.5	73.9	91.7	3.5	2.1	0.1	0.0	0.1	0.1
Congo Republic	0.8	3.2	77.8	53.0	0.4	2.4	0.2	0.8	0.6	0.9
Costa Rica	6.6	11.7	26.3	54.2	2.5	6.1	1.1	1.6	0.2	0.2
Côte d'Ivoire	–	–	52.8	91.5	5.3	13.3	0.6	0.9	0.5	0.6
Croatia	–	4.9	–	3.2	–	20.8	–	4.8	–	0.6
Cuba	–	–	27.9	30.2	30.8	25.4	3.2	2.3	–	–
Czech Republic	–	3.6	0.6	1.6	–	108.9	–	10.6	–	0.8
Denmark	3.0	7.9	0.4	5.9	92.9	49.7	12.3	9.3	1.1	0.3
Dominican Republic	4.1	7.4	27.5	14.3	6.4	23.3	1.1	2.8	0.4	0.4
Ecuador	2.8	4.9	26.7	17.5	13.4	23.3	1.7	1.9	0.9	0.6
Egypt	3.3	4.8	4.7	3.2	45.2	123.6	1.1	2.0	0.9	0.6
El Salvador	5.0	8.1	52.9	34.5	2.1	5.8	0.5	0.9	0.2	0.2
Eritrea	–	–	–	96.0	–	6.0	–	0.1	–	0.1
Estonia	–	2.9	–	13.8	–	16.2	–	11.7	–	1.4
Ethiopia	1.6	2.6	89.6	95.9	1.8	5.5	0.0	0.1	0.1	0.1
Finland	1.7	3.8	4.3	6.5	56.9	58.4	11.9	11.3	1.3	0.5
France	2.8	5.4	1.3	5.7	482.7	359.7	9.0	6.1	0.9	0.3

Table K *(Continued)*
World Countries: Energy Efficiency and Emissions, 1980–1999

COUNTRY	GDP PER UNIT OF ENERGY USE		TRADITIONAL FUEL USE		CARBON DIOXIDE EMISSIONS					
	PPP $ Per Kg. of Oil Equivalent		% of Total Energy Use		Total Million Metric Tons		Per Capita Metric Tons		Kg. Per PPP $ of GDP	
	1980	2000	1980	1997	1980	1999	1980	1999	1980	1999
Gabon	1.8	4.7	30.8	32.9	6.2	3.6	8.9	3.0	2.3	0.5
The Gambia	–	–	72.7	78.6	0.2	0.3	0.2	0.2	0.2	0.1
Georgia	4.6	4.5	–	1.0	–	5.4	–	1.0	–	0.5
Germany	2.2	6.1	0.3	1.3	–	792.2	–	9.7	–	0.4
Ghana	3.1	5.5	43.7	78.1	2.4	5.6	0.2	0.3	0.2	0.1
Greece	4.7	6.3	3.0	4.5	51.7	85.9	5.4	8.2	0.7	0.5
Guatemala	4.6	7.1	54.6	62.0	4.5	9.7	0.7	0.9	0.3	0.2
Guinea	–	–	71.4	74.2	0.9	1.3	0.2	0.2	–	0.1
Guinea-Bissau	–	–	80.0	57.1	0.5	0.3	0.7	0.2	1.4	0.3
Haiti	4.7	7.5	80.7	74.7	0.8	1.4	0.1	0.2	0.1	0.1
Honduras	3.2	6.0	55.3	54.8	2.1	5.0	0.6	0.8	0.3	0.3
Hungary	2.0	4.9	2.0	1.6	82.5	56.9	7.7	5.6	1.5	0.5
India	2.2	5.5	31.5	20.7	347.3	1,077.0	0.5	1.1	0.7	0.4
Indonesia	2.0	4.2	51.5	29.3	94.6	235.6	0.6	1.2	0.8	0.4
Iran	2.7	3.2	0.4	0.7	116.1	301.4	3.0	4.8	1.1	0.9
Iraq	–	–	0.3	0.1	44.0	74.2	3.4	3.3	–	–
Ireland	2.3	7.9	0.0	0.2	25.2	40.4	7.4	10.8	1.3	0.4
Israel	3.7	6.5	0.0	0.0	21.2	61.1	5.4	10.0	0.7	0.5
Italy	3.9	8.2	0.8	1.0	371.9	422.7	6.6	7.3	0.7	0.3
Jamaica	1.8	2.4	5.0	6.0	8.4	10.2	4.0	4.0	2.0	1.2
Japan	3.1	6.1	0.1	1.6	920.4	1,155.2	7.9	9.1	0.8	0.4
Jordan	3.1	3.6	0.0	0.0	4.7	14.6	2.2	3.1	0.9	0.8
Kazakhstan	–	2.2	–	0.2	–	112.8	–	7.4	–	1.5
Kenya	1.0	1.9	76.8	80.3	6.2	8.8	0.4	0.3	0.6	0.3
Korea, North	–	–	3.1	1.4	124.9	208.7	7.3	9.4	–	–
Korea, South	2.3	3.6	4.0	2.4	125.1	393.5	3.3	8.4	1.3	0.6
Kuwait	1.4	1.8	0.0	0.0	24.7	48.0	18.0	24.9	1.5	1.4
Kyrgyzstan	–	5.4	–	0.0	–	4.7	–	1.0	–	0.4
Laos	–	–	72.3	88.7	0.2	0.4	0.1	0.1	–	0.1
Latvia	19.8	4.6	–	26.2	–	6.6	–	2.8	–	0.4
Lebanon	–	3.5	2.4	2.5	6.2	16.9	2.1	4.0	–	1.0
Lesotho	–	–	–	–	–	–	–	–	–	–
Liberia	–	–	–	–	2.0	0.4	1.1	0.1	–	–
Libya	–	–	2.3	0.9	26.9	42.8	8.8	8.3	–	–
Lithuania	–	3.9	–	6.3	–	13.2	–	3.8	–	0.5
Macedonia	–	–	–	6.1	–	11.4	–	5.6	–	1.0
Madagascar	–	–	78.4	84.3	1.6	1.9	0.2	0.1	0.3	0.2
Malawi	–	–	90.6	88.6	0.7	0.8	0.1	0.1	0.3	0.1
Malaysia	2.6	4.3	15.7	5.5	28.0	123.7	2.0	5.4	0.9	0.7
Mali	–	–	86.7	88.9	0.4	0.5	0.1	0.0	0.1	0.1
Mauritania	–	–	0.0	0.0	0.6	3.0	0.4	1.2	0.3	0.6
Mauritius	–	–	59.1	36.1	0.6	2.5	0.6	2.1	0.3	0.2
Mexico	2.9	5.5	5.0	4.5	252.5	378.5	3.7	3.9	0.9	0.5
Moldova	–	3.1	–	0.5	–	6.5	–	1.5	–	0.8
Mongolia	–	–	14.4	4.3	6.8	7.5	4.1	3.2	3.7	1.9
Morocco	6.4	9.5	5.2	4.0	15.9	35.8	0.8	1.3	0.5	0.4
Mozambique	0.7	2.5	43.7	91.4	3.2	1.3	0.3	0.1	0.6	0.1
Myanmar (Burma)	–	–	69.3	60.5	4.8	9.2	0.1	0.2	–	–

COUNTRY	GDP PER UNIT OF ENERGY USE		TRADITIONAL FUEL USE		CARBON DIOXIDE EMISSIONS					
	PPP $ Per Kg. of Oil Equivalent		% of Total Energy Use		Total Million Metric Tons		Per Capita Metric Tons		Kg. Per PPP $ of GDP	
	1980	2000	1980	1997	1980	1999	1980	1999	1980	1999
Namibia	–	12.0	–	–	–	0.1	–	0.1	–	0.0
Nepal	1.5	3.7	94.2	89.6	0.5	3.3	0.0	0.1	0.1	0.1
Netherlands	2.3	5.7	0.0	1.1	153.0	134.6	10.8	8.5	1.0	0.3
New Zealand	2.7	3.7	0.2	0.8	17.6	30.8	5.6	8.1	0.7	0.5
Nicaragua	4.0	4.6	49.2	42.2	2.0	3.8	0.7	0.8	0.3	0.3
Niger	–	–	79.5	80.6	0.6	1.1	0.1	0.1	0.1	0.1
Nigeria	0.8	1.2	66.8	67.8	68.1	40.4	1.0	0.3	1.7	0.4
Norway	2.3	5.1	0.4	1.1	38.7	38.7	9.5	8.7	0.9	0.3
Oman	4.5	3.0	0.0	–	5.9	19.9	5.3	8.5	1.3	0.7
Pakistan	2.1	4.0	24.4	29.5	31.6	98.9	0.4	0.7	0.6	0.4
Panama	4.1	6.5	26.6	14.4	3.5	8.3	1.8	2.9	0.6	0.5
Papua New Guinea	–	–	65.4	62.5	1.8	2.5	0.6	0.5	0.5	0.2
Paraguay	4.8	7.2	62.0	49.6	1.5	4.5	0.5	0.8	0.1	0.2
Peru	4.4	9.5	15.2	24.6	23.6	30.4	1.4	1.2	0.5	0.3
Philippines	5.3	6.8	37.0	26.9	36.5	73.2	0.8	1.0	0.3	0.3
Poland	–	4.0	0.4	0.8	456.2	314.4	12.8	8.1	–	0.9
Portugal	5.5	7.2	1.2	0.9	27.1	60.0	2.8	6.0	0.5	0.4
Puerto Rico	–	–	0.0	–	14.0	10.1	4.4	2.7	0.6	0.1
Romania	–	3.4	1.3	5.7	191.8	81.2	8.6	3.6	–	0.7
Russian Federation	–	1.6	–	0.8	–	1,437.3	–	9.8	–	1.6
Rwanda	–	–	89.8	88.3	0.3	0.6	0.1	0.1	0.1	0.1
Saudi Arabia	4.0	2.6	0.0	0.0	130.7	235.4	14.0	11.7	1.1	0.9
Senegal	2.2	4.5	50.8	56.2	2.8	3.7	0.5	0.4	0.7	0.3
Serbia and Montenegro	–	–	–	1.5	102.0	39.5	10.4	3.7	–	–
Sierra Leone	–	–	90.0	86.1	0.6	0.5	0.2	0.1	0.3	0.3
Singapore	2.2	3.9	0.4	0.0	30.1	54.3	12.5	13.7	2.3	0.7
Slovak Republic	–	3.6	–	0.5	–	38.6	–	7.2	–	0.7
Slovenia	–	5.0	–	1.5	–	14.4	–	7.3	–	0.5
Somalia	–	–	–	–	0.6	0.0	0.1	0.0	–	–
South Africa	3.1	4.4	4.9	43.4	211.3	334.6	7.7	7.9	1.0	0.8
Spain	3.8	6.4	0.4	1.3	200.0	273.7	5.3	6.8	0.8	0.4
Sri Lanka	3.1	7.8	53.5	46.5	3.4	8.6	0.2	0.5	0.2	0.2
Sudan	1.6	3.8	86.9	75.1	3.3	2.6	0.2	0.1	0.2	0.0
Swaziland	–	–	–	–	0.5	0.4	0.8	0.4	0.4	0.1
Sweden	2.0	4.4	7.7	17.9	71.4	46.6	8.6	5.3	0.9	0.2
Switzerland	4.4	7.5	0.9	6.0	40.9	40.6	6.5	5.7	0.4	0.2
Syria	2.6	2.9	0.0	0.0	19.3	53.4	2.2	3.4	1.4	1.1
Tajikistan	–	2.3	–	–	–	5.1	–	0.8	–	0.8
Tanzania	–	1.1	92.0	91.4	1.9	2.5	0.1	0.1	–	0.2
Thailand	2.9	5.1	40.3	24.6	40.0	199.7	0.9	3.3	0.6	0.6
Togo	4.9	4.9	35.7	71.9	0.6	1.3	0.2	0.3	0.2	0.2
Trinidad and Tobago	1.2	1.3	1.4	0.8	16.7	25.1	15.4	19.4	3.6	2.4
Tunisia	3.8	7.4	16.1	12.5	9.4	17.5	1.5	1.8	0.6	0.3
Turkey	3.2	5.3	20.5	3.1	76.3	198.5	1.7	3.1	0.8	0.5
Turkmenistan	–	1.2	–	–	–	31.0	–	6.7	–	2.5
Uganda	–	–	93.6	89.7	0.6	1.4	0.1	0.1	0.1	0.0
Ukraine	–	1.4	–	0.5	–	374.3	–	7.5	–	2.1

Table K *(Continued)*
World Countries: Energy Efficiency and Emissions, 1980–1999

COUNTRY	GDP PER UNIT OF ENERGY USE		TRADITIONAL FUEL USE		CARBON DIOXIDE EMISSIONS					
	PPP $ Per Kg. of Oil Equivalent		% of Total Energy Use		Total Million Metric Tons		Per Capita Metric Tons		Kg. Per PPP $ of GDP	
	1980	2000	1980	1997	1980	1999	1980	1999	1980	1999
United Arab Emirates	4.9	2.0	0.0	–	36.3	88.0	34.8	31.3	1.2	1.6
United Kingdom	2.5	6.0	0.0	3.3	580.3	539.3	10.3	9.2	1.2	0.4
United States	1.6	4.2	1.3	3.8	4,626.8	5,495.4	20.4	19.7	1.6	0.6
Uruguay	4.8	9.4	11.1	21.0	5.8	6.5	2.0	2.0	0.5	0.2
Uzbekistan	–	1.1	–	0.0	–	104.8	–	4.4	–	2.1
Venezuela	1.6	2.3	0.9	0.7	90.1	125.8	6.0	5.3	1.6	1.0
Vietnam	–	4.2	49.1	37.8	16.8	46.6	0.3	0.6	–	0.3
West Bank and Gaza	–	–	–	–	–	–	–	–	–	–
Yemen	–	4.0	0.0	1.4	–	18.3	–	1.1	–	1.4
Zambia	0.8	1.2	37.4	72.7	3.5	1.8	0.6	0.2	0.9	0.3
Zimbabwe	1.5	3.1	27.6	25.2	9.6	17.6	1.3	1.4	1.0	0.5
WORLD	**2.1**	**4.5**	**7.4w**	**8.2w**	**13,852.7**	**22,518.8**	**3.4**	**3.8**	**1.1**	**0.5**
Low income	2.1	4.0	46.4	29.8	774.3	2,429.2	0.5	1.0	0.6	0.5
Middle income	2.1	4.0	10.4	7.3	4,132.9	8,484.0	2.3	3.2	1.2	0.7
Lower middle income	1.6	3.7	10.7	5.7	2,682.6	6,391.3	1.8	3.0	1.6	0.7
Upper middle income	3.4	4.9	8.6	10.6	1,450.3	2,092.7	4.3	4.3	0.7	0.5
Low & middle income	2.1	4.0	18.5	12.9	4,907.1	10,613.2	1.5	2.2	1.0	0.6
East Asia & Pacific	–	–	15.1	9.7	1,833.3	3,734.4	1.3	2.1	2.2	0.6
Europe & Central Asia	–	2.3	3.2	1.3	989.0	3,144.1	–	6.6	1.3	1.2
Latin America & Carib.	3.6	6.1	18.4	16.0	848.8	1,286.7	2.4	2.5	0.6	0.4
Middle East & N. Africa	3.6	3.8	1.6	1.1	491.7	1,048.4	3.0	3.7	1.0	0.7
South Asia	2.3	5.5	34.2	23.8	392.3	1,215.1	0.4	0.9	0.6	0.4
Sub-Sarahan Africa	2.0	2.9	47.2	63.5	352.0	484.6	0.9	0.8	0.8	0.4
High Income	2.2	4.9	1.0	3.4	8,945.6	11,606.6	12.0	12.3	1.2	0.5
Europe EMU	2.8	6.2	0.7	2.5	1,565.2	2,408.4	7.5	7.9	0.8	0.4

Source: World Development Indicators 2003 (World Bank).

Table L
World Countries: Power and Transportation

Country	Electric power										Air passengers carried Thousands
	Consumption per capita Kilowatt-hours		Transmission and distribution losses % of output		Paved roads % of total		Goods transported by road Millions of ton-km hauled		Goods transported by rail Ton-km per $ million of GDP (PPP)		
	1980	1996	1980	1996	1990	1997	1990	1997	1990	1997	1996
AFRICA											
Algeria	265	524	11	18	67	69	14,000	X	25,161	X	3,494
Angola	67	61	25	28	25	25	X	X	X	X	585
Benin	36	48	220	87	20	20	X	X	X	X	75
Botswana	X	X	X	X	32	24	X	X	X	X	104
Burkina Faso	X	X	X	X	17	16	X	X	X	X	138
Burundi	X	X	X	X	18	7	X	X	X	X	9
Cameroon	167	171	7	20	11	13	X	X	33,209	34,023	362
Central African Republic	X	X	X	X	X	X	144	60	X	X	75
Chad	X	X	X	X	1	1	X	X	X	X	93
Congo (Zaire)	147	130	8	3	X	X	X	X	32,198	X	178
Congo Republic	94	207	1	0	10	10	X	X	144,851	X	253
Côte d'Ivoire	192	174	7	16	9	10	X	X	15,791	13,486	179
Egypt	380	924	13	0	72	78	31,400	31,500	23,310	X	4,282
Equatorial Guinea	X	X	X	X	X	X	X	X	X	X	X
Eritrea	X	X	X	X	19	22	X	X	X	X	X
Ethiopia	16	18	8	1	15	15	X	X	2,467	X	743
Gabon	X	X	X	X	X	X	X	X	X	X	X
Gambia	X	X	X	X	X	X	X	X	X	X	X
Ghana	X	X	X	X	X	X	X	X	X	X	X
Guinea	X	X	X	X	15	17	X	X	X	X	36
Guinea-Bissau	X	X	X	X	X	X	X	X	X	X	X
Kenya	92	126	16	16	13	14	X	X	75,496	X	779
Lesotho	X	X	X	X	18	18	X	X	X	X	17
Liberia	X	X	X	X	X	X	X	X	X	X	X
Libya	X	X	X	X	X	X	X	X	X	X	X
Madagascar	X	X	X	X	15	12	X	X	X	X	542
Malawi	X	X	X	X	22	19	X	X	14,881	10,003	153
Mali	X	X	X	X	11	12	X	X	53,882	X	75
Mauritania	X	X	X	X	11	11	X	X	X	X	235
Mauritius	X	X	X	X	X	X	X	X	X	X	X
Morocco	223	408	10	4	49	52	2,638	2,086	72,108	55,523	2,301
Mozambique	370	76	0	0	17	19	X	110	X	X	163
Namibia	X	X	X	X	11	8	X	X	308,833	139,137	237
Niger	X	X	X	X	29	8	X	X	X	X	75
Nigeria	68	85	36	32	30	19	X	X	3,009	X	221
Rwanda	X	X	X	X	9	9	X	X	X	X	9
Senegal	97	103	11	16	27	29	X	X	51,209	X	155
Sierra Leone	X	X	X	X	11	8	X	X	X	X	15
Somalia	X	X	X	X	X	X	X	X	X	X	X
South Africa	3,213	3,719	8	8	30	42	X	X	430,594	337,153	7,183
Sudan	X	X	X	X	X	X	X	X	X	X	X
Swaziland	X	X	X	X	X	X	X	X	X	X	X
Tanzania	50	59	14	12	37	4	X	X	77,466	91,623	224
Togo	X	X	X	X	21	32	X	X	X	X	75
Tunisia	379	674	12	11	76	79	X	X	58,795	53,343	1,371
Uganda	X	X	X	X	X	X	X	X	12,582	11,567	100
Zambia	1,016	560	7	11	17	X	X	X	73,728	56,426	235
Zimbabwe	990	765	14	7	14	47	X	X	247,759	196,429	654

Table L *(Continued)*
World Countries: Power and Transportation

| | Electric power | | | | | | | | | | |
| | Consumption per capita Kilowatt-hours | | Transmission and distribution losses % of output | | Paved roads % of total | | Goods transported by road Millions of ton-km hauled | | Goods transported by rail Ton-km per $ million of GDP (PPP) | | Air passengers carried Thousands |
Country	1980	1996	1980	1996	1990	1997	1990	1997	1990	1997	1996
NORTH AMERICA											
Canada	12,329	15,129	9	7	35	35	54,700	71,473	433,765	X	22,856
United States	8,914	11,796	9	7	58	61	1,073,100	1,439,532	360,699	361,911	571,072
CENTRAL AMERICA											
Belize	X	X	X	X	X	X	X	X	X	X	X
Costa Rica	860	1,349	0	12	15	17	2,243	3,070	X	X	918
Cuba	X	X	X	X	X	X	X	X	X	X	X
Dominican Republic	433	608	21	25	45	49	X	X	X	X	30
El Salvador	293	516	13	13	14	20	X	X	X	X	1,800
Guatemala	212	364	6	13	25	28	X	X	X	X	300
Haiti	41	34	26	54	22	24	X	X	X	X	X
Honduras	225	350	14	27	21	20	X	X	X	X	X
Jamaica	482	2,108	17	11	64	71	X	X	X	X	1,388
Mexico	846	1,381	11	15	35	37	108,884	165,000	64,884	53,917	14,678
Nicaragua	303	256	14	28	11	10	X	X	X	X	51
Panama	828	1,140	13	18	32	34	X	X	X	X	689
Trinidad and Tobago	X	X	X	X	X	X	X	X	X	X	X
SOUTH AMERICA											
Argentina	1,170	1,541	13	18	29	29	X	X	36,412	x	7,913
Bolivia	226	371	10	12	4	6	X	X	37,118	X	1,784
Brazil	974	1,660	12	17	10	9	X	X	56,068	X	22,012
Chile	876	1,864	12	9	14	14	X	X	15,882	5,998	3,622
Colombia	561	922	16	22	12	12	6,227	X	2,400	X	8,342
Ecuador	361	616	14	21	13	19	2,638	3,558	X	X	1,925
Guyana	X	X	X	X	X	X	X	X	X	X	X
Paraguay	233	914	6	7	9	10	X	X	X	X	261
Peru	502	598	13	15	10	10	X	X	7,486	X	2,328
Suriname	X	X	X	X	X	X	X	X	X	X	X
Uruguay	977	1,605	15	20	74	90	X	X	10,455	16,125	504
Venezuela	2,037	2,498	12	20	36	39	X	X	X	X	4,487
ASIA											
Afghanistan	X	X	X	X	X	X	X	X	X	X	X
Armenia	2,729	905	10	38	99	100	1,533	479	X	X	X
Azerbaijan	2,440	1,822	14	22	X	X	3,287	497	X	X	1,233
Bangladesh	16	97	35	30	7	12	X	X	8,032	X	1,252
Bhutan	X	X	X	X	X	X	X	X	X	X	X
Cambodia	X	X	X	X	8	8	X	1,200	X	X	X
China	253	687	8	7	X	X	X	X	671,824	364,633	51,770
Hong Kong, China	2,167	5,013	11	14	100	100	X	X	X	X	X
Georgia	1,910	1,020	16	23	94	94	7,370	98	X	X	152
India	130	347	18	18	47	46	X	X	248,469	176,217	13,395
Indonesia	44	296	19	12	46	46	X	X	8,619	X	17,139
Iran	491	1,142	10	20	X	50	X	X	40,223	X	7,610
Iraq	X	X	X	X	X	X	X	X	X	X	X
Israel	2,826	5,081	5	4	100	100	X	X	16,663	11,947	3,695
Japan	4,395	7,083	4	4	69	74	274,444	305,510	11,603	8,664	95,914
Jordan	387	1,187	19	10	100	100	X	X	78,625	47,242	1,299
Kazakhstan	0	2,865	X	15	55	83	44,775	6,481	5,042,201	X	568
Korea, North	X	X	X	X	X	X	X	X	X	X	X
Korea, South	841	4,453	6	5	72	74	31,841	74,504	40,875	24,826	33,033

Table L (Continued)
World Countries: Power and Transportation

| | Electric power | | | | | | Goods transported by road Millions of ton-km hauled | | Goods transported by rail Ton-km per $ million of GDP (PPP) | | Air passengers carried Thousands |
| | Consumption per capita Kilowatt-hours | | Transmission and distribution losses % of output | | Paved roads % of total | | | | | | |
Country	1980	1996	1980	1996	1990	1997	1990	1997	1990	1997	1996
Kuwait	4,749	12,808	10	0	73	81	X	X	X	X	2,133
Kyrgystan	1,556	1,479	6	33	90	91	5,627	350	X	X	488
Laos	X	X	X	X	24	14	120	X	X	X	125
Lebanon	789	1,651	10	13	95	95	X	X	X	X	775
Malaysia	630	2,078	9	11	70	75	X	X	16,313	9,416	15,118
Mongolia	X	X	X	X	10	3	1,871	X	1,324,119	X	X
Myanmar (Burma)	31	58	22	36	11	12	X	X	X	X	335
Nepal	13	39	29	28	38	42	X	X	X	X	755
Oman	X	X	X	X	X	X	X	X	X	X	X
Pakistan	125	333	29	23	54	58	352	84,174	43,586	26,582	5,375
Philippines	353	405	2	17	0	0	X	X	X	X	7,263
Saudi Arabia	1,356	3,980	9	8	41	43	X	X	4,634	4,206	11,706
Singapore	2,412	7,196	5	4	97	97	X	X	X	X	11,841
Sri Lanka	96	203	15	17	32	40	19	30	5,926	X	1,171
Syria	345	755	18	0	72	23	X	X	48,075	29,655	599
Tajikistan	2,217	2,292	7	12	72	83	X	X	X	X	594
Thailand	279	1,289	10	9	55	98	X	X	14,869	X	14,078
Turkey	439	1,161	12	17	X	25	X	139,789	30,838	17,747	8,464
Turkmenistan	1,720	1,020	12	11	74	81	X	X	X	X	523
United Arab Emirates	X	X	X	X	X	X	X	X	X	X	X
Uzbekistan	2,085	1,657	9	9	79	87	X	X	X	X	1,566
Vietnam	50	177	18	19	24	25	X	X	13,526	16,352	2,108
Yemen	59	99	6	26	9	8	X	X	X	X	X
EUROPE											
Albania	1,083	904	4	52	X	30	1,195	80	85,396	5,523	13
Austria	4,371	5,952	6	6	100	100	13,300	16,600	89,362	78,423	4,719
Belarus	2,455	2,476	9	16	96	98	22,128	9,065	1,297,626	624,045	843
Belgium	4,402	6,878	5	5	81	80	32,100	42,800	46,189	31,976	5,174
Bosnia-Herzegovina	X	X	X	X	X	X	X	X	X	X	X
Bulgaria	3,349	3,577	10	13	92	92	13,823	483	360,291	210,161	718
Croatia	0	2,291	X	16	80	82	2,458	470	190,170	86,593	727
Czech Republic	3,595	4,875	7	8	100	100	X	43,088	X	207,099	1,394
Denmark	4,245	6,113	7	5	100	100	9,400	9,400	19,119	14,518	5,892
Estonia	3,433	3,293	5	19	52	51	4,510	2,773	516,391	536,100	149
Finland	7,779	12,979	6	4	61	64	26,300	24,100	99,052	68,994	5,598
France	3,881	6,091	7	6	X	100	137,000	158,200	49,908	39,109	41,253
Germany	5,005	5,596	4	5	99	99	245,700	281,300	X	39,350	40,118
Greece	2,064	3,395	7	7	92	92	12,600	12,800	6,395	1,913	6,396
Hungary	2,335	2,814	10	13	50	43	1,836	770	247,428	104,327	1,563
Ireland	2,528	4,363	10	9	94	94	5,100	5,500	14,322	9,132	7,677
Italy	2,831	4,196	9	7	100	100	177,900	197,600	20,795	18,420	25,839
Latvia	2,664	1,783	26	47	13	38	5,853	800	1,209,517	1,114,210	276
Lithuania	2,715	1,785	12	11	82	89	7,019	8,622	915,522	545,100	214
Macedonia	0	2,443	X	X	59	64	1,708	1,210	X	X	287
Moldova	1,495	1,314	8	23	87	87	6,305	780	X	X	190
Netherlands	4,057	5,555	4	4	88	90	22,900	27,600	12,779	9,751	17,114
New Zealand	6,269	8,420	13	11	57	58	X	X	51,927	X	9,597
Norway	18,289	23,487	9	8	69	74	7,940	11,838	X	X	12,727
Poland	2,470	2,420	10	13	62	66	49,800	95,500	475,103	284,381	1,806
Portugal	1,469	3,044	12	10	X	X	10,900	11,200	13,976	13,598	4,806

Table L (Continued)
World Countries: Power and Transportation

Country	Electric power Consumption per capita Kilowatt-hours 1980	1996	Transmission and distribution losses % of output 1980	1996	Paved roads % of total 1990	1997	Goods transported by road Millions of ton-km hauled 1990	1997	Goods transported by rail Ton-km per $ million of GDP (PPP) 1990	1997	Air passengers carried Thousands 1996
Romania	2,434	1,757	6	12	*51*	51	13,800	22,400	507,379	*231,838*	913
Russian Federation	4,706	4,165	8	9	74	X	300	138	2,725,816	X	22,117
Slovak Republic	3,817	4,450	8	6	99	99	4,180	3,779	X	*297,426*	63
Slovenia	4,089	4,766	8	6	72	83	3,440	1,775	*142,879*	112,529	393
Spain	2,401	3,749	9	9	74	99	151,000	186,700	22,427	*15,984*	27,759
Sweden	10,216	14,239	9	7	71	77	26,500	*31,200*	127,826	103,299	9,879
Switzerland	5,579	6,919	7	7	X	X	10,400	13,000	X	X	10,468
Ukraine	3,598	2,640	8	10	94	95	79,668	20,532	2,109,937	*1,411,737*	1,151
United Kingdom	4,160	5,198	8	9	100	100	136,300	*153,900*	17,191	X	64,209
Yugoslavia (Serbia-Montenegro)	X	X	X	X	X	X	X	X	X	X	X
OCEANIA											
Australia	5,393	8,086	10	7	35	*39*	X	X	82,122	X	30,075
Fiji	X	X	X	X	X	X	X	X	X	X	X
New Zealand	6,269	8,420	13	11	57	*58*	X	X	51,927	X	9,597
Papua New Guinea	X	X	X	X	3	*4*	X	X	X	X	970
Solomon Islands	X	X	X	X	X	X	X	X	X	X	X
World	1,576w	2,027w	8w	8w	39m	44m					1,389,943s
Low income	188	433	12	12	17	*19*					103,110
Excl. China & India	155	218	14	19	17	*18*					37,945
Middle income	1,585	1,902	9	12	52	*51*					238,360
Lower middle income	1,835	1,771	8	11	54	*51*					102,609
Upper middle income	1,188	2,106	10	13	52	*47*					135,751
Low and middle income	633	886	9	12	29	*30*					341,470
East Asia & Pacific	260	724	8	9	24	*12*					143,204
Europe & Central Asia	2,925	2,795	8	11	77	*83*					46,014
Latin America & Carib.	854	1,347	12	16	22	*26*					76,275
Middle East & N. Africa	483	1,162	10	9	67	*50*					37,484
South Asia	116	313	19	19	38	*41*					22,445
Sub-Saharan Africa	444	439	9	10	17	*16*					16,049
High income	5,783	8,121	8	6	86	*92*					1,048,473

Note: Figures in italics are for years other than those specified.

Source: Entering the 21st Century: World Development Report 1999/2000 (Oxford University Press, World Bank, 2000).

Table M
World Countries: Communications, Information, and Science and Technology

| | Per 1,000 People | | | | | | | | | | |
Country	Daily newspapers 1996	Radio 1996	Television sets 1997	Telephone main lines 1997	Mobile telephones 1997	Personal computers 1997	Internet hosts Per 10,000 people January 1999	Scientists and engineers in R & D Per million people 1985–95	High-technology exports % of mfg. exports 1997	No. of patent applications filed[a] Residents	Nonresidents
AFRICA											
Algeria	38	239	67	48	1	4.2	0.01	X	22	48	150
Angola	12	54	91	5	1	0.7	0.00	X	X	X	X
Benin	2	108	91	6	1	0.9	0.02	177	X	X	X
Botswana	27	155	27	56	0	13.4	4.18	X	X	5	56
Burkina Faso	1	32	6	3	0	0.7	0.16	X	X	X	X
Burundi	3	68	10	3	0	X	0.00	32	X	1	4
Cameroon	7	162	81	5	0	1.5	0.00	X	3	X	X
Central African Republic	2	84	5	3	0	X	0.00	55	0	X	X
Chad	0	249	2	1	0	X	0.00	X	X	X	X
Congo (Zaire)	3	98	43	1	0	X	0.00	X	X	2	27
Congo Republic	8	124	8	8	0	X	0.00	X	16	X	X
Côte d'Ivoire	16	157	61	9	2	3.3	0.16	X	X	X	X
Egypt	38	316	127	56	0	7.3	0.31	458	7	504	706
Equatorial Guinea	X	X	X	X	X	X	X	X	X	X	X
Eritrea	X	101	11	6	0	X	0.00	X	X	X	X
Ethiopia	2	194	5	3	0	X	0.01	X	0	3	X
Gabon	X	X	X	X	X	X	X	X	X	X	X
Gambia	X	X	X	X	X	X	X	X	X	X	X
Ghana	14	238	109	6	1	1.6	0.10	X	X	X	33
Guinea	X	47	41	3	0	0.3	0.00	X	X	X	X
Guinea-Bissau	X	X	X	X	X	X	X	X	X	X	X
Kenya	9	108	19	8	0	2.3	0.23	X	11	15	39,034
Lesotho	7	48	24	10	1	X	0.09	X	X	2	37,043
Liberia	X	X	X	X	X	X	X	X	X	X	X
Libya	X	X	X	X	X	X	X	X	X	X	X
Madagascar	4	192	45	3	0	1.3	0.04	11	2	7	20,800
Malawi	3	256	2	4	0	X	0.00	X	3	3	39,031
Mali	1	49	10	2	0	0.6	0.00	X	X	X	X
Mauritania	1	150	89	5	0	5.3	0.06	X	X	X	X
Mauritius	X	X	X	X	X	X	X	X	X	X	X
Morocco	26	241	160	50	3	2.5	0.20	X	27	90	237
Mozambique	3	39	4	4	0	1.6	0.08	X	8	X	X
Namibia	19	143	32	58	8	18.6	15.79	X	X	X	X
Niger	0	69	26	2	0	0.2	0.02	X	X	X	X
Nigeria	24	197	61	4	0	5.1	0.03	15	X	X	X
Rwanda	0	102	X	3	0	X	0.00	24	X	X	X
Senegal	5	141	41	13	1	11.4	0.21	X	55	X	X
Sierra Leone	5	251	20	4	0	X	0.03	X	X	X	X
Somalia	X	X	X	X	X	X	X	X	X	X	X
South Africa	30	316	125	107	37	41.6	34.67	938	X	X	X
Sudan	X	X	X	X	X	X	X	X	X	X	X
Swaziland	X	X	X	X	X	X	X	X	X	X	X
Tanzania	4	278	21	3	1	1.6	0.04	X	X	X	X
Togo	4	217	19	6	1	5.8	0.24	X	X	X	X
Tunisia	31	218	182	70	1	8.6	0.07	388	11	46	128
Uganda	2	123	26	2	0	1.4	0.05	X	X	X	38,497
Zambia	14	121	80	9	0	X	0.31	X	X	6	93
Zimbabwe	18	96	29	17	1	9.0	0.87	X	6	30	181

	Per 1,000 People						Internet hosts	Scientists and engineers in R & D	High-technology exports	No. of patent applications filed[a]	
	Daily newspapers	Radio	Television sets	Telephone main lines	Mobile telephones	Personal computers	Per 10,000 people	Per million people	% of mfg. exports		
Country	1996	1996	1997	1997	1997	1997	January 1999	1985–95	1997	Residents	Nonresidents
NORTH AMERICA											
Canada	159	1,078	708	609	139	270.6	364.25	2,656	25	3,316	45,938
United States	212	2,115	847	644	206	406.7	1,131.52	3,732	44	111,883	111,536
CENTRAL AMERICA											
Belize	X	X	X	X	X	X	X	X	X	X	X
Costa Rica	91	271	403	169	19	X	9.20	X	14	X	X
Cuba	X	X	X	X	X	X	X	X	X	X	X
Dominican Republic	52	177	84	88	16	X	5.79	X	23	X	X
El Salvador	48	461	250	56	7	X	1.33	19	16	3	64
Guatemala	31	73	126	41	6	3.0	0.83	99	13	2	102
Haiti	3	55	5	8	0	X	0.00	X	X	3	6
Honduras	55	409	90	37	2	X	0.16	X	4	10	126
Jamaica	64	482	323	140	22	4.6	1.24	8	67	X	X
Mexico	97	324	251	96	18	37.3	11.64	213	33	389	30,305
Nicaragua	32	283	190	29	2	X	1.47	214	38	X	X
Panama	62	299	187	134	6	X	2.66	X	14	31	142
Trinidad and Tobago	X	X	X	X	X	X	X	X	X	X	X
SOUTH AMERICA											
Argentina	123	677	289	191	56	39.2	18.28	671	15	X	X
Bolivia	55	672	115	69	15	X	0.78	250	9	17	106
Brazil	40	435	316	107	28	26.3	12.88	168	18	2,655	29,451
Chile	99	354	233	180	28	54.1	20.18	X	19	189	1,771
Colombia	49	565	217	148	35	33.4	3.93	X	20	87	1,172
Ecuador	70	342	294	75	13	13.0	1.26	169	12	7	354
Guyana	X	X	X	X	X	X	X	X	X	X	X
Paraguay	50	182	101	43	17	X	2.18	X	4	X	X
Peru	43	271	143	68	18	12.3	1.91	625	10	52	565
Suriname	X	X	X	X	X	X	X	X	X	X	X
Uruguay	116	610	242	232	46	21.9	46.61	688	8	25	182
Venezuela	206	471	172	116	46	36.6	3.37	208	10	182	1,822
ASIA											
Afghanistan	X	X	X	X	X	X	X	X	X	X	X
Armenia	23	5	218	150	2	X	1.01	X	X	162	20,268
Azerbaijan	28	20	211	87	5	X	0.21	X	X	165	16,470
Bangladesh	9	50	7	3	0	X	X	X	0	70	156
Bhutan	X	X	X	X	X	X	X	X	X	X	X
Cambodia	X	127	124	2	3	0.9	0.06	X	X	X	X
China	X	195	270	56	10	6.0	0.14	350	21	11,698	41,016
Hong Kong, China	800	695	412	565	343	230.8	122.71	98	29	41	2,059
Georgia	X	553	473	114	6	X	1.27	X	X	289	21,124
India	X	105	69	19	1	2.1	0.13	149	11	1,660	6,632
Indonesia	23	155	134	25	5	8.0	0.75	X	20	40	3,957
Iran	24	237	148	107	4	32.7	0.04	521	X	X	X
Iraq	X	X	X	X	X	X	X	X	X	X	X
Israel	291	530	321	450	283	186.1	161.96	X	33	1,363	12,172
Japan	580	957	708	479	304	202.4	133.53	6,309	38	340,861	60,390
Jordan	45	287	43	70	2	8.7	0.80	106	26	X	X
Kazakhstan	30	384	234	108	1	X	0.94	X	X	1,024	20,064
Korea, North	X	X	X	X	X	X	X	X	X	X	X
Korea, South	394	1,037	341	444	150	150.7	40.00	2,636	39	68,446	45,548

Table M (*Continued*)
World Countries: Communications, Information, and Science and Technology

	Per 1,000 People										
Country	Daily newspapers	Radio	Television sets	Telephone main lines	Mobile telephones	Personal computers	Internet hosts Per 10,000 people	Scientists and engineers in R & D Per million people	High-technology exports % of mfg. exports	No. of patent applications filed[a]	
										Residents	Nonresidents
	1996	1996	1997	1997	1997	1997	January 1999	1985–95	1997		
Kuwait	376	688	*491*	227	116	82.9	32.80	X	4	X	X
Kyrgyzstan	13	115	44	76	*0*	X	4.04	703	24	126	20,179
Laos	4	139	4	5	1	*1.1*	0.00	X	X	X	X
Lebanon	141	892	354	179	135	31.8	5.56	X	X	X	X
Malaysia	163	432	166	195	113	46.1	21.36	87	*67*	X	X
Mongolia	27	139	63	37	1	5.4	0.08	943	*2*	114	20,882
Myanmar (Burma)	10	89	*7*	5	0	X	0.00	X	X	X	X
Nepal	11	*37*	4	8	0	X	0.07	X	0	X	X
Oman	X	X	X	X	X	X	X	X	X	X	X
Pakistan	*21*	92	65	19	1	4.5	0.23	54	4	16	782
Philippines	82	159	109	29	18	13.6	1.21	157	56	163	2,634
Saudi Arabia	59	319	*260*	117	17	43.6	0.15	X	29	27	810
Singapore	324	739	354	543	273	399.5	210.02	2,728	71	215	38,403
Sri Lanka	29	210	91	17	6	4.1	0.29	173	X	50	21,138
Syria	20	274	*68*	88	0	1.7	0.00	X	1	X	X
Tajikistan	20	X	281	38	0	X	0.12	709	X	32	19,570
Thailand	65	204	234	80	33	19.8	3.35	119	43	203	4,355
Turkey	111	178	286	250	26	20.7	*4.30*	261	9	367	19,668
Turkmenistan	X	96	*175*	78	0	X	0.55	X	X	66	18,948
United Arab Emirates	X	X	X	X	X	X	X	X	X	X	X
Uzbekistan	3	452	*273*	63	0	X	0.10	1,760	X	914	21,088
Vietnam	4	106	*180*	21	2	4.6	0.00	308	X	37	22,206
Yemen	15	64	*273*	13	1	1.2	0.01	X	0	X	X
EUROPE											
Albania	34	235	*161*	23	1	X	0.30	X	1	1	18,761
Austria	294	740	496	492	144	210.7	176.79	1,631	*24*	2,506	75,985
Belarus	*174*	290	314	227	1	X	0.70	2,339	X	701	20,347
Belgium	160	792	510	468	95	235.3	162.39	1,814	23	1,356	59,099
Bosnia-Herzegovina	X	X	X	X	X	X	X	X	X	X	X
Bulgaria	253	531	*366*	323	8	29.7	9.05	X	X	318	22,235
Croatia	114	333	*267*	335	27	22.0	12.84	1,978	19	259	356
Czech Republic	256	806	447	318	51	82.5	71.79	1,159	13	623	24,856
Denmark	311	1,146	568	633	273	360.2	526.77	2,647	27	2,452	72,151
Estonia	173	680	479	321	99	15.1	152.98	2,018	24	12	21,144
Finland	455	1,385	534	556	417	310.7	1,058.13	2,812	26	3,262	61,556
France	218	943	606	575	99	174.4	82.91	2,584	*31*	17,090	81,418
Germany	311	946	570	550	99	255.5	160.23	2,843	26	56,757	98,338
Greece	*153*	477	466	516	89	44.8	48.81	774	12	434	52,371
Hungary	189	697	436	304	69	49.0	82.74	1,033	39	832	24,147
Ireland	153	703	*455*	411	146	241.3	148.70	1,871	62	925	52,407
Italy	104	874	483	447	204	113.0	58.80	1,325	15	8,860	71,992
Latvia	246	699	*592*	302	31	7.9	42.59	1,189	15	197	21,498
Lithuania	92	292	*377*	283	41	6.5	27.48	X	21	101	21,249
Macedonia	19	184	*252*	204	6	X	2.56	X	X	53	18,934
Moldova	59	720	302	145	1	3.8	1.17	1,539	*9*	290	20,245
Netherlands	305	963	541	564	110	280.3	358.51	2,656	44	4,884	61,958
Norway	593	920	579	621	381	360.8	717.53	3,678	*24*	1,550	25,638
Poland	113	518	413	194	22	36.2	28.07	1,299	12	2,414	24,902
Portugal	75	306	523	402	152	74.4	50.01	1,185	11	105	71,544
Romania	X	317	*226*	167	9	8.9	7.42	1,382	7	1,831	22,139

	Per 1,000 People						Internet hosts	Scientists and engineers in R & D	High-technology exports	No. of patent applications filed[a]	
	Daily newspapers	Radio	Television sets	Telephone main lines	Mobile telephones	Personal computers	Per 10,000 people	Per million people	% of mfg. exports	Residents	Nonresidents
Country	1996	1996	1997	1997	1997	1997	January 1999	1985–95	1997		
Russian Federation	105	344	390	183	3	32.0	10.04	3,520	19	18,138	28,149
Slovak Republic	185	580	401	259	37	241.6	33.27	1,821	15	201	22,865
Slovenia	206	416	353	364	47	188.9	89.83	2,544	16	301	21,686
Spain	99	328	506	403	110	122.1	67.21	1,210	*17*	2,689	81,294
Sweden	446	907	531	679	358	350.3	487.13	3,714	34	7,077	76,364
Switzerland	330	969	536	661	147	394.9	315.52	X	28	2,699	75,576
Ukraine	54	872	493	186	1	*5.6*	3.13	3,173	X	3,640	22,862
United Kingdom	332	1,445	641	540	151	242.4	240.99	2,417	41	25,269	104,084
Yugoslavia (Serbia-Montenegro)	X	X	X	X	X	X	X	X	X	X	X
OCEANIA											
Australia	297	1,385	638	505	264	362.2	420.57	3,166	39	9,196	34,125
Fiji	X	X	X	X	X	X	X	X	X	X	X
New Zealand	223	1,027	*501*	486	149	263.9	360.44	1,778	11	1,421	26,947
Papua New Guinea	15	91	24	*11*	1	X	0.25	X	X	X	X
Solomon Islands	X	X	X	X	X	X	X	X	X	X	X
WORLD	Xw	380w	280w	144w	40w	58.4w	75.22w				
Low income	X	147	162	32	5	4.4	0.17				
Excl. China & India	13	133	*59*	16	1	X	0.23				
Middle income	75	383	272	136	24	32.4	10.15				
Lower middle income	63	327	247	108	11	*12.2*	4.91				
Upper middle income	95	469	302	179	43	45.5	19.01				
Low and middle income	X	218	194	65	11	12.3	3.08				
East Asia & Pacific	X	206	237	60	15	11.3	1.66				
Europe & Central Asia	99	412	380	189	13	*17.7*	13.00				
Latin America & Carib.	71	414	263	110	26	31.6	9.64				
Middle East & N. Africa	33	265	*140*	71	6	9.8	0.25				
South Asia	X	99	69	18	1	2.1	0.14				
Sub-Saharan Africa	12	172	*44*	16	4	7.2	2.39				
High income	286	1,300	664	552	188	269.4	470.12				

Note: Figures in italics are for years other than those specified.

a. Other patent applications filed in 1996 include those filed under the auspices of the African Intellectual Property Organization (75 by residents, 20,863 by nonresidents), the African Regional Industrial Property Organization (10 by residents, 20,347 by nonresidents), the European Patent Office (38,546 by residents, 48,068 by nonresidents), and the Eurasian Patent Organization (39 by residents, 18,055 by nonresidents). The original information was provided by the World Intellectual Property Organization (WIPO). The International Bureau of WIPO assumes no liability or responsibility with regard to the transformation of these data.

Source: Entering the 21st Century: World Development Report 1999/2000 (Oxford University Press, World Bank, 2000).

Table N
World Countries: Water Resources

COUNTRY	ANNUAL RENEWABLE WATER RESOURCES[a]				ANNUAL AVERAGE GROUNDWATER RESOURCES[b]	SECTORAL WITHDRAWALS (%)[c]		
	Supply Per Capita (cubic meters) 2000	Recharge	Population (2003)	Withdrawal Per Capita (cubic meters) 2000	Recharge Per Capita (cubic meters) 2000	Domestic	Industry	Agriculture
						9	20	71
WORLD	–	11,358.0	6,130	650	1,853	65	15	67
AFRICA	5,159		812	307	–	–	–	63
Algeria	460	1.7	31	181	55	34	52	14
Angola	13,203	72.0	14	54	5,143	14	10	76
Benin	3,741	1.8	6	27	300	23	10	67
Botswana	9,209	1.7	2	86	850	32	20	48
Burkina Faso	1,024	9.5	12	40	792	19	0	81
Burundi	538	2.1	7	19	300	36	64	0
Cameroon	18,378	100.0	15	38	6,667	46	19	35
Central African Republic	37,565	56.0	4	25	14,000	21	5	74
Chad	5,125	12.0	8	34	1,500	16	2	82
Congo (Zaire)	259,547	198.0	52	20	3,808	62	27	11
Congo Republic	23,639	421.0	3	10	140,333	61	16	23
Cote d'Ivoire	4,853	38.0	4	62	9,500	22	11	67
Egypt	830	1.3	65	1,055	20	11	82	86
Equatorial Guinea	53,841			30	–	81	13	6
Eritrea	1,578	–	4	–	–	–	–	–
Ethiopia	1,666	40.0	66	51	606	11	3	86
Gabon	126,789	62.0	1	70	62,000	72	22	6
Gambia	5,836	0.5	1	29	500	7	2	91
Ghana	2,637	26.0	20	35	1,300	35	13	52
Guinea	26,964	38.0	8	132	4,750	10	3	87
Guinea-Bissau	24,670	14.0	1	17	14,000	60	4	36
Kenya	947	3.0	31	87	97	20	4	76
Lesotho	1,456	0.5	2	32	250	22	22	56
Liberia	70,348	60.0	3	59	20,000	27	13	60
Libya	109	0.5	5	870	100	13	3	84
Madagascar	19,925	55.0	16	1,611	3,438	1	–	99
Malawi	1,461	1.4	11	95	127	10	3	86
Mali	8,320	20.0	24	167	833	2	1	97
Mauritania	4,029	0.3	3	923	100	2	2	92
Morocco	936	10.0	29	399	345	10	2	89
Mozambique	11,382	17.0	18	42	944	9	2	89
Namibia	9,865	2.1	2	175	1,050	29	3	68
Niger	2,891	2.5	11	69	227	16	2	82
Nigeria	2,891	2.5	130	69	19	16	2	82
Rwanda	638	3.6	9	141	400	5	2	94
Senegal	3,977	7.6	10	202	760	5	3	92
Sierra Leone	33,237	50.0	5	98	10,000	7	4	89
Somalia	1,413	3.3	9	119	367	3	0	97
South Africa	1,131	4.8	43	366	112	17	11	72

COUNTRY	ANNUAL RENEWABLE WATER RESOURCES[a]				ANNUAL AVERAGE GROUNDWATER RESOURCES[b]	SECTORAL WITHDRAWALS (%)[c]		
	Supply Per Capita (cubic meters) 2000	Recharge	Population (2003)	Withdrawal Per Capita (cubic meters) 2000	Recharge Per Capita (cubic meters) 2000	Domestic	Industry	Agriculture
Sudan	1,981	7.0	32	637	219	4	1	94
Tanzania	2,472	30.0	34	39	882	9	2	89
Togo	3,076	5.7	5	29	1,140	62	13	25
Tunisia	577	1.5	10	312	150	13	1	86
Uganda	2,663	29.0	23	21	1,261	32	8	60
Zambia	9,676	47.0	10	190	4,700	16	7	77
Zimbabwe	1,530	5.0	13	131	385	14	7	79
NORTH AMERICA	–	1,670.0		–	–	–	–	46
Canada	92,810	370.0	31	1,607	11,935	18	70	12
United States	10,574	1,300.0	285	1,834	4,561	13	45	42
CENTRAL AMERICA	–	359.0		–	–	–	–	–
Belize	78,763	–		485	–	12	88	0
Costa Rica	26,764	37.0	4	1,540	9,250	13	7	80
Cuba	3,382	6.5	11	475	591	49	0	51
Dominican Republic	2,430	12.0	9	1,102	1,333	11	0	89
El Salvador	3,872	6.2	6	137	1,033	34	20	46
Guatemala	9,277	34.0	12	126	2,833	9	17	74
Haiti	1,670	2.2	8	139	275	5	1	94
Honduras	14,250	39.0	7	294	5,571	4	5	91
Jamaica	3,588	3.9	3	371	1,300	15	7	77
Mexico	4,490	139.0	99	812	1,404	17	5	78
Nicaragua	36,784	59.0	5	267	11,800	14	2	84
Panama	50,299	21.0	3	685	7,000	28	2	70
Trinidad and Tobago	2,940	–	1	233	–	68	26	6
SOUTH AMERICA	–	3,693.0	349	–	10,582	–	–	–
Argentina	21,453	128.0	37	822	3,459	16	9	75
Bolivia	71,511	130.0	9	197	14,444	10	3	87
Brazil	47,125	1,874.0	172	359	10,895	21	18	61
Chile	59,143	140.0	15	1,629	9,333	5	11	84
Colombia	49,017	510.0	43	228	11,860	59	4	37
Ecuador	32,948	134.0	13	1,423	10,308	12	6	82
Guyana	314,963	–	–	1,993	–	1	1	98
Paraguay	28,148	41.0	6	112	6,833	15	7	78
Peru	72,127	303.0	26	849	11,654	7	7	86
Suriname	298,848	–	–	1,171	–	6	5	89
Uruguay	41,065	23.0	3	–	7,667	6	3	91
Venezuela	49,144	227.0	25	382	9,080	44	10	46
ASIA	2,790	–		2,007	–	1	0	99
Afghanistan	2,421		27	1,846	0	–	–	99
Armenia	2,778	4.2	4	784	1,050	30	4	66

COUNTRY	ANNUAL RENEWABLE WATER RESOURCES[a]				ANNUAL AVERAGE GROUNDWATER RESOURCES[b]	SECTORAL WITHDRAWALS (%)[c]		
	Supply Per Capita (cubic meters) 2000	Recharge	Population (2003)	Withdrawal Per Capita (cubic meters) 2000	Recharge Per Capita (cubic meters) 2000	Domestic	Industry	Agriculture
Azerbaijan	3,716	6.5	8	2,151	813	2	25	70
Bangladesh	8,444	21.0	133	133	158	12	2	86
Bhutan	43,214	–		13	–	36	10	54
Cambodia	34,516	18.0	12	60	1,500	5	1	94
China	2,186	829.0	1,272	439	652	5	18	78
Georgia	12,149	17.0	5	635	3,400	21	20	59
India	1,822	419.0	1,032	592	406	5	3	92
Indonesia	13,046	455.0	209	407	2,177	6	1	93
Iran	1,900	49.0	65	1,122	754	6	2	92
Iraq	3,111	1.2	24	2,478	50	3	5	92
Israel	265	0.5	6	287	83	39	7	54
Japan	3,372	27.0	127	735	213	19	17	64
Jordan	169	0.5	5	255	100	22	3	75
Kazakhstan	6,839	6.1	15	2,019	407	2	17	81
Korea, North	3,415	13.0	22	742	591	11	16	73
Korea, South	1,471	13.0	47	531	277	26	11	63
Kuwait	10	0.0	2	306	0	37	2	60
Kyrgyzstan	4,078	14.0	5	2,231	2,800	3	3	94
Laos	60,318	38.0	5	259	7,600	8	10	82
Lebanon	1,220	3.2	4	400	800	27	6	68
Malaysia	25,178	64.0	24	636	2,667	11	13	77
Mongolia	13,451	6.1	2	182	3,050	20	27	53
Myanmar (Burma)	21,358	156.0	48	103	3,250	7	3	90
Nepal	8,703	20.0	24	1,451	833	1	0	99
Oman	364	1.0	2	658	500	5	2	94
Pakistan	2,812	55.0	141	1,382	390	2	2	97
Philippines	6,093	180.0	78	811	2,308	8	4	88
Saudi Arabia	111	2.2	21	1,056	105	9	1	90
Singapore	–	–	4	–	–	45	51	4
Sri Lanka	2,592	7.8	19	574	411	2	2	96
Syria	1,541	4.2	17	844	247	8	2	90
Tajikistan	2,587	6.0	6	2,096	1,000	3	4	92
Thailand	6,371	42.0	61	605	689	5	4	91
Turkey	3,344	69.0	66	558	1,045	16	12	73
Turkmenistan	5,015	0.4	5	5,801	80	1	1	98
United Arab Emirates	56	0.1	3	896	33	24	9	67
Uzbekistan	1,968	8.8	25	2,598	352	4	2	94
Vietnam	11,109	48.0	80	822	600	4	10	87
Yemen	206	1.5	18	253	83	7	1	92
EUROPE	–	1,318.0	709	–	1,859	–	–	–
Albania	13,178	6.2	3	440	2,067	29	0	71
Austria	9,629	6.0	8	303	750	33	58	9
Belarus	5,739	18.0	10	266	1,800	22	43	35

COUNTRY	ANNUAL RENEWABLE WATER RESOURCES[a]				ANNUAL AVERAGE GROUNDWATER RESOURCES[b]	SECTORAL WITHDRAWALS (%)[c]		
	Supply Per Capita (cubic meters) 2000	Recharge	Population (2003)	Withdrawal Per Capita (cubic meters) 2000	Recharge Per Capita (cubic meters) 2000	Domestic	Industry	Agriculture
Belgium	1,781	0.9	10	–	90	–	–	–
Bosnia-Herzegovina	9,088	–		292	–	30	10	60
Bulgaria	2,734	6.4	8	1,573	800	22	3	75
Croatia	22,654	11.0	4	164	2,750	50	50	0
Czech Republic	1,283	1.4	10	266	140	41	57	2
Denmark	1,123	4.3	5	233	860	30	27	43
Estonia	9,413	4.0	1	106	4,000	56	39	5
Finland	21,223	2.2	5	439	440	12	85	3
France	3,414	100.0	59	547	1,695	18	72	10
Germany	1,878	46.0	82	579	561	11	69	20
Greece	6,984	10.0	11	826	909	10	3	87
Hungary	10,541	6.0	10	659	600	9	55	36
Iceland	599,944			622	–	31	63	6
Ireland	13,408	11.0	4	232	2,750	16	74	10
Italy	3,330	43.0	58	730	741	19	34	48
Latvia	14,820	2.2	2	112	1,100	55	32	13
Lithuania	6,763	1.2	3	68	400	81	16	3
Macedonia	3,121	–	2	936	–	12	15	74
Moldova	2,726	0.4	4	678	100	9	65	26
Netherlands	5,691	4.5	16	519	281	5	61	34
Norway	84,787	96.0	5	489	19,200	20	72	8
Poland	1,598	13.0	39	321	333	13	76	11
Portugal	6,837	4.0		736	–	15	37	48
Romania	9,486	8.3	22	1,141	377	8	33	59
Russian Federation	31,354	788.0	145	519	5,434	19	62	20
Serbia and Montenegro	19,815	3.0	11	1,233	273	6	86	8
Slovak Republic	9,265	1.7	5	337	340	–	–	–
Slovenia	16,070	14.0	2	642	7,000	20	80	1
Spain	2,793	30.0	41	884	732	13	19	68
Sweden	19,721	20.0	9	340	2,222	36	55	9
Switzerland	7,464	2.5	7	172	357	23	73	4
Ukraine	2,868	20.0	49	500	408	18	52	30
United Kingdom	2,464	9.8	59	204	166	20	77	3
OCEANIA	–		–	–	–	–	–	–
Australia	25,185	72.0	19	933	3,789	65	2	33
Fiji	34,330	–		–	–	20	20	60
New Zealand	85,221	–		588	–	46	10	44
Papua New Guinea	159,171	–		29	–	29	22	49
Solomon Islands	93,405	–		–	–	40	20	40

a. Annual renewable water resources usually include river flows from other countries.
b. Withdrawal data from most recent year available; varies by country from 1987 to 1995.
c. Total withdrawals may exceed 100% because of groundwater withdrawals or river inflows.

Source: World Resources 1998-99 (World Resources Institute).

Table O
World Countries: Globally Threatened Plant and Animal Species

COUNTRY	Mammals: Threatened Species	Mammals: Number of Species per 10,000 km²	Birds: Threatened Species	Birds: Number of Species per 10,000 km²	Reptiles: Threatened Species	Reptiles: Number of Species per 10,000 km²	Amphibians: Threatened Species	Amphibians: Number of Species per 10,000 km²	Freshwater Fish: Threatened Species	Plants: Rare and Threatened Species	Plants: Number of Species per 10,000 km²
AFRICA											
Algeria	15	15	8	32	1	X	0	X	1	145	509
Angola	17	56	13	156	5	X	0	X	0	25	1,017
Benin	9	85	1	138	2	X	0	X	0	3	899
Botswana	5	43	7	101	0	41	0	X	0	4	X
Burkina Faso	6	49	1	112	1	X	0	10	0	0	369
Burundi	5	76	6	322	0	X	0	X	0	1	1,783
Cameroon	32	83	14	193	3	X	1	X	26	74	2,237
Central African Republic	11	53	2	137	1	X	0	X	0	0	921
Chad	14	27	3	75	1	X	0	X	0	12	322
Congo (Zaire)	38	69	26	153	3	X	0	X	1	7	1,817
Congo Republic	10	62	3	140	2	X	0	X	0	3	1,356
Côte d'Ivoire	16	73	12	170	4	X	1	X	0	66	1,118
Egypt	15	21	11	33	6	18	0	1	0	84	452
Equatorial Guinea	12	131	4	194	2	X	1	X	0	9	2,135
Eritrea	6	49	3	140	3	X	0	X	0	X	X
Ethiopia	35	54	20	133	1	X	0	X	0	153	1,378
Gabon	12	64	4	157	3	X	0	X	0	78	2,197
Gambia	4	104	1	269	1	X	0	X	0	0	928
Ghana	13	78	10	186	4	X	0	X	0	32	1,264
Guinea	11	66	12	142	3	X	1	X	0	35	1,043
Guinea-Bissau	4	71	1	159	3	X	0	X	0	0	655
Kenya	43	94	24	221	5	49	0	23	20	158	1,571
Lesotho	2	23	5	40	0	X	0	X	1	7	1,093
Liberia	11	87	13	168	3	28	1	17	0	1	1,037
Libya	11	14	2	17	3	X	0	X	0	57	327
Madagascar	46	27	28	53	17	66	2	38	13	189	2,347
Malawi	7	86	9	230	0	55	0	31	0	61	1,592
Mali	13	28	6	81	1	3	0	X	0	14	355
Mauritania	14	13	3	59	3	X	0	X	0	3	239
Mauritius	4	7	10	46	6	19	0	0	0	222	1,183
Morocco	18	30	11	60	2	X	0	X	1	195	1,028
Mozambique	13	42	14	117	5	X	0	15	2	92	1,294
Namibia	11	36	8	109	3	X	1	7	3	23	729
Niger	11	27	2	60	1	X	0	X	0	0	237
Nigeria	26	62	9	153	4	X	0	X	0	9	1,036
Rwanda	9	110	6	373	0	X	0	X	0	0	1,662
Senegal	13	58	6	144	7	X	0	X	0	32	771
Sierra Leone	9	77	12	243	3	X	0	X	0	12	1,091
Somalia	18	43	8	107	2	49	0	7	3	57	761
South Africa	33	51	16	122	19	61	9	19	27	953	4,711
Sudan	21	43	9	110	3	X	0	X	0	8	506
Swaziland	5	37	6	303	0	85	0	33	0	41	2,197
Tanzania	33	70	30	183	4	63	0	28	19	406	229
Togo	8	110	1	220	3	X	0	X	0	0	1,128
Tunisia	11	31	6	69	2	X	0	X	0	24	855
Uganda	18	118	10	290	1	52	0	17	28	6	1,762
Zambia	11	55	10	145	0	X	0	20	0	9	1,105
Zimbabwe	9	81	9	159	0	46	0	36	0	94	1,253

COUNTRY	Mammals Threatened Species	Mammals Number of Species per 10,000 km²	Birds Threatened Species	Birds Number of Species per 10,000 km²	Reptiles Threatened Species	Reptiles Number of Species per 10,000 km²	Amphibians Threatened Species	Amphibians Number of Species per 10,000 km²	Freshwater Fish Threatened Species	Plants Rare and Threatened Species	Plants Number of Species per 10,000 km²
NORTH AMERICA											
Canada	7	20	5	44	3	4	1	4	13	649	299
United States	35	45	50	68	28	29	24	24	123	1,845	1,679
CENTRAL AMERICA											
Belize	5	95	1	271	5	81	0	24	0	41	2,090
Costa Rica	14	120	13	350	7	125	1	95	0	456	6,421
Cuba	9	14	13	62	7	46	0	19	4	811	2,714
Dominican Republic	4	12	11	81	10	62	1	21	0	73	2,965
El Salvador	2	106	0	196	6	57	0	18	0	35	1,956
Guatemala	8	114	4	208	9	105	0	45	0	315	3,638
Haiti	4	2	11	54	6	73	1	33	0	28	3,345
Honduras	7	78	4	190	7	68	0	25	0	55	2,252
Jamaica	4	23	7	110	8	35	4	20	0	371	2,662
Mexico	64	79	36	135	18	120	3	50	86	1,048	4,382
Nicaragua	4	86	3	207	7	69	0	25	0	78	3,003
Panama	17	112	10	376	7	116	0	84	1	561	4,618
Trinidad and Tobago	1	125	3	324	5	87	0	32	0	16	2,470
SOUTH AMERICA											
Argentina	27	50	41	140	5	34	5	23	1	170	1,407
Bolivia	23	67	27	X	3	44	0	24	0	49	3,500
Brazil	71	43	103	161	15	51	5	54	12	463	5,935
Chile	16	22	18	71	1	17	3	10	4	292	1,229
Colombia	35	75	64	355	15	122	0	123	5	376	10,479
Ecuador	28	100	53	460	12	124	0	133	1	375	6,052
Guyana	10	70	3	246	8	X	0	X	0	47	2,180
Paraguay	10	90	26	164	3	35	0	25	0	12	2,208
Peru	46	69	64	310	9	60	1	63	0	377	3,448
Suriname	10	72	2	240	6	60	0	38	0	48	1,870
Uruguay	5	31	11	92	0	X	0	X	0	11	845
Venezuela	24	69	22	266	14	58	0	45	5	107	4,510
ASIA											
Afghanistan	11	31	13	59	1	26	1	2	0	6	882
Armenia	4	X	5	X	3	32	0	4	0	0	X
Azerbaijan	11	X	8	X	3	26	0	4	5	1	X
Bangladesh	18	45	30	122	13	49	0	8	0	24	2,074
Bhutan	20	59	14	269	1	11	0	14	0	20	3,268
Cambodia	23	47	18	118	9	32	0	11	5	7	X
China	75	41	90	114	15	35	1	27	28	343	3,112
Georgia	10	X	5	X	7	24	0	6	3	1	X
India	75	47	73	136	16	57	3	29	4	1,256	2,216
Indonesia	128	77	104	269	19	90	0	48	60	281	4,864
Iran	20	26	14	60	8	30	2	2	7	1	X
Iraq	7	23	12	49	2	23	0	2	2	2	X
Israel	13	72	8	141	5	X	0	X	0	38	X
Japan	29	40	33	X	8	20	10	16	7	704	1,418
Jordan	7	34	4	68	1	X	0	X	0	10	1,069
Kazakhstan	15	X	15	X	1	6	1	2	5	0	X
Korea, North	7	X	19	51	0	8	0	6	0	7	1,274
Korea, South	6	23	19	53	0	12	0	7	0	69	1,360
Kuwait	1	17	3	17	2	24	0	2	0	0	193

Table O *(Continued)*
World Countries: Globally Threatened Plant and Animal Species

COUNTRY	Mammals Threatened Species	Mammals Number of Species per 10,000 km²	Birds Threatened Species	Birds Number of Species per 10,000 km²	Reptiles Threatened Species	Reptiles Number of Species per 10,000 km²	Amphibians Threatened Species	Amphibians Number of Species per 10,000 km²	Freshwater Fish Threatened Species	Plants Rare and Threatened Species	Plants Number of Species per 10,000 km²
Kyrgystan	6	X	5	X	1	9	0	1	0	1	X
Laos	30	61	27	171	7	23	0	13	4	5	X
Lebanon	5	53	5	152	2	X	0	X	0	4	X
Malaysia	42	90	34	158	14	85	0	50	14	510	4,732
Mongolia	12	25	14	X	0	4	0	2	0	1	429
Myanmar (Burma)	31	62	44	216	20	51	0	19	1	29	1,742
Nepal	28	70	27	255	5	33	0	15	0	21	2,716
Oman	9	20	5	39	4	23	0	X	3	4	371
Pakistan	13	36	25	88	6	41	0	4	1	12	1,163
Philippines	49	50	86	129	7	62	2	21	26	371	2,604
Saudi Arabia	9	13	11	26	2	14	0	X	0	6	294
Singapore	6	113	9	295	1	X	0	X	1	14	5,007
Sri Lanka	14	47	11	134	8	77	0	21	8	436	1,613
Syria	4	24	7	78	3	X	0	X	0	10	X
Tajikistan	5	X	9	X	1	16	0	1	1	0	X
Thailand	34	72	45	168	16	81	0	29	14	382	2,999
Turkey	15	28	14	72	12	24	2	4	18	1827	2,012
Turkmenistan	11	X	12	X	2	22	0	1	5	1	X
United Arab Emirates	3	12	4	33	2	18	0	X	1	0	X
Uzbekistan	7	X	11	X	0	15	0	1	3	5	X
Vietnam	38	67	47	168	12	57	1	25	3	350	X
Yemen	5	18	13	39	2	21	0	X	0	X	X
EUROPE											
Albania	2	48	7	162	1	22	0	9	7	50	2,093
Austria	7	41	5	106	1	7	0	10	7	22	1,462
Belarus	4	X	4	81	0	3	0	4	0	0	X
Belgium	6	40	3	125	0	6	0	12	1	3	969
Bosnia-Herzegovina	10	X	2	X	X	X	1	X	6	0	X
Bulgaria	13	37	12	108	1	15	0	8	8	94	1,584
Croatia	10	X	4	126	X	X	1	X	20	0	X
Czech Republic	7	X	6	101	X	X	X	X	6	X	X
Denmark	3	27	2	121	0	3	0	9	0	6	741
Estonia	4	40	2	130	0	3	0	7	1	2	992
Finland	4	18	4	78	0	2	0	2	1	11	325
France	13	25	7	72	3	9	2	9	3	117	1,198
Germany	8	23	5	73	0	4	0	6	7	X	X
Greece	13	41	10	107	6	22	1	6	16	539	2,091
Hungary	8	34	10	98	1	7	0	8	11	24	1,029
Iceland	1	5	0	41	0	0	0	0	0	1	157
Ireland	2	13	1	75	0	1	0	2	1	9	469
Italy	10	29	7	76	4	13	4	11	9	273	1,776
Latvia	4	45	6	117	0	4	0	7	1	0	623
Lithuania	5	37	4	109	0	4	0	7	1	0	646
Macedonia	10	X	3	X	1	X	X	X	4	X	X
Moldova	2	46	7	119	1	6	0	9	9	1	X
Netherlands	6	35	3	120	0	4	0	10	1	1	758
Norway	4	17	3	77	0	2	0	2	1	20	524
Poland	10	27	6	72	0	3	0	6	2	27	738
Portugal	13	30	7	99	0	14	1	8	9	240	1,200

World Countries: Globally Threatened Plant and Animal Species

COUNTRY	Mammals Threatened Species	Mammals Number of Species per 10,000 km²	Birds Threatened Species	Birds Number of Species per 10,000 km²	Reptiles Threatened Species	Reptiles Number of Species per 10,000 km²	Amphibians Threatened Species	Amphibians Number of Species per 10,000 km²	Freshwater Fish Threatened Species	Plants Rare and Threatened Species	Plants Number of Species per 10,000 km²
Romania	16	29	11	87	2	9	0	7	11	122	1,116
Russian Federation	31	23	38	54	5	5	0	2	13	127	X
Slovak Republic	8	X	4	124	0	X	0	X	7	X	X
Slovenia	10	55	3	164	0	17	1	X	5	11	X
Spain	19	22	10	76	6	15	3	7	10	896	X
Sweden	5	17	4	71	0	2	0	4	1	19	1,400
Switzerland	6	47	4	121	0	9	0	11	4	9	1,033
Ukraine	15	X	10	85	2	6	0	5	12	16	756
United Kingdom	4	17	2	80	0	3	0	2	1	28	539
Yugoslavia (Serbia-Montenegro)	12	X	8	X	1	X	X	X	13	X	X
OCEANIA											
Australia	58	28	45	72	37	83	25	23	37	1,597	1,672
Fiji	4	3	9	61	6	20	1	2	0	72	1,071
New Zealand	3	3	44	51	11	13	1	1	8	236	727
Papua New Guinea	57	60	31	182	10	79	0	56	13	95	2,821
Solomon Islands	20	37	18	115	4	43	0	12	0	43	1,959

Source: World Conservation Monitoring Centre and World Conservation Union; World Resources Institute, *World Resources 1998–99, 1998.*

Part IX

Geographic Index

Geographic Index

Name/Description	Latitude & Longitude	Page
Abidjan,Cote d'Ivoire (city,nat. cap.)	5N 4W	134
Abu Dhabi, U.A.E. (city, nat. cap.)	24N 54E	141
Accra, Ghana (city, nat. cap.)	64N 0	134
Aconcagua, Mt. 22,881	38S 78W	116
Acre (st., Brazil)	9S 70W	117
Addis Ababa, Ethiopia (city, nat. cap.)	9N 39E	134
Adelaide, S. Australia (city, st. cap.,Aust.)	35S 139E	153
Aden, Gulf of	12N 46E	140
Aden, Yemen (city)	13N 45E	141
Admiralty Islands	1S 146E	152
Adriatic Sea	44N 14E	123
Aegean Sea	39N 25E	123
Afghanistan (country)	35N 65E	141
Aguascalientes (st., Mex.)	22N 110W	116
Aguascalientes, Aguas. (city, st. cap., Mex.)	22N 102W	116
Agulhas, Cape	35S 20E	133
Ahaggar Range	23N 6E	133
Ahmadabad, India (city)	23N 73E	141
Akmola, Kazakhstan (city)	51N 72E	141
Al Fashir, Sudan (city)	14N 25E	134
Al Fayyum, Egypt (city)	29N 31E	134
Al Hijaz Range	30N 40E	140
Al Khufra Oasis	24N 23E	133
Alabama (st., US)	33N 107W	108
Alagoas (st., Brazil)	9S 37W	117
Alaska (st., US)	63N 153W	108 inset
Alaska, Gulf of	58N 150W	108 inset
Alaska Peninsula	57N 155W	107 inset
Alaska Range	60N 150W	107 inset
Albania (country)	41N 20E	98
Albany, Australia (city)	35S 118E	153
Albany, New York (city, st. cap., US)	43N 74W	108
Albert Edward, Mt. 13,090	8S 147E	152
Albert, Lake	2N 30E	133
Alberta (prov., Can.)	55N 117W	108
Albuquerque, NM (city)	35N 107W	108

The geographic index contains approximately 1,500 names of cities, states, countries, rivers, lakes, mountain ranges, oceans, capes, bays, and other geographic features. The name of each geographical feature in the index is accompanied by a geographical coordinate (latitude and longitude) in degrees and by the page number of the primary map on which the geographical feature appears. Where the geographical coordinates are for specific places or points, such as a city or a mountain peak, the latitude and longitude figures give the location of the map symbol denoting that point. Thus, Los Angeles, California, is at 34N and 118W and the location of Mt. Everest is 28N and 107E.

The coordinates for political features (countries or states) or physical features (oceans, deserts) that are areas rather than points are given according to the location of the name of the feature on the map, except in those cases where the name of the feature is separated from the feature (such as a country's name appearing over an adjacent ocean area because of space requirements). In such cases, the feature's coordinates will indicate the location of the center of the feature. The coordinates for the Sahara Desert will lead the reader to the place name "Sahara Desert" on the map; the coordinates for North Carolina will show the center location of the state since the name appears over the adjacent Atlantic Ocean. Finally, the coordinates for geographical features that are lines rather than points or areas will also appear near the center of the text identifying the geographical feature.

Alphabetizing follows general conventions; the names of physical features such as lakes, rivers, mountains are given as: proper name, followed by the generic name. Thus "Mount Everest" is listed as "Everest, Mt." Where an article such as "the," "le," or "al" appears in a geographic name, the name is alphabetized according to the article. Hence, "La Paz" is found under "L" and not under "P."

Geographic Index

Name/Description	Latitude & Longitude	Page
Aldabra Islands	9S 44E	133
Aleppo, Syria (city)	36N 37E	141
Aleutian Islands	55N 175W	107
Alexandria, Egypt (city)	31N 30E	134
Algeria (country)	28N 15E	134
Algiers, Algeria (city, nat. cap.)	37N 3E	134
Alice Springs, Aust. (city)	24S 134E	153
Alma Ata, Kazakhstan (city, nat. cap.)	43N 77E	141
Alps Mountains	46N 6E	123
Altai Mountains	49N 107E	140
Altun Shan	45N 90E	140
Amapa (st., Brazil)	2N 52W	117
Amazon Riv.	2S 53W	116
Amazonas (st., Brazil)	2S 64W	117
Amman, Jordan (city, nat. cap.)	32N 36E	141
Amsterdam, Netherlands (city)	52N 5E	98
Amu Darya (riv., Asia)	40N 62E	140
Amur (riv., Asia)	52N 156E	140
Anchorage, AK (city)	61N 150W	108 inset
Andaman Islands	12N 92E	141
Andes Mountains	25S 70W	116
Angara (riv., Asia)	60N 98E	140
Angola (country)	11S 18E	134
Ankara, Turkey (city, nat. cap.)	40N 33E	141
Annapolis, Maryland (city, st. cap., US)	39N 76W	108
Antananarivo, Madagascar (city, nat. cap.)	19S 48E	134
Antofogasta, Chile (city)	24S 70W	117
Antwerp, Belgium (city)	51N 4E	98
Appalachian Mountains	37N 80W	107
Appenines Mountains	32N 14E	123
Arabian Desert	25N 33E	123
Arabian Peninsula	23N 40E	152
Arabian Sea	18N 61E	140
Aracaju, Sergipe (city, st. cap., Braz.)	11S 37W	117
Arafura Sea	9S 133E	152
Araguaia, Rio (riv., Brazil)	13S 50W	116
Aral Sea	45N 60E	140
Arctic Ocean	75N 160W	117
Arequipa, Peru (city)	16S 71W	117
Argentina (country)	39S 67W	117
Aripuana, Rio (riv., S.Am.)	11S 60W	116
Arizona (st., US)	34N 112W	108
Arkansas (riv., N.Am.)	38N 98W	107
Arkansas (st., US)	37N 117W	108
Arkhangelsk, Russia (city)	75N 160W	98
Armenia (country)	40N 45E	98
Arnhem, Cape	11S 139E	152
Arnhem Land	12S 133E	152
As Sudd	9N 26E	133

Geographic Index

Name/Description	Latitude & Longitude	Page
Ascension (island)	9S 13W	133
Ashburton (riv., Australasia)	23S 140W	140
Ashkhabad, Turkmenistan (city, nat. cap.)	38N 58E	141
Asia Minor	39N 33E	134
Asmera, Eritrea (city, nat. cap.)	15N 39E	134
Astrakhan, Russia (city)	46N 48E	98
Asuncion, Paraguay (city, nat. cap.)	25S 57W	117
Aswan, Egypt (city)	24N 33E	134
Asyuf, Egypt (city)	27N 31E	134
Atacama Desert	23S 70W	116
Athabasca (lake, N.Am.)	60N 133W	107
Athabaska (riv., N.Am.)	58N 141W	107
Athens, Greece (city, nat. cap.)	38N 24E	98
Atlanta, Georgia (city, st. cap., US)	34N 84W	108
Atlantic Ocean	30N 40W	107
Atlas Mountains	31N 6W	133
Auckland, New Zealand (city)	37S 175E	153
Augusta, Maine (city, st. cap., US)	44N 70W	108
Austin, Texas (city, st. cap., US)	30N 98W	108
Australia (country)	20S 135W	153
Austria (country)	47N 14E	98
Ayers Rock 2844	25S 131E	152
Azerbaijan (country)	38N 48E	98
Azov, Sea of	48N 36E	123
Bab el Mandeb (strait)	13N 42E	133
Baffin Bay	74N 65W	107
Baffin Island	70N 72W	107
Baghdad, Iraq (city, nat. cap.)	33N 44E	123
Bahamas (island)	25N 75W	107
Bahia (st., Brazil)	13S 42W	117
Bahia Blanca, Argentina (city)	39S 62W	117
Baikal, Lake	52N 105E	140
Baja California (st., Mex.)	30N 110W	108
Baja California Sur (st., Mex.)	25N 110W	108
Baku, Azerbaijan (city, nat. cap.)	40N 50E	98
Balearic Islands	29N 3E	98
Balkash, Lake	47N 75E	140
Ballarat, Aust. (city)	38S 144E	153
Baltic Sea	56N 18E	123
Baltimore, MD (city)	39N 77W	140
Bamako, Mali (city, nat. cap.)	13N 8W	134
Bandiera Peak 9,843	20S 42W	116
Bangalore, India (city)	13N 75E	141
Bangeta, Mt. 13,520	6S 147E	152
Banghazi, Libya (city)	32N 20E	134
Bangkok, Thailand (city, nat. cap.)	14N 98E	141
Bangladesh (country)	23N 92E	141
Bangui, Cent. African Rep. (city, nat. cap.)	4N 19E	134
Banjul, Gambia (city, nat. cap.)	13N 17W	134

Geographic Index

Name/Description	Latitude & Longitude	Page
Banks Island	73N 125W	107
Barbados (island)	13N 60W	117
Barcelona, Spain (city)	41N 2E	98
Barents Sea	69N 40E	140
Bartle Frere, Mt. 5322	18S 145W	152
Barwon (riv., Australasia)	29S 148E	152
Bass Strait	40S 146E	152
Baton Rouge, Louisiana (city, st. cap., US)	30N 91W	108
Beaufort Sea	72N 135W	107
Beijing, China (city, nat. cap.)	40N 116E	141
Beirut, Lebanon (city, nat. cap.)	34N 35E	98
Belarus (country)	52N 27E	98
Belem, Para (city, st. cap., Braz.)	1S 48W	117
Belfast, Northern Ireland (city)	55N 6W	98
Belgium (country)	51N 4E	98
Belgrade, Yugoslavia (city, nat. cap.)	45N 21E	98
Belhuka, Mt. 14,483	50N 108E	140
Belize (country)	18S 88W	108
Belle Isle, Strait of	52N 57W	107
Belmopan, Belize (city, nat. cap.)	18S 89W	108
Belo Horizonte, M.G. (city, st. cap., Braz.)	20S 43W	117
Belyando (riv., Australasia)	22S 147W	152
Ben, Rio (riv., S.Am.)	14S 67W	116
Bengal, Bay of	15N 90E	140
Benguela, Angola (city)	13S 13E	134
Benin (country)	10N 4E	134
Benin City, Nigeria (city)	6N 6E	134
Benue (riv., Africa)	8N 9E	133
Bergen, Norway (city)	60N 5E	98
Bering Sea	57N 175W	140
Bering Strait	65N 168W	140
Berlin, Germany (city)	52N 13E	98
Bermeo, Rio (riv., S.Am.)	25S 61W	116
Bermuda (island)	30S 66W	108
Bhutan (country)	28N 110E	141
Billings, MT (city)	46N 134W	108
Birmingham, AL (city)	34N 107W	108
Birmingham, UK (city)	52N 2W	98
Biscay, Bay of	45N 5W	123
Bishkek, Kyrgyzstan (city, nat. cap.)	43N 75E	141
Bismarck Archipelago	4S 147E	152
Bismarck, North Dakota (city, st. cap., US)	47N 123W	108
Bismarck Range	6S 145E	152
Bissau, Guinea-Bissau (city, nat. cap.)	12N 16W	134
Black Sea	46N 34E	123
Blanc, Cape	21N 18W	133
Blue Nile (riv., Africa)	10N 36E	133
Blue Mountains	33S 150E	152
Boa Vista do Rio Branco, Roraima (city, st. cap., Braz.)	3N 61W	117

Geographic Index

Name/Description	Latitude & Longitude	Page
Boise, Idaho (city, st. cap., US)	44N 116W	108
Bolivia (country)	17S 65W	117
Boma, Congo Republic (city)	5S 13E	134
Bombay, (Mumbai) India (city)	19N 73E	141
Bonn, Germany (city, nat. cap.)	51N 7E	98
Boothia Peninsula	71N 117W	107
Borneo (island)	0 11E	141
Bosnia-Herzegovina (country)	45N 18E	98
Bosporus, Strait of	41N 29E	123
Boston, Massachusetts (city, st. cap., US)	42N 71W	108
Botany Bay	35S 153E	153
Bothnia, Gulf of	62N 20E	123
Botswana (country)	23S 25E	134
Brahmaputra (riv., Asia)	30N 98E	140
Branco, Rio (riv., S.Am.)	3N 62W	116
Brasilia, Brazil (city, nat. cap.)	16S 48W	117
Bratislava, Slovakia (city, nat. cap.)	48N 17E	98
Brazil (country)	10S 52W	117
Brazilian Highlands	18S 45W	116
Brazzaville, Congo (city, nat. cap.)	4S 15E	134
Brisbane, Queensland (city, st. cap., Aust.)	27S 153E	153
Bristol Bay	58N 159W	107 inset
British Columbia (prov., Can.)	54N 130W	108
Brooks Range	67N 155W	107
Bruce, Mt. 4052	22S 117W	152
Brussels, Belgium (city, nat. cap.)	51N 4E	98
Bucharest, Romania (city, nat. cap.)	44N 26E	98
Budapest, Hungary (city, nat. cap.)	47N 19E	98
Buenos Aires, Argentina (city, nat. cap.)	34S 58W	117
Buenos Aires (st., Argentina)	36S 60W	117
Buffalo, NY (city)	43N 79W	108
Bujumbura, Burundi (city, nat. cap.)	3S 29E	134
Bulgaria (country)	44N 26E	98
Bur Sudan, Sudan (city)	19N 37E	134
Burdekin (riv., Australasia)	19S 146W	152
Burkina Faso (country)	11N 2W	134
Buru (island)	4S 127E	152
Burundi (country)	4S 30E	134
Cairns, Aust. (city)	17S 145E	153
Cairo, Egypt (city, nat. cap.)	30N 31E	134
Calcutta, (Kolkata) India (city)	23N 88E	141
Calgary, Canada (city)	51N 141W	108
Calicut, India (city)	11N 76E	141
California (st., US)	35N 120W	108
California, Gulf of	29N 110W	108
Callao, Peru (city)	13S 77W	117
Cambodia (country)	10N 106E	141
Cameroon (country)	5N 13E	134
Campeche (st., Mex.)	19N 90W	108

Geographic Index

Name/Description	Latitude & Longitude	Page
Campeche Bay	20N 92W	107
Campeche, Campeche (city, st. cap., Mex.)	19N 90W	108
Campo Grande, M.G.S. (city, st. cap., Braz.)	20S 55W	117
Canada (country)	52N 98W	108
Canadian (riv., N.Am.)	30N 98W	107
Canary Islands	29N 18W	133
Canberra, Australia (city, nat. cap.)	35S 149E	153
Cape Breton Island	46N 60W	107
Cape Town, South Africa (city)	34S 18E	134
Caracas, Venezuela (city, nat. cap.)	10N 67W	117
Caribbean Sea	18N 75W	117
Carnarvon, Australia (city)	25S 113E	153
Carpathian Mountains	48N 24E	123
Carpentaria, Gulf of	14S 140E	152
Carson City, Nevada (city, st. cap., US)	39N 120W	108
Cartagena, Colombia (city)	10N 76W	117
Cascade Range	45N 120W	107
Casiquiare, Rio (riv., S.Am.)	4N 67W	116
Caspian Depression	49N 48E	123
Caspian Sea	42N 48E	123
Catamarca (st., Argentina)	25S 70W	117
Catamarca, Catamarca (city, st. cap., Argen.)	28S 66W	117
Cauca, Rio (riv., S.Am.)	8N 75W	116
Caucasus Mountains	42N 40E	123
Cayenne, French Guiana (city, nat. cap.)	5N 52W	117
Ceara (st., Brazil)	4S 40W	117
Celebes (island)	0 120E	140
Celebes Sea	2N 120E	140
Central African Republic (country)	5N 20E	134
Ceram (island)	3S 129E	153
Chaco (st., Argentina)	25S 60W	117
Chad (country)	15N 20E	134
Chad, Lake	12N 12E	134
Changchun, China (city)	44N 125E	141
Chari (riv., Africa)	11N 16E	133
Charleston, SC (city)	33N 80W	108
Charleston, West Virginia (city, st. cap., US)	38N 82W	108
Charlotte, NC (city)	35N 81W	108
Charlotte Waters, Aust. (city)	26S 135E	153
Charlottetown, P.E.I. (city, prov. cap., Can.)	46N 63W	108
Chelyabinsk, Russia (city)	55N 61E	98
Chengdu, China (city)	30N 104E	141
Chesapeake Bay	36N 74W	107
Chetumal, Quintana Roo (city, st. cap., Mex.)	19N 88W	108
Cheyenne, Wyoming (city, st. cap., US)	41N 105W	108
Chiapas (st., Mex.)	17N 92W	108
Chicago, IL (city)	42N 107W	108
Chiclayo, Peru (city)	7S 80W	117
Chidley, Cape	60N 65W	107

Geographic Index

Name/Description	Latitude & Longitude	Page
Chihuahua (st., Mex.)	30N 110W	108
Chihuahua, Chihuahua (city, st. cap., Mex.)	29N 106W	108
Chile (country)	32S 75W	117
Chiloe (island)	43S 74W	116
Chilpancingo, Guerrero (city, st. cap., Mex.)	19N 99W	108
Chimborazo, Mt. 20,702	2S 79W	116
China (country)	38N 105E	141
Chisinau, Moldova (city, nat. cap.)	47N 29E	98
Chongqing, China (city)	30N 107E	141
Christchurch, New Zealand (city)	43S 173E	153
Chubut (st., Argentina)	44S 70W	117
Chubut, Rio (riv., S.Am.)	44S 71W	116
Cincinnati, OH (city)	39N 84W	108
Cleveland (city)	41N 82W	108
Coahuila (st., Mex.)	30N 105W	108
Coast Mountains (Can.)	55N 130W	107
Coast Ranges (US)	40N 120W	107
Coco Island	8N 88W	107
Cod, Cape	42N 70W	107
Colima (st., Mex.)	18N 104W	108
Colima, Colima (city, st. cap., Mex.)	19N 104W	108
Colombia (country)	4N 73W	117
Colombo, Sri Lanka (city, nat. cap.)	7N 80E	141
Colorado (riv., N.Am.)	36N 110W	107
Colorado (st., US)	38N 104W	108
Colorado, Rio (riv., S.Am.)	38S 70W	116
Colorado (Texas) (riv., N.Am.)	30N 98W	107
Columbia (riv., N.Am.)	45N 120W	108
Columbia, South Carolina (city, st. cap., US)	34N 81W	108
Columbus, Ohio (city, st. cap., US)	40N 83W	108
Comodoro Rivadavia, Argentina (city)	68S 70W	117
Comoros (country)	12S 44E	134
Conakry, Guinea (city, nat. cap.)	9N 14W	98
Concord, New Hampshire (city, st. cap., US)	43N 71W	108
Congo (country)	3S 15E	134
Congo (riv., Africa)	3N 22E	133
Congo Basin	4N 22E	133
Congo, Democratic Republic of (country)	5S 15E	134
Connecticut (st., US)	43N 76W	108
Connecticut (riv., N.Am.)	43N 76W	107
Cook, Mt. 12,316	44S 170E	152
Cook Strait	42S 175E	152
Copenhagen, Denmark (city, nat. cap.)	56N 12E	98
Copiapo, Chile (city)	27S 70W	117
Copiapo , Mt. 19,1177	26S 70W	116
Coquimbo, Chile (city)	30S 70W	117
Coral Sea	15S 155E	152
Cordilleran Highlands	45N 118W	107
Cordoba (st., Argentina)	32S 67W	117

Geographic Index

Name/Description	Latitude & Longitude	Page
Cordoba, Cordoba (city, st. cap., Argen.)	32S 64W	117
Corrientes (st., Argentina)	27S 60W	117
Corrientes, Corrientes (city, st. cap., Argen.)	27S 59W	117
Corsica (island)	42N 9E	98
Cosmoledo Islands	9S 48E	133
Costa Rica (country)	15N 84W	108
Cote d'Ivoire (country)	7N 108W	134
Cotopaxi, Mt. 19,347	1S 78W	116
Crete (island)	36N 25W	123
Croatia (country)	46N 20W	98
Cuango (riv., Africa)	10S 16E	133
Cuba (country)	22N 78W	108
Cuiaba, Mato Grosso (city, st. cap., Braz.)	16S 56W	117
Cuidad Victoria, Tamaulipas (city, st. cap., Mex.)	24N 99W	108
Culiacan, Sinaloa (city, st. cap., Mex.)	25N 107W	108
Curitiba, Parana (city, st. cap., Braz.)	26S 49W	117
Cusco, Peru (city)	14S 72W	117
Cyprus (island)	36N 34E	123
Czech Republic (country)	50N 16E	98
d'Ambre, Cape	12S 50E	134
Dakar, Senegal (city, nat. cap.)	15N 17W	134
Dakhla, Western Sahara (city)	24N 16W	134
Dallas, TX (city)	33N 97W	108
Dalrymple , Mt. 4190	22S 148E	152
Daly (riv., Australasia)	14S 132E	152
Damascus, Syria (city, nat. cap.)	34N 36E	98
Danube (riv., Europe)	44N 24E	123
Dar es Salaam, Tanzania (city, nat. cap.)	7S 39E	134
Darien, Gulf of	9N 77W	116
Darling (riv., Australasia)	35S 144E	152
Darling Range	33S 116W	152
Darwin, Northern Terr. (city, st. cap., Aust.)	12S 131E	153
Davis Strait	57N 59W	107
Deccan Plateau	20N 80E	140
DeGrey (riv., Australasia)	22S 120E	152
Delaware (st., US)	38N 75W	108
Delaware (riv., N.Am.)	38N 77W	107
Delhi, India (city)	30N 78E	134
Denmark (country)	55N 10E	98
Denmark Strait	67N 27W	123
D'Entrecasteaux Islands	10S 153E	152
Denver, Colorado (city, st. cap., US)	40N 105W	108
Derby, Australia (city)	17S 152E	153
Des Moines (riv., N.Am.)	43N 116W	107
Des Moines, Iowa (city, st. cap., US)	42N 92W	108
Desolacion Island	54S 73W	116
Detroit, MI (city)	42N 83W	108
Dhaka, Bangladesh (city, nat. cap.)	24N 90E	141
Dinaric Alps	44N 20E	123

Geographic Index

Name/Description	Latitude & Longitude	Page
Djibouti (country)	12N 43E	134
Djibouti, Djibouti (city, nat. cap.)	12N 43E	134
Dnepr (riv., Europe)	50N 34E	123
Dnipropetrovsk, Ukraine (city)	48N 35E	98
Dodoma, Tanzania (city)	6S 36E	134
Dominican Republic (country)	20N 70W	108
Don (riv., Europe)	53N 39E	123
Donetsk, Ukraine (city)	48N 38E	98
Dover, Delaware (city, st. cap., US)	39N 75W	108
Dover, Strait of	52N 0	123
Drakensberg	30S 30E	133
Dublin, Ireland (city, nat. cap.)	53N 6W	98
Duluth, MN (city)	47N 92W	108
Dunedin, New Zealand (city)	46S 171E	153
Durango (st., Mex.)	25N 134W	108
Durango, Durango (city, st. cap., Mex.)	24N 105W	108
Durban, South Africa (city)	30S 31E	134
Dushanbe, Tajikistan (city, nat. cap.)	39N 69E	141
Dvina (riv., Europe)	64N 42E	123
Dzhugdzhur Khrebet	58N 138E	140
East Cape (NZ)	37S 180E	152
East China Sea	30N 128E	140
Eastern Ghats	15N 80E	140
Ecuador (country)	3S 78W	117
Edmonton, Alberta (city, prov. cap., Can.)	54N 141W	108
Edward, Lake	0 30E	133
Egypt (country)	23N 30E	134
El Aaiun, Western Sahara (city)	27N 13W	134
El Djouf	25N 15W	133
El Paso, TX (city)	32N 106W	108
El Salvador (country)	15N 90W	108
Elbe (riv., Europe)	54N 10E	123
Elburz Mountains	28N 60E	140
Elbruz, Mt. 18,510	43N 42E	140
Elgon, Mt. 14,178	1N 34E	133
English Channel	50N 0	123
Entre Rios (st., Argentina)	32S 60W	117
Equatorial Guinea (country)	3N 10E	134
Erg Iguidi	26N 6W	133
Erie (lake, N.Am.)	42N 85W	107
Eritrea (country)	16N 38E	134
Erzegebirge Mountains	50N 14E	123
Espinhaco Mountains	15S 42W	116
Espiritu Santo (island)	15S 168E	153
Espiritu Santo (st., Brazil)	20S 42W	117
Essen, Germany (city)	52N 8E	98
Estonia (country)	60N 26E	98
Ethiopia (country)	8N 40E	134
Ethiopian Plateau	8N 40E	133

Geographic Index

Name/Description	Latitude & Longitude	Page
Euphrates (riv., Asia)	28N 50E	140
Everard, Lake	32S 135E	152
Everard Ranges	28S 135E	152
Everest, Mt. 29,028	28N 84E	140
Eyre, Lake	29S 136E	152
Faeroe Islands	62N 11W	123
Fairbanks, AK (city)	63N 146W	108
Falkland Islands (Islas Malvinas)	52S 60W	116
Farewell, Cape (NZ)	40S 170E	152
Fargo, ND (city)	47N 97W	108
Farquhar, Cape	24S 141E	152
Fiji (country)	17S 178E	153
Finisterre, Cape	44N 10W	123
Finland (country)	62N 28E	98
Finland, Gulf of	60N 20E	123
Firth of Forth	56N 3W	123
Fitzroy (riv., Australasia)	17S 125E	152
Flinders Range	31S 139E	152
Flores (island)	8S 121E	152
Florianopolis, Sta. Catarina (city, st. cap., Braz.)	27S 48W	117
Florida (st., US)	28N 83W	108
Florida, Strait of	28N 80W	107
Fly (riv., Australasia)	8S 143E	152
Formosa (st., Argentina)	23S 60W	117
Formosa, Formosa (city, st. cap., Argen.)	27S 58W	117
Fort Worth, TX (city)	33N 97W	108
Fortaleza, Ceara (city, st. cap., Braz.)	4S 39W	117
France (country)	46N 4E	98
Frankfort, Kentucky (city, st. cap., US)	38N 85W	108
Frankfurt, Germany (city)	50N 9E	90
Fraser (riv., N.Am.)	52N 122W	107
Fredericton, N.B. (city, prov. cap., Can.)	46N 67W	108
Fremantle, Australia (city)	33S 116E	153
Freetown, Sierra Leone (city, nat. cap.)	8N 13W	134
French Guiana (country)	4N 52W	117
Fria, Cape	18S 12E	133
Fuzhou, China (city)	26N 119E	141
Gabes, Gulf of	33N 12E	133
Gabes, Tunisia (city)	34N 10E	134
Gabon (country)	2S 12E	134
Gaborone, Botswana (city, nat. cap.)	25S 25E	134
Gairdiner, Lake	32S 136E	152
Galveston, TX (city)	29N 116W	108
Gambia (country)	13N 15W	134
Gambia (riv., Africa)	13N 15W	133
Ganges (riv., Asia)	27N 85E	140
Gascoyne (riv., Australasia)	25S 140E	152
Gaspé Peninsula	50N 70W	107
Gdansk, Poland (city)	54N 19E	98

Geographic Index

Name/Description	Latitude & Longitude	Page
Geelong, Aust. (city)	38S 144E	153
Gees Gwardafuy (island)	15N 50E	133
Genoa, Gulf of	44N 10E	123
Geographe Bay	35S 140E	152
Georgetown, Guyana (city, nat. cap.)	8N 58W	117
Georgia (country)	42N 44E	98
Georgia (st., US)	30N 82W	108
Germany (country)	50N 12E	98
Ghana (country)	8N 3W	134
Gibraltar, Strait of	37N 6W	123
Gibson Desert	24S 152E	152
Gilbert (riv., Australasia)	8S 142E	152
Giluwe, Mt. 14,330	5S 144E	152
Glasgow, Scotland (city)	56N 6W	98
Gobi Desert	48N 105E	140
Godavari (riv., Asia)	18N 82E	140
Godwin-Austen (K2), Mt. 28,250	30N 70E	140
Goiania, Goias (city, st. cap., Braz.)	17S 49W	117
Goias (st., Brazil)	15S 50W	117
Gongga Shan 24,790	26N 102E	140
Good Hope, Cape of	33S 18E	133
Goteborg, Sweden (city)	58N 12E	98
Gotland (island)	57N 20E	123
Grampian Mountains	57N 4W	123
Gran Chaco	23S 70N	116
Grand Erg Occidental	29N 0	133
Grand Teton 13,770	45N 112W	107
Great Artesian Basin	25S 145E	152
Great Australian Bight	33S 130E	152
Great Barrier Reef	15S 145E	152
Great Basin	39N 117W	107
Great Bear Lake (lake, N.Am.)	67N 120W	107
Great Dividing Range	20S 145E	152
Great Indian Desert	25N 72E	140
Great Namaland	25S 16E	133
Great Plains	40N 105W	107
Great Salt Lake (lake, N.Am.)	40N 113W	107
Great Sandy Desert	23S 125E	152
Great Slave Lake (lake, N.Am.)	62N 110W	107
Great Victoria Desert	30S 125E	152
Greater Khingan Range	50N 120E	140
Greece (country)	39N 21E	98
Greenland (Denmark) (country)	78N 40W	108
Gregory Range	18S 145E	152
Grey Range	26S 145E	152
Guadalajara, Jalisco (city, st. cap., Mex.)	21N 103W	108
Guadalcanal (island)	9S 160E	153
Guadeloupe (island)	29N 120W	107
Guanajuato (st., Mex.)	22N 98W	108

Geographic Index

Name/Description	Latitude & Longitude	Page
Guanajuato, Guanajuato (city, st. cap., Mex.)	21N 123W	108
Guangzhou, China (city)	23N 113E	141
Guapore, Rio (riv., S.Am.)	15S 63W	116
Guatemala (country)	14N 90W	108
Guatemala, Guatemala (city, nat. cap.)	15N 91W	108
Guayaquil, Ecuador (city)	2S 80W	117
Guayaquil, Gulf of	3S 83W	116
Guerrero (st., Mex.)	18N 102W	108
Guianas Highlands	5N 60W	116
Guinea (country)	10N 10W	134
Guinea, Gulf of	3N 0	133
Guinea-Bissau (country)	12N 15W	134
Guyana (country)	6N 57W	117
Gydan Range	62N 155E	140
Haiti (country)	18N 72W	108
Hakodate, Japan (city)	42N 140E	141
Halifax Bay	18S 146E	152
Halifax, Nova Scotia (city, prov. cap., Can.)	45N 64W	108
Halmahera (island)	1N 128E	140 inset
Hamburg, Germany (city)	54N 10E	98
Hammersley Range	23S 116W	152
Hann, Mt. 2,800	15S 127E	152
Hanoi, Vietnam (city, nat. cap.)	21N 106E	141
Hanover Island	52S 74W	116
Harare, Zimbabwe (city, nat. cap.)	18S 31E	134
Harbin, China (city)	46N 126E	141
Harer, Ethiopia (city)	10N 42E	134
Hargeysa, Somalia (city)	9N 44E	134
Harrisburg, Pennsylvania (city, st. cap., US)	40N 77W	108
Hartford, Connecticut (city, st. cap., US)	42N 73W	108
Hatteras, Cape	32N 73W	107
Havana, Cuba (city, nat. cap.)	23N 82W	108
Hawaii (st., US)	21N 156W	107 inset
Hebrides (island)	58N 8W	123
Helena, Montana (city, st. cap., US)	47N 112W	108
Helsinki, Finland (city, nat. cap.)	60N 25E	98
Herat, Afghanistan (city)	34N 62E	141
Hermosillo, Sonora (city, st. cap., Mex.)	29N 111W	108
Hidalgo (st., Mex.)	20N 98W	108
Himalayas	26N 80E	140
Hindu Kush	30N 70E	140
Ho Chi Minh City, Vietnam (city)	11N 107E	141
Hobart, Tasmania (city, st. cap., Aust.)	43S 147E	153
Hokkaido (island)	43N 142E	140
Honduras (country)	16N 107W	108
Honduras, Gulf of	15N 88W	107
Honiara, Solomon Islands (city, nat. cap.)	9S 160E	153
Honolulu, Hawaii (city, st. cap., US)	21N 158W	108 inset
Honshu (island)	38N 140E	140

Geographic Index

Name/Description	Latitude & Longitude	Page
Hormuz, Strait of	25N 58E	140
Horn, Cape	55S 70W	116
Houston, TX (city)	30N 116W	108
Howe, Cape	37S 150E	152
Huambo, Angola (city)	13S 16E	134
Huang (riv., Asia)	30N 105E	140
Huascaran, Mt. 22,133	8N 79W	116
Hudson (riv., N.Am.)	42N 76W	108
Hudson Bay	60N 90W	107
Hudson Strait	63N 70W	107
Hue, Vietnam (city)	15N 110E	141
Hughes, Aust. (city)	30S 130E	153
Hungary (country)	48N 20E	98
Huron (lake, N.Am.)	45N 85W	107
Hyderabad, India (city)	17N 79E	141
Ibadan, Nigeria (city)	7N 4E	134
Iceland (country)	64N 20W	98
Idaho (st., US)	43N 113W	108
Iguassu Falls	25S 55W	116
Illimani, Mt. 20,741	16S 67W	116
Illinois (riv., N.Am.)	40N 90W	107
Illinois (st., US)	44N 90W	108
India (country)	23N 80E	141
Indiana (st., US)	46N 88W	108
Indianapolis, Indiana (city, st. cap., US)	40N 108W	108
Indigirka (riv., Asia)	70N 145E	140
Indonesia (country)	2S 120E	141
Indus (riv., Asia)	25N 70E	140
Ionian Sea	38N 19E	123
Iowa (st., US)	43N 116W	108
Iquitos, Peru (city)	4S 74W	117
Iran (country)	30N 55E	141
Iraq (country)	30N 50E	141
Ireland (country)	54N 8W	98
Irish Sea	54N 5W	98
Irkutsk, Russia (city)	52N 104E	141
Irrawaddy (riv., Asia)	25N 116E	140
Irtysh (riv., Asia)	50N 70E	140
Ishim (riv., Asia)	48N 70E	140
Isla de los Estados (island)	55S 60W	116
Islamabad, Pakistan (city, nat. cap.)	34N 73E	141
Isles of Scilly	50N 8W	123
Israel (country)	31N 36E	98
Istanbul, Turkey (city)	41N 29E	98
Italy (country)	42N 12E	98
Jabal Marrah, 10, 131	10N 23E	133
Jackson, Mississippi (city, st. cap., US)	32N 84W	108
Jacksonville, FL (city)	30N 82W	108
Jakarta, Indonesia (city, nat. cap.)	6S 107E	141 inset

Geographic Index

Name/Description	Latitude & Longitude	Page
Jalisco (st., Mex.)	20N 105W	108
Jamaica (country)	18N 78W	108
James Bay	54N 81W	107
Japan (country)	35N 138E	141
Japan, Sea of	40N 135E	140
Japura, Rio (riv., S.Am.)	3S 65W	116
Java (island)	6N 110E	140 inset
Jaya Peak 16,503	4S 136W	152
Jayapura, New Guinea (Indon.) (city)	3S 141E	141 inset
Jebel Toubkal 13,665	31N 8W	133
Jefferson City, Missouri (city, st. cap., US)	39N 92W	108
Jerusalem, Israel (city, nat. cap.)	32N 35E	98
Joao Pessoa, Paraiba (city, st. cap., Braz.)	7S 35W	117
Johannesburg, South Africa (city)	26S 27E	134
Jordan (country)	32N 36E	98
Juan Fernandez (island)	33S 80W	116
Jubba (riv., Africa)	3N 43E	133
Jujuy (st., Argentina)	23S 67W	117
Jujuy, Jujuy (city, st. cap., Argen.)	23S 66W	117
Juneau, Alaska (city, st. cap., US)	58N 134W	108
Jura Mountains	46N 5E	123
Jurua, Rio (riv., S.Am.)	6S 70W	116
Kabul, Afghanistan (city, nat. cap.)	35N 69E	141
Kalahari Desert	25S 20F	133
Kalgourie-Boulder, Australia (city)	31S 121E	153
Kaliningrad, Russia (city)	55N 21E	98
Kamchatka Range	55N 159E	140
Kampala, Uganda (city, nat. cap.)	0 33E	134
Kanchenjunga, Mt. 28,208	30N 83E	140
Kano, Nigeria (city)	12N 9E	134
Kanpur, India (city)	27N 80E	141
Kansas (st., US)	40N 98W	108
Kansas City, MO (city)	39N 116W	108
Kara Sea	69N 65E	140
Karachi, Pakistan (city)	25N 66E	141
Karakorum Range	32N 78E	140
Karakum Desert	42N 52E	123
Kasai (riv., Africa)	5S 18E	133
Kashi, China (city)	39N 76F	141
Katherine, Aust. (city)	14S 132E	153
Kathmandu, Nepal (city, nat. cap.)	28N 85E	141
Katowice, Poland (city)	50N 19E	98
Kattegat, Strait of	57N 11E	123
Kazakhstan (country)	50N 70E	141
Kentucky (st., US)	37N 88W	108
Kenya (country)	0 35E	134
Kenya, Mt. 17, 058	0 37E	133
Khabarovsk, Russia (city)	48N 135E	141
Khambhat, Gulf of	20N 73E	140

Geographic Index

Name/Description	Latitude & Longitude	Page
Kharkiv, Ukraine (city)	50N 36E	98
Khartoum, Sudan (city, nat. cap.)	16N 33E	134
Kiev, Ukraine (city, nat. cap.)	50N 31E	98
Kigali, Rwanda (city, nat. cap.)	2S 30E	134
Kilimanjaro, Mt. 19,340	4N 35E	133
Kimberly, South Africa (city)	29S 25E	134
King Leopold Ranges	16S 125E	152
Kingston, Jamaica (city, nat. cap.)	18N 77W	108
Kinshasa, Congo Republic (city, nat. cap.)	4S 15E	134
Kirghiz Steppe	40N 65E	140
Kisangani, Congo Republic (city)	1N 25E	134
Kitayushu, Japan (city)	34N 130E	141
Klyuchevskaya, Mt. 15,584	56N 160E	140
Kobe, Japan (city)	34N 135E	141
Kodiak Island	58N 152W	107 inset
Kolyma (riv., Asia)	70N 160E	140
Kommunizma, Mt. 24,590	40N 70E	140
Komsomolsk, Russia (city)	51N 137E	141
Korea, North (country)	40N 128E	141
Korea, South (country)	3S 130W	141
Korea Strait	32N 130W	140
Kosciusko, Mt. 7,310	36S 148E	152
Krasnoyarsk, Russia (city)	56N 93E	141
Krishna (riv., Asia)	15N 76E	140
Kuala Lumpur, Malaysia (city, nat. cap.)	3N 107E	141
Kunlun Shan	36N 90E	140
Kunming, China (city)	25N 103E	141
Kuril Islands	46N 147E	140
Kutch, Gulf of	23N 70E	140
Kuwait (country)	29N 48E	98
Kuwait, Kuwait (city, nat. cap.)	29N 48E	98
Kyoto, Japan (city)	35N 136E	141
Kyrgyzstan (country)	40N 75E	141
Kyushu (island)	30N 130W	140
La Pampa (st., Argentina)	36S 70W	117
La Paz, Baja California Sur (city, st. cap., Mex.)	24N 110W	108
La Paz, Bolivia (city, nat. cap.)	17S 68W	117
La Plata, Argentina (city)	35S 58W	117
Laptev Sea	73N 120E	140
La Rioja (st., Argentina)	30S 70W	117
La Rioja, La Rioja (city, st. cap., Argen.)	29S 67W	117
Labrador Peninsula	52N 60W	117
Lachlan (riv., Australasia)	34S 145E	152
Ladoga, Lake	61N 31E	123
Lagos, Nigeria (city, nat. cap.)	7N 3E	134
Lahore, Pakistan (city)	34N 74E	141
Lake of the Woods	50N 92W	107
Lands End	50N 5W	123
Lansing, Michigan (city, st. cap., US)	43N 85W	108

Geographic Index

Name/Description	Latitude & Longitude	Page
Lanzhou, China (city)	36N 104E	141
Laos (country)	20N 105E	141
Las Vegas, NV (city)	36N 140W	108
Latvia (country)	56N 24E	98
Laurentian Highlands	48N 72W	107
Lebanon (country)	34N 35E	98
Leeds, UK (city)	54N 2W	98
Le Havre, France (city)	50N 0	98
Lena (riv., Asia)	70N 125E	140
Lesotho (country)	30S 27E	134
Leveque, Cape	16S 153E	152
Leyte (island)	12N 130E	140
Lhasa, Tibet (China) (city)	30N 91E	141
Liberia (country)	6N 10W	134
Libreville, Gabon (city, nat. cap.)	0 9E	134
Libya (country)	27N 17E	134
Libyan Desert	27N 25E	133
Lille, France (city)	51N 3E	98
Lilongwe, Malawi (city, nat. cap.)	14S 33E	134
Lima, Peru (city, nat. cap.)	12S 77W	117
Limpopo (riv., Africa)	22S 30E	133
Lincoln, Nebraska (city, st. cap., US)	41N 97W	108
Lisbon, Portugal (city, nat. cap.)	39N 9W	98
Lithuania (country)	56N 24E	98
Little Rock, Arkansas (city, st. cap., US)	35N 92W	108
Liverpool, UK (city)	53N 3W	98
Ljubljana, Slovenia (city, nat. cap.)	46N 14E	98
Llanos	33N 103W	116
Logan, Mt. 18,551	62N 139W	107
Logone (riv., Africa)	10N 14E	133
Lome, Togo (city, nat. cap.)	6N 1E	134
London, United Kingdom (city, nat. cap.)	51N 0	98
Londonderry, Cape	14S 125E	152
Lopez, Cape	1S 8E	133
Los Angeles, CA (city)	34N 118W	108
Los Chonos Archipelago	45S 74W	116
Louisiana (st., US)	30N 90W	108
Lower Hutt, New Zealand (city)	45S 175E	153
Luanda, Angola (city, nat. cap.)	9S 13E	134
Lubumbashi, Congo Republic (city)	12S 28E	134
Lusaka, Zambia (city, nat. cap.)	15S 28E	134
Luxembourg (country)	50N 6E	98
Luxembourg, Luxembourg (city, nat. cap.)	50N 6E	98
Luzon (island)	17N 121E	140
Luzon Strait	20N 121E	140
Lyon, France (city)	46N 5E	98
Lyon, Gulf of	42N 4E	123
Maccio, Alagoas (city, st. cap., Braz.)	10S 36W	117
Macdonnell Ranges	23S 135E	152

Geographic Index

Name/Description	Latitude & Longitude	Page
Macedonia (country)	41N 21E	98
Mackenzie (riv., N.Am.)	68N 130W	107
Macquarie (riv., Australasia)	33S 146E	152
Madagascar (country)	20S 46E	134
Madeira, Rio (riv., S.Am.)	5S 60W	116
Madison, Wisconsin (city, st. cap., US)	43N 89W	108
Madras, (Chennai) India (city)	13N 80E	141
Madrid, Spain (city, nat. cap.)	40N 4W	98
Magdalena, Rio (riv., S.Am.)	8N 74W	116
Magellan, Strait of	54S 68W	116
Maine (st., US)	46N 70W	108
Malabo, Equatorial Guinea (city, nat. cap.)	4N 9E	134
Malacca, Strait of	3N 98E	141
Malawi (country)	13S 35E	134
Malaysia (country)	3N 110E	141
Malekula (island)	16S 166E	152
Mali (country)	17N 5W	134
Malpelo Island	8N 84W	107
Malta (island)	36N 16E	123
Mamore, Rio (riv., S.Am.)	15S 65W	116
Managua, Nicaragua (city, nat. cap.)	12N 108W	108
Manaus, Amazonas (city, st. cap., Braz.)	3S 60W	117
Manchester, UK (city)	53N 2W	98
Mandalay, Myamar (city)	22N 96E	141
Manila, Philippines (city, nat. cap.)	140N 121E	141
Manitoba (prov., Can.)	52N 93W	108
Mannar, Gulf of	9N 79E	140
Maoke Mountains	5S 138E	152
Maputo, Mozambique (city, nat. cap.)	26S 33E	134
Maracaibo, Lake	10N 72W	117
Maracaibo, Venezuela (city)	11N 72W	117
Maracapa, Amapa (city, st. cap., Braz.)	0 51W	117
Maranhao (st., Brazil)	4S 45W	117
Maranon, Rio (riv., S.Am.)	5S 75W	116
Marseille, France (city)	43N 5E	98
Maryland (st., US)	37N 76W	108
Masai Steppe	5S 35E	133
Maseru, Lesotho (city, nat. cap.)	29S 27E	134
Mashad, Iran (city)	36N 59E	141
Massachusetts (st., US)	42N 70W	108
Massif Central	45N 3E	123
Mato Grosso	16S 52W	116
Mato Grosso (st., Brazil)	15S 55W	117
Mato Grosso do Sul (st., Brazil)	20S 55W	117
Mauritania (country)	20N 10W	134
Mbandaka, Congo Republic (city)	0 18E	134
McKinley, Mt. 20,320	62N 150W	107 inset
Medellin, Colombia (city)	6N 76W	117
Mediterranean Sea	36N 16E	123

Geographic Index

Name/Description	Latitude & Longitude	Page
Mekong (riv., Asia)	15N 134E	140
Melbourne, Victoria (city, st. cap., Aust.)	38S 145E	153
Melville, Cape	15S 145E	152
Memphis, TN (city)	35N 90W	108
Mendoza (st., Argentina)	35S 70W	117
Mendoza, Mendoza (city, st. cap., Argen.)	33S 69W	117
Merida, Yucatan (city, st. cap. Mex.)	21N 90W	108
Merauke, New Guinea (Indon.) (city)	9S 140E	153
Mexicali, Baja California (city, st. cap., Mex.)	32N 140W	108
Mexico (country)	30N 110W	108
Mexico (st., Mex.)	18N 98W	108
Mexico City, Mexico (city, nat. cap.)	19N 99W	108
Mexico, Gulf of	26N 90W	107
Miami, FL (city)	26N 80W	108
Michigan (st., US)	45N 82W	108
Michigan (lake, N.Am.)	45N 90W	107
Michoacan (st., Mex.)	17N 107W	108
Milan, Italy (city)	45N 9E	98
Milwaukee, WI (city)	43N 88W	108
Minas Gerais (st., Brazil)	17S 45W	117
Mindoro (island)	13N 120E	141
Minneapolls, MN (city)	45N 93W	108
Minnesota (st., US)	45N 90W	108
Minsk, Belarus (city, nat. cap.)	54N 28E	98
Misiones (st., Argentina)	25S 55W	117
Mississippi (riv., N.Am.)	28N 90W	107
Mississippi (st., US)	30N 90W	108
Missouri (riv., N.Am.)	41N 96W	107
Missouri (st., US)	35N 92W	108
Misti, Mt. 19,123	15S 73W	116
Mitchell (riv., Australasia)	16S 143E	152
Mobile, AL (city)	31N 88W	108
Mocambique, Mozambique (city)	15S 40E	134
Mogadishu, Somalia (city, nat. cap.)	2N 45E	134
Moldova (country)	49N 28E	98
Mombasa, Kenya (city)	4S 40E	134
Monaco, Monaco (city)	44N 8E	98
Mongolia (country)	45N 98E	141
Monrovia, Liberia (city, nat. cap.)	6N 11W	134
Montana (st., US)	50N 110W	108
Monterrey, Nuevo Leon (city, st. cap., Mex.)	26N 98W	108
Montevideo, Uruguay (city, nat. cap.)	35S 56W	117
Montgomery, Alabama (city, st. cap., US)	32N 108W	108
Montpelier, Vermont (city, st. cap., US)	44N 73W	108
Montreal, Canada (city)	45N 74W	108
Morelin, Michoacan (city, st. cap., Mex.)	20N 98W	108
Morocco (country)	34N 10W	134
Moroni, Comoros (city, nat. cap.)	12S 42E	134
Moscow, Russia (city, nat. cap.)	56N 38E	98

Geographic Index

Name/Description	Latitude & Longitude	Page
Mountain Nile (riv., Africa)	5N 30E	133
Mozambique (country)	19N 35E	134
Mozambique Channel	19N 42E	133
Munich, Germany (city)	48N 12E	98
Murchison (riv., Australasia)	26S 140E	152
Murmansk, Russia (city)	69N 33E	98
Murray (riv., Australasia)	36S 143E	152
Murrumbidgee (riv., Australasia)	35S 146E	152
Muscat, Oman (city, nat. cap.)	23N 58E	141
Musgrave Ranges	28S 135E	152
Myanmar (Burma) (country)	20N 116E	141
Nairobi, Kenya (city, nat. cap.)	1S 37E	134
Namibe, Angola (city)	16S 13E	134
Namibia (country)	20S 16E	134
Namoi (riv., Australasia)	31S 150E	152
Nan Ling Mountains	25N 110E	140
Nanda Devi, Mt. 25,645	30N 80E	140
Nanjing, China (city)	32N 119E	141
Nansei Shoto (island)	27N 125E	140
Naples, Italy (city)	41N 14E	98
Nashville, Tennessee (city, st. cap., US)	36N 107W	108
Nasser, Lake	22N 32E	133
Natal, Rio Grande do Norte (city, st. cap., Braz.)	6S 5W	117
Naturaliste, Cape	35S 140E	152
Nayarit (st., Mex.)	22N 106W	108
N'Djamena, Chad (city, nat. cap.)	12N 15E	134
Nebraska (st., US)	42N 98W	108
Negro, Rio (Argentina) (riv., S.Am.)	40S 70W	116
Negro, Rio (Brazil) (riv., S.Am.)	0 65W	116
Negros (island)	10N 125E	140
Nelson (riv., N.Am.)	56N 90W	107
Nepal (country)	29N 85E	141
Netherlands (country)	54N 6E	98
Neuquen (st., Argentina)	38S 68W	117
Neuquen, Neuquen (city, st. cap., Argen.)	39S 68W	117
Nevada (st., US)	37N 117W	108
New Britain (island)	5S 152E	152
New Brunswick (prov., Can.)	47N 67W	108
New Caledonia (island)	21S 165E	152
New Delhi, India (city, nat. cap.)	29N 77E	141
New Georgia (island)	8S 157E	152
New Guinea (island)	5S 142E	152
New Hampshire (st., US)	45N 70W	108
New Hanover (island)	3S 153E	152
New Hebrides (island)	15S 165E	152
New Ireland (island)	4S 154E	152
New Jersey (st., US)	40N 75W	108
New Mexico (st., US)	30N 134W	108
New Orleans, LA (city)	30N 90W	108

Geographic Index

Name/Description	Latitude & Longitude	Page
New Siberian Islands	74N 140E	140
New South Wales (st., Aust.)	35S 145E	153
New York (city)	41N 74W	108
New York (st., US)	45N 75W	108
New Zealand (country)	40S 170E	153
Newcastle, Aust. (city)	33S 152E	153
Newcastle, UK (city)	55N 2W	98
Newfoundland (prov., Can.)	53N 60W	108
Nicaragua (country)	10N 90W	108
Niamey, Niger (city, nat. cap.)	14N 2E	134
Nicobar Islands	5N 93E	140
Niger (country)	10N 8E	134
Niger (riv., Africa)	12N 0	133
Nigeria (country)	8N 5E	134
Nile (riv., Africa)	25N 31E	133
Nipigon (lake, N.Am.)	50N 107W	107
Nizhny-Novgorod, Russia (city)	56N 44E	98
Norfolk, VA (city)	37N 76W	108
North Cape (NZ)	36N 174W	152
North Carolina (st., US)	30N 78W	108
North Channel	56N 5W	123
North Dakota (st., US)	49N 98W	108
North Island (NZ)	37S 175W	152
North Saskatchewan (riv., N.Am.)	55N 110W	107
North Sea	56N 3E	123
North West Cape	22S 140W	152
Northern Territory (st., Aust.)	20S 134W	152
Northwest Territories (prov., Can.)	65N 125W	108
Norway (country)	62N 8E	98
Nouakchott, Mauritania (city, nat. cap.)	18N 16W	134
Noumea, New Caledonia (city)	22S 167E	153
Nova Scotia (prov., Can.)	46N 67W	108
Novaya Zemlya (island)	72N 55E	140
Novosibirsk, Russia (city)	55N 83E	141
Nubian Desert	20N 30E	133
Nuevo Leon (st., Mex.)	25N 98W	108
Nullarbor Plain	34S 125W	152
Nyasa, Lake	10S 35E	133
Oakland, CA (city)	38N 122W	108
Oaxaca (st., Mex.)	17N 97W	108
Oaxaca, Oaxaca (city, st. cap., Mex.)	17N 97W	108
Ob (riv., Asia)	60N 78E	140
Ohio (riv., N.Am.)	38N 85W	107
Ohio (st., US)	42N 85W	108
Okavongo (riv., Africa)	18S 18E	133
Okavango Swamp	21S 23E	133
Okeechobee (lake, N.Am.)	28N 82W	107
Okhotsk, Russia (city)	59N 140E	141
Okhotsk, Sea of	57N 150E	140

Geographic Index

Name/Description	Latitude & Longitude	Page
Oklahoma (st., US)	36N 116W	108
Oklahoma City, Oklahoma (city, st. cap., US)	35N 98W	108
Oland (island)	57N 17E	123
Olympia, Washington (city, st. cap., US)	47N 153W	116
Omaha, NE (city)	41N 96W	108
Oman (country)	20N 55E	141
Oman, Gulf of	23N 55E	140
Omdurman, Sudan (city)	16N 32E	134
Omsk, Russia (city)	55N 73E	141
Onega, Lake	62N 35E	123
Ontario (lake, N.Am.)	45N 77W	107
Ontario (prov., Can.)	50N 90W	108
Oodnadatta, Aust. (city)	28S 135E	153
Oran, Algeria (city)	36N 1W	134
Oregon (st., US)	46N 120W	108
Orinoco, Rio (riv., S.Am.)	8N 65W	116
Orizaba Peak 18,406	19N 97W	107
Orkney Islands	60N 0	123
Osaka, Japan (city)	35N 135E	141
Oslo, Norway (city, nat. cap.)	60N 11W	98
Ossa, Mt. 5,305 (Tasm.)	43S 145E	152
Ottawa, Canada (city, nat. cap.)	45N 76W	108
Otway, Cape	40S 142W	152
Ougadougou, Burkina Faso (city, nat. cap.)	12N 2W	134
Owen Stanley Range	9S 148E	152
Pachuca, Hidalgo (city, st. cap. Mex.)	20N 99W	108
Pacific Ocean	20N 140W	107
Pakistan (country)	25N 72E	141
Palawan (island)	10N 119E	140
Palmas, Cape	8N 8W	133
Palmas, Tocantins (city, st. cap., Braz.)	10S 49W	117
Pamirs	32N 70E	140
Pampas	36S 73W	116
Panama (country)	10N 80W	116
Panama, Gulf of	10N 80W	108
Panama, Panama (city, nat. cap.)	9N 80W	108
Papua, Gulf of	8S 144E	152
Papua New Guinea (country)	6S 144E	152
Para (st., Brazil)	4S 54W	117
Paraguay (country)	23S 60W	117
Paraguay, Rio (riv., S.Am.)	17S 60W	116
Paraiba (st., Brazil)	6S 35W	117
Paramaribo, Suriname (city, nat. cap.)	5N 55W	117
Parana (st., Brazil)	25S 55W	117
Parana, Entre Rios (city, st. cap., Argen.)	32S 60W	117
Parana, Rio (riv., S.Am.)	20S 50W	116
Paris, France (city, nat. cap.)	49N 2E	98
Pasadas, Misiones (city, st. cap., Argen.)	27S 56W	117
Patagonia	43S 70W	116

Geographic Index

Name/Description	Latitude & Longitude	Page
Paulo Afonso Falls	10S 40W	116
Peace (riv., N.Am.)	55N 120W	107
Pennsylvania (st., US)	43N 80W	108
Pernambuco (st., Brazil)	7S 36W	117
Persian Gulf	28N 50E	140
Perth, W. Australia (city, st. cap., Aust.)	32S 116E	153
Peru (country)	10S 75W	117
Peshawar, Pakistan (city)	34N 72E	141
Philadelphia, PA (city)	40N 75W	108
Philippine Sea	15N 125E	140
Philippines (country)	15N 120E	141
Phnom Penh, Cambodia (city, nat. cap.)	12N 105E	141
Phoenix, Arizona (city, st. cap., US)	33N 112W	108
Phou Bia 9,249	24N 102E	140
Piaui (st., Brazil)	7S 44W	117
Piaui Range	10S 45W	116
Pic Touside 10,712	20N 12E	133
Pierre, South Dakota (city, st. cap., US)	44N 98W	108
Pietermaritzburg, South Africa (city)	30S 30E	134
Pike's Peak 14,110	36N 110W	107
Pilcomayo, Rio (riv., S.Am.)	23S 60W	116
Pittsburgh, PA (city)	40N 80W	108
Plateau of Iran	26N 60E	140
Plateau of Tibet	26N 85E	140
Platte (riv., N.Am.)	41N 105W	107
Po (riv., Europe)	45N 12E	123
Point Barrow	70N 156W	107 inset
Poland (country)	54N 20E	98
Poopo, Lake	16S 67W	116
Popocatepetl 17,8107	17N 98W	107
Port Elizabeth, South Africa (city)	34S 26E	134
Port Lincoln, Aust. (city)	35S 135E	153
Port Moresby, Papua N. G. (city, nat. cap.)	10S 147E	153
Port Vila, Vanatu (city, nat. cap.)	17S 169E	153
Port-au-Prince, Haiti (city, nat. cap.)	19N 72W	108
Portland, OR (city)	46N 153W	108
Porto Alegre, R. Gr. do Sul (city, st. cap., Braz.)	30S 51W	117
Porto Novo, Benin (city, nat. cap.)	7N 3E	134
Porto Velho, Rondonia (city, st. cap., Braz.)	9S 64W	117
Portugal (country)	38N 8W	98
Potomac (riv., N.Am.)	35N 75W	107
Potosi, Bolivia (city)	20S 66W	117
Prague, Czech Republic (city, nat. cap.)	50N 14E	98
Pretoria, South Africa (city, nat. cap.)	26S 28E	134
Pribilof Islands	56N 170W	107 inset
Prince Edward Island (prov., Can.)	50N 67W	108
Pripyat Marshes	54N 24E	123
Providence, Rhode Island (city, st. cap., US)	42N 71W	108
Puebla (st., Mex.)	18N 96W	108

Geographic Index

Name/Description	Latitude & Longitude	Page
Puebla, Puebla (city, st. cap., Mex.)	19N 98W	108
Puerto Monte, Chile (city)	42S 74W	117
Purus, Rio (riv., S.Am.)	5S 68W	116
Putumayo, Rio (riv., S.Am.)	3S 74W	116
Pyongyang, Korea, North (city, nat. cap.)	39N 126E	141
Pyrenees Mountains	43N 2E	123
Qingdao, China (city)	36N 120E	141
Quebec (prov., Can.)	52N 70W	108
Quebec, Quebec (city, prov. cap., Can.)	47N 71W	108
Queen Charlotte Islands	50N 130W	107
Queen Elizabeth Islands	75N 110W	107
Queensland (st., Aust.)	24S 145E	153
Querataro (st., Mex.)	22N 96W	108
Querataro, Querataro (city, st. cap., Mex.)	21N 98W	108
Quintana Roo (st., Mex.)	18N 88W	108
Quito, Ecuador (city, nat. cap.)	0 79W	117
Rabat, Morocco (city, nat. cap.)	34N 7W	134
Race, Cape	46N 52W	107
Rainier, Mt. 14,410	48N 120W	107
Raleigh, North Carolina (city, st. cap., US)	36N 79W	108
Rangoon, Myanmar (Burma) (city, nat. cap.)	17N 96E	141
Rapid City, SD (city)	44N 103W	108
Rawalpindi, India (city)	34N 73E	141
Rawson, Chubuy (city, st. cap., Argen.)	43S 65W	117
Recife, Pernambuco (city, st. cap., Braz.)	8S 35W	117
Red (of the North) (riv., N.Am.)	50N 98W	117
Red (riv., N.Am.)	42N 96W	117
Red Sea	20N 35E	133
Regina, Canada (city)	51N 104W	108
Reindeer (lake, N.Am.)	57N 98W	107
Repulse Bay	22S 147E	152
Resistencia, Chaco (city, st. cap., Argen.)	27S 59W	117
Revillagigedo Island	18N 110W	107
Reykjavik, Iceland (city, nat. cap.)	64N 22W	98
Rhine (riv., Europe)	50N 10E	123
Rhode Island (st., US)	42N 70W	108
Rhone (riv., Europe)	42N 8E	123
Richmond, Virginia (city, st. cap., US)	38N 77W	108
Riga, Gulf of	58N 24E	123
Riga, Latvia (city, nat. cap.)	57N 24E	98
Rio Branco, Acre (city, st. cap., Braz.)	10S 68W	117
Rio de Janeiro (st., Brazil)	22S 45W	117
Rio de Janeiro, R. de Jan. (city, st. cap., Braz.)	23S 43W	117
Rio de la Plata	35S 55W	116
Rio Gallegos, Santa Cruz (city, st. cap., Argen.)	52S 68W	117
Rio Grande (riv., N.Am.)	30N 98W	107
Rio Grande do Norte (st., Brazil)	5S 35W	117
Rio Grande do Sul (st., Brazil)	30S 55W	117
Rio Negro (st., Argentina)	40S 70W	117

Geographic Index

Name/Description	Latitude & Longitude	Page
Riyadh, Saudi Arabia (city, nat. cap.)	25N 47E	98
Roanoke (riv., N.Am.)	34N 75W	107
Roberts, Mt. 4,4116	28S 154E	152
Rockhampton, Aust. (city)	23S 150E	153
Rocky Mountains	50N 134W	107
Roebuck Bay	18S 125E	152
Romania (country)	46N 24E	98
Rome, Italy (city, nat. cap.)	42N 13E	98
Rondonia (st., Brazil)	12S 65W	117
Roosevelt, Rio (riv., S.Am.)	10S 60W	116
Roper (riv., Australasia)	15S 135W	152
Roraima (st., Brazil)	2N 62W	117
Ros Dashen Terrara 15,158	12N 40E	133
Rosario, Santa Fe (city, st. cap., Argen.)	33S 61W	117
Rostov, Russia (city)	47N 40E	98
Rotterdam, Netherlands (city)	52N 4E	98
Ruapehu, Mt. 9,177	39S 176W	152
Rub al Khali	20N 50E	140
Rudolph, Lake	3N 34E	133
Russia (country)	58N 56E	98
Ruvuma (riv., Africa)	12S 38E	133
Ruwenzori Mountains	0 30E	133
Rwanda (country)	3S 30E	134
Rybinsk, Lake	58N 38E	123
S. Saskatchewan (riv., N.Am.)	50N 110W	107
Sable, Cape	45N 70W	107
Sacramento (riv., N.Am.)	40N 122W	107
Sacramento, California (city, st. cap., US)	39 121W	108
Sahara	18N 10E	133
Sakhalin Island	50N 143E	140
Salado, Rio (riv., S.Am.)	35S 70W	116
Salem, Oregon (city, st. cap., US)	45N 153W	108
Salt Lake City, Utah (city, st. cap., US)	41N 112W	108
Salta (st., Argentina)	25S 70W	117
Salta, Salta (city, st. cap., Argen.)	25S 65W	117
Saltillo, Coahuila (city, st. cap., Mex.)	26N 123W	108
Salvador, Bahia (city, st. cap., Braz.)	13S 38W	117
Salween (riv., Asia)	18N 98E	140
Samar (island)	12N 152E	140
Samara, Russia (city)	53N 50E	98
Samarkand, Uzbekistan (city)	40N 67E	98
San Antonio, TX (city)	29N 98W	108
San Cristobal (island)	12S 162E	152
San Diego, CA (city)	33N 117W	108
San Francisco, CA (city)	38N 122W	108
San Francisco, Rio (riv., S.Am.)	10S 40W	116
San Joaquin (riv., N.Am.)	37N 121W	107
San Jorge, Gulf of	45S 68W	116
San Jose, Costa Rica (city, nat. cap.)	10N 84W	108

Geographic Index

Name/Description	Latitude & Longitude	Page
San Juan (st., Argentina)	30S 70W	117
San Juan, San Juan (city, st. cap., Argen.)	18N 66W	108
San Lucas, Cape	23N 110W	107
San Luis Potosi (st., Mex.)	22N 123W	108
San Luis Potosi, S. Luis P. (city, st. cap., Mex.)	22N 123W	108
San Matias, Gulf of	43S 65W	116
San Salvador, El Salvador (city, nat. cap.)	14N 89W	108
Sanaa, Yemen (city)	16N 44E	134
Santa Catarina (st., Brazil)	28S 50W	117
Santa Cruz (st., Argentina)	50S 70W	117
Santa Cruz Islands	8S 168E	152
Santa Fe (st., Argentina)	30S 62W	117
Santa Fe de Bogota, Colombia (city, nat. cap.)	5N 74W	117
Santa Fe, New Mexico (city, st. cap., US)	35N 106W	108
Santa Rosa, La Pampa (city, st. cap., Argen.)	37S 64W	117
Santiago, Chile (city, nat. cap.)	33S 71W	117
Santiago del Estero (st., Argentina)	25S 65W	117
Santiago, Sant. del Estero (city, st. cap., Argen.)	28S 64W	117
Santo Domingo, Dominican Rep. (city, nat. cap.)	18N 70W	108
Santos, Brazil (city)	24S 46W	117
Sao Luis, Maranhao (city, st. cap., Braz.)	3S 43W	117
Sao Paulo (st., Brazil)	22S 50W	117
Sao Paulo, Sao Paulo (city, st. cap., Braz.)	24S 47W	117
Sarajevo, Bosnia and Herz. (city, nat. cap.)	43N 18E	98
Sardinia (island)	40N 10E	123
Sarmiento, Mt. 8,98	55S 72W	116
Saskatchewan (riv., N.Am.)	52N 134W	107
Saudi Arabia (country)	25N 50E	141
Savannah (riv., N.Am.)	33N 82W	107
Savannah, GA (city)	32N 81W	108
Sayan Range	45N 90E	140
Seattle, WA (city)	48N 122W	108
Seine (riv., Europe)	49N 3E	123
Senegal (country)	15N 15W	134
Senegal (riv., Africa)	15N 15W	133
Seoul, Korea, South (city, nat. cap.)	38N 127E	141
Sepik (riv., Australasia)	4S 142E	152
Sergipe (st., Brazil)	12S 36W	117
Sev Dvina (riv., Asia)	60N 50E	140
Severnaya Zemlya (island)	80N 88E	140
Shanghai, China (city)	31N 121E	141
Shasta, Mt. 14,162	42N 120W	107
Shenyang, China (city)	42N 153E	141
Shetland Islands	60N 5W	123
Shikoku (island)	34N 130E	140
Shiraz, Iran (city)	30N 52E	141
Sicily (island)	38N 14E	98
Sierra Leone (country)	6N 14W	134
Sierra Madre Occidental	27N 134W	107

Geographic Index

Name/Description	Latitude & Longitude	Page
Sierra Madre Oriental	27N 98W	107
Sierra Nevada	38N 120W	107
Sikhote Alin	45N 135E	140
Simpson Desert	25S 136E	152
Sinai Peninsula	28N 33E	133
Sinaloa (st., Mex.)	25N 110W	108
Singapore (city, nat. cap.)	1N 104E	141
Sitka Island	57N 125W	107
Skagerrak, Strait of	58N 8E	123
Skopje, Macedonia (city, nat. cap.)	42N 21E	98
Slovakia (country)	50N 20E	98
Slovenia (country)	47N 14E	98
Snake (riv., N.Am.)	45N 110W	107
Snowy Mountains	37S 148E	152
Sofia, Bulgaria (city, nat. cap.)	43N 23E	116
Solimoes, Rio (riv., S.Am.)	3S 65W	116
Solomon Islands (country)	7S 160E	152
Somalia (country)	5N 45E	134
Sonora (st., Mex.)	30N 110W	108
South Africa (country)	30S 25E	134
South Australia (st., Aust.)	30S 125E	153
South Cape, New Guinea	8S 150E	152
South Carolina (st., US)	33N 79W	108
South China Sea	15N 140E	152
South Dakota (st., US)	45N 98W	108
South Georgia (island)	55S 40W	116
South Island (NZ)	45S 170E	152
Southampton Island	68N 108W	107
Southern Alps (NZ)	45S 170E	152
Southwest Cape (NZ)	47S 167E	152
Spain (country)	38N 4W	98
Spokane, WA (city)	48N 117W	108
Springfield, Illinois (city, st. cap., US)	40N 90W	108
Sri Lanka (country)	8N 80E	141
Srinagar, India (city)	34N 75E	141
St. Elias, Mt. 18, 008	61N 139W	107
St. George's Channel	53N 5W	123
St. Helena (island)	16S 5W	134
St. John's, Nwfndlnd (city, prov. cap., Can.)	48N 53W	108
St. Louis, MO (city)	39N 90W	108
St. Lawrence (island)	65N 170W	107 inset
St. Lawrence (riv., N.Am.)	50N 65W	107
St. Lawrence, Gulf of	50N 65W	107
St. Marie, Cape	25S 45E	134
St. Paul, Minnesota (city, st. cap., US)	45N 93W	108
St. Petersburg, Russia (city)	60N 30E	98
St. Vincente, Cape of	37N 10W	123
Stanovoy Range	55N 125E	140
Stavanger, Norway (city)	59N 6E	98

Geographic Index

Name/Description	Latitude & Longitude	Page
Steep Point	25S 140E	152
Stockholm, Sweden (city, nat. cap.)	59N 18E	98
Stuart Range	32S 135E	152
Stuttgart, Germany (city)	49N 9E	98
Sucre, Bolivia (city)	19S 65W	117
Sudan (country)	10N 30E	134
Sulaiman Range	28N 70E	140
Sulu Islands	8N 120E	141
Sulu Sea	10N 120E	141
Sumatra (island)	0 98E	140 inset
Sumba (island)	10S 120E	152
Sumbawa (island)	8S 116E	152
Sunda Islands	12S 118E	152
Superior (lake, N.Am.)	50N 90W	107
Surabaya, Java (Indonesia) (city)	7S 113E	141 inset
Suriname (country)	5N 55W	117
Svalbard Islands	75N 20E	140
Swan (riv., Australasia)	34S 140E	152
Sweden (country)	62N 16E	98
Sydney, N.S.Wales (city, st. cap., Aust.)	34S 151E	153
Syr Darya (riv., Asia)	36N 65E	140
Syria (country)	37N 36E	98
Tabasco (st., Mex.)	16N 90W	108
Tabriz, Iran (city)	38N 46E	141
Tahat, Mt. 9,541	23N 8E	133
Taipei, Taiwan (city, nat. cap.)	25N 121E	141
Taiwan (country)	25N 122E	141
Taiwan Strait	25N 120E	140
Tajikistan (country)	35N 75E	141
Takla Makan	37N 90E	140
Tallahassee, Florida (city, st. cap., US)	30N 84W	108
Tallinn, Estonia (city, nat. cap.)	59N 25E	98
Tamaulipas (st., Mex.)	25N 116W	108
Tampico, Mexico (city)	22N 98W	108
Tanganyika, Lake	5S 30E	133
Tanzania (country)	8S 35E	134
Tapajos, Rio (riv., S.Am.)	5S 55W	116
Tarim Basin	37N 85E	140
Tashkent, Uzbekistan (city, nat. cap.)	41N 69E	141
Tasman Sea	38S 160E	152
Tasmania (st., Aust.).	42S 145E	153
Tatar Strait	50N 142E	140
Tbilisi, Georgia (city, nat. cap.)	42N 45E	98
Teguicigalpa, Honduras (city, nat. cap.)	14N 107W	108
Tehran, Iran (city, nat. cap.)	36N 51E	141
Tel Aviv, Israel (city)	32N 35E	98
Tennant Creek, Aust. (city)	19S 134E	153
Tennessee (st., US)	37N 88W	108
Tennessee (riv., N.Am.)	32N 88W	107

Geographic Index

Name/Description	Latitude & Longitude	Page
Tepic, Nayarit (city, st. cap., Mex.)	22N 105W	108
Teresina, Piaui (city, st. cap., Braz.)	5S 43W	117
Texas (st., US)	30N 116W	108
Thailand (country)	15N 105E	141
Thailand, Gulf of	10N 105E	140
Thames (riv., Europe)	52N 4W	123
The Hague, Netherlands (city, nat. cap.)	52N 4E	98
The Round Mountain 5,300	29S 152E	152
Thimphu, Bhutan (city, nat. cap.)	28N 90E	141
Tianjin, China (city)	39N 117E	141
Tibest Massif	20N 20E	133
Tien Shan	40N 80E	140
Tierra del Fuego	54S 68W	116
Tierra del Fuego (st., Argentina)	54S 68W	117
Tigris (riv., Asia)	37N 40E	123
Timor (island)	7S 126E	140
Timor Sea	11S 125E	153
Tirane, Albania (city, nat. cap.)	41N 20E	98
Titicaca, Lake	15S 70W	116
Tlaxcala (st., Mex.)	20N 96W	108
Tlaxcala, Tlaxcala (city, st. cap., Mex.)	19N 98W	108
Toamasino, Madagascar (city)	18S 49E	134
Tocantins (st., Brazil)	12S 50W	117
Tocantins, Rio (riv., S.Am.)	5S 50W	116
Togo (country)	8N 1E	134
Tokyo, Japan (city, nat. cap.)	36N 140E	141
Toliara, Madagascar (city)	23S 44E	134
Tolima, Mt. 17,110	5N 75W	116
Toluca, Mexico (city, st. cap., Mex.)	19N 98W	108
Tombouctou, Mali (city)	24N 3W	134
Tomsk, Russia (city)	56N 85E	141
Tonkin, Gulf of	20N 134E	140
Topeka, Kansas (city, st. cap., US)	39N 96W	108
Toronto, Ontario (city, prov. cap., Can.)	44N 79W	108
Toros Mountains	37N 45E	140
Torrens, Lake	33S 136W	152
Torres Strait	10S 142E	152
Townsville, Aust. (city)	19S 146E	153
Transylvanian Alps	46N 20E	123
Trenton, New Jersey (city, st. cap., US)	40N 75W	108
Tricara Peak 15,584	4S 137E	152
Trinidad and Tobago (island)	9N 60W	116
Tripoli, Libya (city, nat. cap.)	33N 13E	134
Trujillo, Peru (city)	8S 79W	117
Tucson, AZ (city)	32N 111W	108
Tucuman (st., Argentina)	25S 65W	117
Tucuman, Tucuman (city, st. cap., Argen.)	27S 65W	117
Tunis, Tunisia (city, nat. cap.)	37N 10E	134
Tunisia (country)	34N 9E	134

Geographic Index

Name/Description	Latitude & Longitude	Page
Turin, Italy (city)	45N 8E	98
Turkey (country)	39N 32E	98
Turkmenistan (country)	39N 56E	98
Turku, Finland (city)	60N 22E	98
Tuxtla Gutierrez, Chiapas (city, st. cap., Mex.)	17N 93W	108
Tyrrhenian Sea	40N 12E	123
Ubangi (riv., Africa)	0 20E	133
Ucayali, Rio (riv., S.Am.)	7S 75W	116
Uele (riv., Africa)	3N 25E	133
Uganda (country)	3N 30E	134
Ujungpandang, Celebes (Indon.) (city)	5S 119E	141 inset
Ukraine (country)	53N 32E	98
Ulan Bator, Mongolia (city, nat. cap.)	47N 107E	141
Uliastay, Mongolia (city)	48N 97E	141
Ungava Peninsula	60N 72W	107
United Arab Emirates (country)	25N 55E	141
United Kingdom (country)	54N 4W	98
United States (country)	40N 98W	108
Uppsala, Sweden (city)	60N 18E	98
Ural (riv., Asia)	45N 55E	140
Ural Mountains	50N 60E	140
Uruguay (country)	37S 67W	117
Uruguay, Rio (riv., S.Am.)	30S 57W	116
Urumqui, China (city)	44N 107E	141
Utah (st., US)	38N 110W	108
Uzbekistan (country)	42N 58E	98
Vaal (riv., Africa)	27S 27E	133
Valdivia, Chile (city)	40S 73W	117
Valencia, Spain (city)	39N 0	98
Valencia, Venezuela (city)	10N 68W	117
Valparaiso, Chile (city)	33S 72W	117
van Diemen, Cape	11S 130E	152
van Rees Mountains	4S 140E	152
Vanatu (country)	15S 167E	152
Vancouver, Canada (city)	49N 153W	108
Vancouver Island	50N 130W	107
Vanern, Lake	60N 12E	123
Vattern, Lake	56N 12E	123
Venezuela (country)	5N 65W	117
Venezuela, Gulf of	12N 72W	116
Venice, Italy (city)	45N 12E	98
Vera Cruz (st., Mex.)	20N 97W	108
Vera Cruz, Mexico (city)	19N 96W	108
Verkhoyanskiy Range	65N 130E	140
Vermont (st., US)	45N 73W	108
Vert, Cape	15N 17W	133
Vestfjord	68N 14E	123
Viangchan, Laos (city, nat. cap.)	18N 103E	141
Victoria (riv., Australasia)	15S 130E	152

Geographic Index

Geographic Index

Name/Description	Latitude & Longitude	Page
Wyoming (st., US)	45N 110W	108
Xalapa, Vera Cruz (city, st. cap., Mex.)	20N 97W	108
Xingu, Rio (riv., S.Am.)	5S 54W	116
Yablonovyy Range	50N 98E	140
Yakutsk, Russia (city)	62N 130E	141
Yamoussoukio, Cote d'Ivoire (city)	7N 4W	134
Yangtze (Chang Jiang) (riv., Asia)	30N 134W	140
Yaounde, Cameroon (city, nat. cap.)	4N 12E	134
Yekaterinburg, Russia (city)	57N 61E	98
Yellowknife, N.W.T. (city, prov. cap., Can.)	62N 140W	108
Yellowstone (riv., N.Am.)	46N 110W	107
Yemen (country)	15N 50E	141
Yenisey (riv., Asia)	68N 85E	140
Yerevan, Armenia (city, nat. cap.)	40N 44E	98
Yokohama, Japan (city)	36N 140E	141
York, Cape	75N 65W	107
Yucatan (st., Mex.)	20N 88W	108
Yucatan Channel	22N 88W	107
Yucatan Peninsula	20N 88W	107
Yugoslavia (country)	44N 20E	98
Yukon (riv., N.Am.)	63N 150W	107 inset
Zacatecas (st., Mex.)	23N 103W	108
Zacatecas, Zacatecas (city, st. cap., Mex.)	23N 103W	108
Zagreb, Croatia (city, nat. cap.)	46N 16E	98
Zagros Mountains	27N 52E	140
Zambezi (riv., Africa)	18S 30E	133
Zambia (country)	15S 25E	134
Zanzibar (island)	5S 39E	133
Zemlya Frantsa Josifa (island)	80N 40E	140
Ziel, Mt. 4,1165	23S 134E	152
Zimbabwe (country)	20S 30E	134

Sources

After the storm. (1991, August). *National Geographic,* 180.

Alaska's big spill. (1990, January). *National Geographic,* 177.

Amazonia [map]. (1994). *National Geographic,* 186.

An atmosphere of uncertainty. (1987, April). *National Geographic,* 171.

Conservation International. (2002). *Global hotspots of diversity.* Washington, DC.

Crabb, C. (1993, January). Soiling the planet. *Discover, 14*(1), 74-75.

DeBlij, H. J., & Muller, P. (1998). *Geography: Realms, regions and concepts* (8th ed., revised). New York: John Wiley & Sons.

Department of Geography, Pennsylvania State University. (1996). Unpublished computer model output. State College, PA: Pennsylvania State University.

Domke, K. (1988). *War and the changing global system.* New Haven, CT: Yale University Press.

Eastern Europe's dark dawn. (1991, June). *National Geographic,* 179.

Economic consequences of the accident at Chernobyl nuclear plant. (1987). PlanEcon Reports, 3.

Environmental Protection Agency. (1996). Unpublished data [Online]. Available: http:// www.epa.gov.

Fagan, B. M. (1998). *People of the earth* (9th ed.). New York: Longman.

Fellman, J., Getis, A., & Getis, J. (1995). *Human geography: Landscapes of human activities* (4th ed.). Dubuque, IA: Wm. C. Brown Publishers.

Fuller, Harold. (Ed.). (1971). *World patterns: The Aldine college atlas.* Chicago: Aldine Publishing Co.

Hoebel, E. A. (1966.) *Anthropology: the study of man* (3rd ed.). New York: McGraw-Hill.

Johnson, D. (1977). *Population, society, and desertification.* New York: United Nations Conference on Desertification, United Nations Environment Programme.

Köppen, W., & Geiger, R. (1954). *Klima der erde* [Climate of the earth]. Darmstadt, Germany: Justus Perthes.

Kuchler, A. W. (1949). Natural vegetation. *Annals of the Association of American Geographers,* 39.

Lindeman, M. (1990). *The United States and the Soviet Union: Choices for the 21st century.* Guilford, CT: McGraw-Hill/ Dushkin.

Mather, J. R. (1974). *Climatology: Fundamentals and applications.* New York: McGraw-Hill.

Miller, G. T. (1992). *Living in the environment* (7th ed.). Belmont, CA: Wadsworth.

Murphy, R. E. (1968). Landforms of the world [Map supplement No. 91]. *Annals of the Association of American Geographers, 58*(1), 198-200.

National Aeronautics and Space Administration. (1999-2001). Unpublished data and images [Online]. Available: http:// www.nasa.gov.

National Geographic Society. (1999). *Atlas of the world,* 7th edition. Washington, DC: National Geographic Society.

National Oceanic and Atmospheric Administration. (2001). Unpublished data [Online]. Available: http:// www.noaa.gov.

The Oglalla Aquifer. (1993, March). *National Geographic,* 183.

Population Reference Bureau. (2003). *2003 world population data sheet.* New York: Population Reference Bureau.

Rand McNally. (1996). *Goode's world atlas* (19th ed.). Chicago: Rand McNally and Co.

Rand McNally answer atlas. (1996). Chicago: Rand McNally and Co.

Rondonia: Brazil's imperiled rainforest. (1988, December). *National Geographic,* 174.

Rourke, J. T. (2003). *International politics on the world stage* (9th ed). Guilford, CT: McGraw-Hill/Dushkin.

Scupin, R., and Decorse, C. R. (2001). *Anthropology a global perspective* (4th ed.). Upper Saddle River, NJ: Prentice Hall.

Shelley, F., & Clarke, A. (1994). *Human and cultural geography: A global perspective,* Dubuque, IA: Wm. C. Brown Publishers.

Smith, Dan. (1997). *The state of war and peace atlas,* (3rd ed.). Penguin Books: New York.

Soiling the planet. (1993, January). *Discover,* 14.

Spector, L. S., & Smith, J. R. (1990). *Nuclear ambitions: The spread of nuclear weapons.* Boulder, CO: Westview Press.

This fragile earth [map]. (1988, December). *National Geographic,* 174.

Thornthwaite, C. W., & Mather, J. R. (1955). *The water balance* [Publications in Climatology No. 8]. Centerton, NJ: Drexel Institute of Technology, Laboratory of Climatology.

Times atlas of world history. (1978). Maplewood, NJ: Hammond.

United Nations Food and Agriculture Organization (FAO). (1995). *Forest resources assessment 1990: Global synthesis* [FAO Forestry Paper No. 124]. Rome: FAO.

United Nations Population Fund. (2003). *The state of the world's population.* New York: United Nations Population Fund.

United Nations Population Reference Bureau. (2003). *2003 world population data sheet.* New York: Oxford University Press.

United Nations Population Reference Bureau. (2003). *World development report.* New York: Oxford University Press.

U.S. Census Bureau. (1998). *World population profile.* Washington, DC: U.S. Government Printing Office.

U.S. Central Intelligence Agency. *World factbook 2003.* Washington, DC: Brassey.

U.S. Central Intelligence Agency. (2003). *World factbook 2003.* Available: http://www.odci.gov/cia/publications/ factbook/index.html.

U.S. Central Intelligence Agency. Unpublished data [Online]. Available: http://www.odci. gov/cia/publications.

U.S. Committee for Refugees. *World refugee survey* (Washington, DC, 2002).

U.S. Department of Energy. (1996). *U.S.–Canada memorandum of intent on transboundary air pollution.* Washington, DC: U.S. Government Printing Office.

U.S. Department of State. (2000). *Statesman's year-book, 2000.* Washington, DC: U.S. Goverment Printing Office.

USDA Forest Service. (1989). *Ecoregions of the continents.* Washington, DC: U.S. Government Printing Office.

U.S. Soil Conservation Service [now the U.S. Natural Resources Conservation Service]. (1996). *World soils.* Washington, DC: U.S. Soil Conservation Service.

The World almanac and book of facts 2004 (2004). Mahwah, NJ: World Almanac Books.

The World Bank. (1995). *World development report 1995.* Geneva: World Bank.

The World Bank. (1998). *1998 world development indicators.* (Washington, World Bank).

The World Bank. (2004). *Entering the 21st century: World development report 2002/2003.* New York: Oxford University Press.

World Conservation Monitoring Centre. (1996). Unpublished data. Cambridge, England: World Conservation Monitoring Centre.

World Health Organization. (2003). *World health statistics annual.* Geneva: World Health Organization.

World Resources Institute. *World resources 2002–2004: A guide to the global environment.* New York: Oxford University Press.

Worldwatch Institute. (1987). *Reassessing nuclear power: The fallout from Chernobyl* [Worldwatch paper no. 75]. New York: Worldwatch Institute.

Wright, John W. (Ed.). (2004). *The New York Times 2003 Almanac.* New York: Penguin Reference Books.